This textbook for engineering students provides an introduction to design for function, using many examples of manufactured artifacts and living organisms to demonstrate fundamental principles.

Nature has evolved numerous solutions to difficult design problems over millions of years. The author shows that the same design principles can be traced in both man-made objects and living organisms, with common themes in mechanisms such as knees and car suspensions, structures such as feathers and ships, and systems such as horses and hydrofoils. The book draws out the central importance of energy, materials and information, and discusses the relation between aesthetics and function. The crucial relation of design to production in artifacts and reproduction and growth in living organisms is dealt with, and the nature and work of designers is discussed, with some useful ideas on how to design and invent.

This second edition has been brought right up to date, with more examples and a new chapter based on case studies from the author's experience which illustrate some of the key ideas. In addition, some exercises have been added to help reinforce important points in the text.

Although this book is intended primarily as a textbook for students of engineering, design and manufacturing technology, it will be of considerable interest to professional engineers, designers and biologists.

Invention and Evolution: 2nd Edition

Invention and Evolution: 2nd Edition

Design in Nature and Engineering

MICHAEL FRENCH

Emeritus Professor of Engineering Design
Lancaster University

CAMBRIDGE
UNIVERSITY PRESS

Published by the Press Syndicate of the University of Cambridge
The Pitt Building, Trumpington Street, Cambridge CB2 1RP
40 West 20th Street, New York, NY 10011-4211, USA
10 Stamford Road, Oakleigh, Melbourne 3166, Australia

First published 1988
Second edition 1994

Printed in Great Britain at the University Press, Cambridge

A catalogue record for this book is available from the British Library

Library of Congress cataloguing in publication data

French, M. J.
Invention and evolution: design in nature and engineering/
Michael French. — 2nd ed.
p. cm.
Includes bibliographical references and index.
ISBN 0–521–46503–6. — ISBN 0–521–46911–2 (pbk.)
1. Engineering design. 2. Structural design. 3. Nature
(Aesthetics) I. Title.
TA174.F75 1994
620′.0042 — dc20   93–48554 CIP

ISBN 0 521 46503 6 hardback
ISBN 0 521 46911 2 paperback

Contents

Preface to the second edition   *page* xv
Preface to the first edition   xvii
Acknowledgements   xix

1 The designed world   1
   1.1 Design   1
   1.2 The intellectual nature of design   3
   1.3 Early design by man: tools and weapons   4
   1.4 Architecture   5
   1.5 The aqueducts of Rome   6
   1.6 The violin   6
   1.7 Naval architecture   8
        1.7.1 The eighteenth-century man-of-war   10
   1.8 Design and drawing   12
   1.9 Design and science   15
   1.10 Elegance in design   15
        1.10.1 The 'Trojan' engine   15
   1.11 Evolution   18
   1.12 Design in nature   19
   1.13 The great inventions of nature   20
   1.14 A limitation of natural evolution as a means of design   22

2 Energy   23
   2.1 Energy   23
   2.2 Mechanics   23
        2.2.1 Forces   23
        2.2.2 Work and energy   25
   2.3 The forms of energy   26

2.4		Conservation of energy	27
	2.4.1	The pendulum	28
	2.4.2	A bouncing ball	29
	2.4.3	The convertibility of heat	29
2.5		The Second Law of Thermodynamics	31
2.6		Power	32
	2.6.1	Energy, power and flight: birds	33
	2.6.2	Energy, power and flight: aircraft	35
	2.6.3	Efficiency	36
2.7		Fish and ships: scale effects	37
2.8		Peak power requirements	39
	2.8.1	Oxygen debt	40
	2.8.2	Peak power in aircraft	41
2.9		Energy conservation and storage	41
	2.9.1	Energy conservation in insect flight	41
	2.9.2	Apparent mass of a vibrating system	43
	2.9.3	Simple vibratory systems	43
	2.9.4	Energy conservation in locomotion – the switch-back railway	44
	2.9.5	Regenerative braking	44
	2.9.6	Peak power requirements in cars: the energy-storing car	45
	2.9.7	The quickness of fleas	47
	2.9.8	The quickness of bows	47
	2.9.9	Missile throwing without energy storage	49
2.10		Systematic invention and design	51
2.11		Energy and design	51
2.12		Specific energy and specific power	52
3		Materials	54
3.1		Materials, design and craftsmanship	54
3.2		Factors affecting choice of material	54
	3.2.1	Tensile strength	56
	3.2.2	Ductility	57
	3.2.3	Brittleness	58
	3.2.4	Compressive and shear strength	60
	3.2.5	Specific strength	62
	3.2.6	Stiffness	64
	3.2.7	Vibration and stiffness	65
	3.2.8	Stiffness of materials	68
	3.2.9	Flexibility	68
3.3		Molecular structure	69
	3.3.1	Polymers	69

3.4   Mechanical structure   70
       3.4.1   Structuring of composites   71
       3.4.2   The composite bow   73
       3.4.3   The comparative strength of composites   74
       3.4.4   Wood   74
3.5   Fluid materials   75
3.6   'Smart' materials   77
3.7   Energy, materials and information   78

4   Mechanism   79
    4.1   Mechanism   79
    4.2   Pivots   79
           4.2.1   A butterfly valve   80
    4.3   Car front suspensions   85
    4.4   The slider-crank chain   88
           4.4.1   Insect flight mechanisms   89
    4.5   The human knee   90
           4.5.1   A model knee   91
           4.5.2   The behaviour of the instantaneous centre   93
    4.6   The wheel   94
    4.7   Degrees of freedom: conformity   95
           4.7.1   Nesting of degrees of freedom   98
    4.8   Flexural pivots   101
    4.9   The use of degrees of freedom to balance forces   101
    4.10   Locks and keys   104
    4.11   The action of enzymes   106

5   Structures   108
    5.1   Elements of structures: struts and ties   108
           5.1.1   Structures built from struts and ties   111
           5.1.2   Components of a force and equilibrium   112
           5.1.3   The suspension bridge   113
           5.1.4   The arch   115
           5.1.5   Masonry arches   118
           5.1.6   Cathedrals   119
           5.1.7   The brontosaurus and the Saltash bridge   119
           5.1.8   The use of ties and struts   120
    5.2   Structural economy   122
    5.3   Shear   125
    5.4   Beams and bending   126
           5.4.1   Bending moments: hogging and sagging   130
           5.4.2   The economical design of beams   131
           5.4.3   Beams that must also resist twisting   132

5.5   Feathers   134
5.6   Hoops, arteries and domes   136
       5.6.1   Ring carrying inward loads   138
       5.6.2   The bicycle wheel   139
5.7   Prestressing   140
       5.7.1   Prestressed concrete bridges   143
5.8   Fluids as structural elements   144

6   Systems   147
    6.1   Process plants   147
    6.2   Service systems   148
    6.3   Interaction of components in systems: matching   149
    6.4   Characteristics of systems   150
           6.4.1   Characteristics: the man/bicycle system 150
           6.4.2   Characteristics   153
           6.4.3   Degrees of freedom   154
           6.4.4   Gearing   155
           6.4.5   Piston-engined ships   157
           6.4.6   Other matching problems   157
    6.5   Matching in flapping-wing flight   158
           6.5.1   The propeller-driven swan   159
    6.6   Other forms of matching   160
           6.6.1   The matching of powder grains   161
           6.6.2   The matching of fish muscles   163
           6.6.3   Matching in human muscles   164
           6.6.4   Matching in bows   164
    6.7   Control systems   166
           6.7.1   Design of control systems   167
           6.7.2   Reliability of systems   168
           6.7.3   More advanced control systems   169
    6.8   Communications   170
           6.8.1   Television bandwidths   171
           6.8.2   Data-processing   173
    6.9   The human brain   175
           6.9.1   Neural networks   175
    6.10   Systems, mechanisms and structures   176
    6.11   Systems design in nature   177

7   The practice, principles and philosophy of design   180
    7.1   Introduction   180
    7.2   Practice   180
           7.2.1   Standardisation   181
           7.2.2   Bearings and pivots   182

	7.2.3	The repertoire of bearings	186
	7.2.4	Cylinders subject to high internal pressures	188
	7.2.5	The evolution of gun-barrels	190
	7.2.6	Containing pressure in string bags	191
	7.2.7	Nemertine worms	192
	7.2.8	Fusion power	193
	7.2.9	Pumps and turbines	194
	7.2.10	A classification of kinetic pumps and motors	196

7.3   The repertoire of known means   198
7.4   Some design principles   199
      7.4.1   Biasing   199
      7.4.2   The principle of least constraint   199
      7.4.3   The regenerative principle   201
      7.4.4   Man and thermal regeneration   203
      7.4.5   Side-ponds   203
      7.4.6   Avoidance of irreversibilities   205
      7.4.7   The cascade principle   206
      7.4.8   Number of stages   209
      7.4.9   Matching   209
7.5   Some design 'philosophies'   211
      7.5.1   The philosophy of wave energy systems   211
      7.5.2   A different philosophy   213
      7.5.3   Suspended roofs   216
      7.5.4   Dams   216
      7.5.5   The gas-cooled nuclear reactor   218
      7.5.6   'Philosophies' of living designs   220

8   Economy, form and beauty   222
    8.1   Economy, form and beauty   222
    8.2   The form of plants   223
          8.2.1   The form of trees   225
          8.2.2   A tree regarded as an engineering design   227
    8.3   The form of animals   229
    8.4   Regular and organic curves   232
    8.5   Function and beauty   235
          8.5.1   Unity and chimeras   237
          8.5.2   Pattern, rhythm and texture   240
          8.5.3   Aesthetics and the experience of design   240
          8.5.4   Insight and intuition   244
    8.6   Style and the means of production   246
    8.7   The aesthetics of mass-production   249
    8.8   The aesthetics of functional design   250
    8.9   The aesthetic basis of design   251

9 Production, reproduction, evolution and design 253
  9.1 Production methods 253
  9.2 Replicative methods: casting and moulding 255
     9.2.1 Replicative processes: blanking, pressing and drawing 256
     9.2.2 Planar replicative processes: printing, photography and microelectronics 257
  9.3 Some other interesting manufacturing methods 260
  9.4 The relation between manufacture and design 264
  9.5 Production in nature: reproduction and growth 264
     9.5.1 The genetic inheritance 266
     9.5.2 Mutations 267
  9.6 Design in nature: evolution 268
     9.6.1 The pollination of flowers 268
     9.6.2 Niches and nichewise evolution 270
     9.6.3 The severity of the design specifications of living creatures 273
     9.6.4 A speculation on hypothetical limitations of the genetic code 274
  9.7 Design by man 276
  9.8 Nichewise evolution in the history of design 276
     9.8.1 Evolution in the design of particular products 277
     9.8.2 Aircraft propulsion 280
     9.8.3 Typewriters 281
  9.9 Genetic engineering 281
  9.10 Non-evolutionary design 282

10 Designing and inventing 283
  10.1 Designing and inventing 283
  10.2 The character of inventors and designers 283
     10.2.1 Measures of creativity 284
     10.2.2 Analysis and synthesis 286
     10.2.3 Mathematicians, physicists and others 286
     10.2.4 Who are the designers and inventors? 287
  10.3 Design in the schools (1) 289
  10.4 Verbal and visual thought 289
  10.5 Design in the schools (2) 292
     10.5.1 Design appreciation 292
     10.5.2 Drawing for thinking 293
  10.6 The higher education of designers 295
  10.7 Some aids to design 295
     10.7.1 Ways of stimulating ideas: maieutics 295
     10.7.2 'Hard' maieutics 298

	10.7.3	Application of analysis of function to the invention of a wave energy converter	300
	10.7.4	Classification	303
	10.7.5	Breaking logical chains	304
	10.7.6	Combinations and separation of functions	305
	10.7.7	The gear-pump	305
10.8	Abstraction as an aid to design		307
	10.8.1	The design repertoire	308
	10.8.2	Standard problems	309
10.9	Further examples and principles		310
	10.9.1	An application of analogy	310
	10.9.2	The steam catapult	312
	10.9.3	Clothes-pegs: specification and classification	314
	10.9.4	Avoiding arbitrary decisions	319
	10.9.5	Clarity of function	320
10.10	Designing and inventing: summary		320
10.11	The engineering designer, society and the future		321

11	Some case studies		322
	11.1	Introduction	322
	11.2	End-balancing of gas turbines	322
		11.2.1 End loads	324
	11.3	A first solution to the end-balancing problem	324
	11.4	Further ideas	326
		11.4.1 Clearances during transients	327
	11.5	Desalination of seawater using ocean temperature differences	328
		11.5.1 Generating the liquid propane	330
	11.6	Construction of the desalination plant	331
	11.7	Raising the cold water	332
	11.8	Ocean transport of liquid natural gas	332
		11.8.1 A naive idea	335
	11.9	Membrane tanks	335
		11.9.1 A solution	336
		11.9.2 Practical forms	337
	11.10	Hybrid tanks	338
	11.11	Nitrogen backhauling	339
	11.12	Air-cooled turbine blades	340
		11.12.1 Performance	343
	11.13	Continuously-variable transmissions (CVTs)	343
		11.13.1 Critique of the 'onion' gear	347
	11.14	Conclusions	348

Questions	350
Answers	355
Suggestions for further reading	358
Index	362

Preface to the second edition

In preparing this second edition I have increased the number of engineering examples, both in the body of the book and in an additional chapter based on my own experiences. I have also seized the opportunity to introduce more up-to-date material, to improve the clarity in some sections and to introduce some more interesting illustrations of principles. I hope and believe the result is a better book, meatier but no less accessible to the lay reader while at the same time enhanced in value to students of engineering.

I wish to thank all those people who have said kind things about the first edition, and especially those who have made helpful suggestions, some of whom may be able to detect traces of their advice in the new version.

Preface to the first edition

Design for function is even more important than design for appearance. It is also a fascinating pursuit which brings delight and challenge to engineers and others who engage in it.

Living organisms are examples of design strictly for function, the product of blind evolutionary forces rather than conscious thought, yet far excelling the products of engineering. When the engineer looks at nature he sees familiar principles of design being followed, often in surprising and elegant ways. Sometimes, as in the case of flight, he is inspired to invention: more commonly, he discovers his ideas are already embodied in some animal or plant.

This book is about design for function, and invention, which is the grandest form of design, in both engineering and nature. It draws out general principles, and shows how similar problems have led to similar solutions, both millions of years ago in the course of evolution and by engineers in our own time. In arctic animals, heat is conserved in an elegant way which we use in making steel and in gas turbines. The helical alignment of muscle fibres in fish puzzled biologists, but is now seen to have much the same role as the helical lay of the strands in a rope. The function of the alula, the small winglet springing from the leading edge of some bird wings, is still not fully understood but is probably basically the same as that of the Handley Page slot which was invented to delay stall in aircraft wings.

The first chapter looks at the history of design and its prehistory in evolution. There follows a discussion of the working media of engineers and nature, energy and materials, and then an account of the three characteristic forms in which design appears, mechanism, structure and systems. Chapter 7 discusses the intellectual resources the designer can call upon in his work, reference to what has been done before, general principles, and sometimes a 'philosophy', a central informing idea from which the rest of a design springs. The relations between economy, form and beauty are explored, together with the way in which some of our aesthetic preferences spring from the functional imperatives of design. Next comes a discussion of the all-important influence on design of the methods of production, which are reproduction and growth in nature. Then the characteristics and education of designers are examined, and finally some aids to design and invention are offered.

Little science is needed to follow the argument, and that little is explained in the text, so that the general reader should find no difficulty. However, the chief appeal should be to engineers, particularly to engineering students, and to school pupils considering engineering as a career, to whom I hope it will convey some idea of the deep intellectual challenge and the sheer fascination of design for function.

Acknowledgements

General acknowledgement

In the course of writing this work, I have received help and inspiration from very many people and sources. To list them all would be too long and invidious oversights might arise, but nevertheless I am deeply grateful for their advice and encouragement, to which this book owes so much.

Line drawings

David Cooper and Bruce Harwood.

Additional drawings, second edition

Audrey Parker
Figure 1.9, Vincent Oh and Azzelarabe Taleb-Bendiab.
CAD figures by the author.

Photographs

Fig. no.	Title	by courtesy of
1.1	Munich Olympic Stadium	Münchner Olympiapark GMBH
1.3	Pont du Gard	Comité Départmental du Tourisme du Gard
1.7	Illustration from *Traité sur l'architecture navale*	F. H. de Chapman

5.10	Flying buttress, Westminster Abbey	Rattee & Kett
5.11	Paddington Station	British Rail (WR)
8.5	Blue tit	Biophoto Associates
8.6	Concorde	Adrian Meredith Photography
8.12	Nike of Samothrace	Musée du Louvre, Musées Nationaux
8.13	Arabian Camel	Biophoto Associates
8.14	Water-nut pliers	
8.15	Gas holder, Salford	
8.17	Space-frame, Gatwick Airport	British Steel Corporaton
8.18	Waterloo Bridge	Greater London Council
8.19	'Bonheur du jour' by Gillow, late 18th century	Lancaster City Council Museum
8.20	Railway carriage, 1902 (Great Northern Railway)	National Railway Museum, York
8.21	Rolls-Royce car	Rolls-Royce Motors

1

The designed world

1.1 Design

Suppose we are asked which of these three is the odd one out, a waterfall, a buttercup and a steam locomotive. One answer would be, the locomotive, because it alone is man-made, another would be, the buttercup, because it alone is alive. But the third possible answer would also be justifiable: the buttercup and the locomotive show evidence of design, but the waterfall does not – its shape simply happens, it has no symmetry, it is not contrived to have any function or to serve any end. The locomotive and the buttercup show symmetry and regularity – several wheels or petals all the same shape, a long regular frame with wheels attached below in pairs, or a long regular stem with leaves arising from it. All the parts of both are adapted to particular purposes, purposes which serve greater ends – in the buttercup, to survive, to grow and propagate its kind, in the locomotive, to haul goods and passengers.

It is easy to recognise that the locomotive is a work of knowledge and cunning and that the relations of its many parts are subtle and exact. It is clear, too, that it is simplicity itself compared with the intricacies of the buttercup. Nevertheless, these two designs, so different in appearance and nature, are informed by the same principles, subject to the same practical necessities and demonstrate similar answers to similar problems. It might be expected, therefore, that the study of design would be a major intellectual interest of man, since it underlies the two sets of things which chiefly engage him, the living world, including himself, and the world he has himself made. But it is not so.

For all its importance, functional design has hardly been studied at all. The most important practitioners of design are engineers, who rarely give any account of their work; when they do, it is generally only an outline, restricted to conclusions and decisions, without much of the thinking behind them. This is only to be expected; a scientist writing a paper gives only the end-product of his thinking, not the route, often unnecessarily tortuous, by which he reached it.

1

Some architects are much less reserved and give accounts of the abstract considerations and principles entering into their work. These are usually concerned much more with the aesthetic aspects than with the functional ones, which are usually rather simple anyway. However, some of the most valuable insights into design are provided by those who have combined architecture and engineering – Nervi, Torroja, Buckminster Fuller, Otto (Fig. 1.1) – and have designed great structures presenting non-trivial functional problems.

Biologists, because they study living things, have always been students of design to that extent. Nowadays some of them are becoming increasingly conscious of organisms as solutions to design problems, and are adopting viewpoints which have much in common with those of engineers. Two recent trends have perhaps increased this tendency. Firstly, careful research often reveals small details of species of animals and plants which contribute in unsuspected ways to their ability to survive, and this is one of the fruits of the study of ecology, the relation of living things to their environment. Thus the increased interest in ecology draws attention to the engineering of living organisms. Secondly, the uncovering of the nature of the genetic 'blueprint' for the organism, the double-helical molecule of DNA, and other advances in molecular biology have shown the extraordinary biochemical design problems to which evolution has found answers.

Fig. 1.1. Olympic stadium roof, Munich.

However, the designers, the engineers and architects, and the interpreters of designs in nature, the biologists, are all very narrow and selective in their approach, concerned with particular problems or organisms or processes. The possibility of studying functional design generally seems to have been largely overlooked.

1.2   The intellectual nature of design

The recognised peaks of human intellectual achievement have always lain in the arts, in Art, and in the sciences, and never in the humble useful arts, which would be disdained for their usefulness, lacking other reasons. But the crown of the useful arts is functional design, and its finest achievements have about them the same elegance, the same apparent fortuity from which there develops a marvellous, deeply satisfying aptness, that marks a beautiful proof in mathematics, say, or a great symphony. Underpinning these great designs, there is a vast and only vaguely apprehended structure of themes and patterns recurrent in the works of both man and nature that have never been explored by scholars or illuminated by academic study. For example, historians have often pointed out the importance in the increase of human wealth and leisure of the *division of labour,** whereby instead of each family growing its own food and grinding its own flour, making its own clothes and building its own house, the tasks are distributed, so that the farmer grows food for many families and buys his clothes and his house from tailors or builders who themselves grow no food. The division of labour is but a special case of a more general principle of functional design, the *separation of functions.* Thus simple single-celled organisms have to provide all their functions in one cell, whereas higher animals and plants have many different kinds of cell for special purposes, carrying sap, extracting water and minerals from the soil, transmitting signals, secreting digestive juices, etc. The early steam engines, following Newcomen's design of 1712, had a cylinder in which the steam did work and in which it was also condensed. Watt's engine, 50 years or so later, separated the functions of working cylinder and steam condenser, so greatly increasing the efficiency.

Themes like the separation of functions recur again and again in functional design, in both the living and the man-made world, often enough in particular patterns. For example, two functions which it is often useful to

* Smith, A. (1776). *The Wealth of Nations.*

separate are those which are most simply described as 'preventing bursting' and 'preventing leaking', and many good designs have been arrived at by making separate provision for these two purposes, which may seem to be nearly the same and hence easily fulfilled by a single means.

1.3 Early design by man: tools and weapons

Some animals produce structures showing design – the waxen combs of bees and the paper nests of wasps, the trapdoor spider's lair and the beaver's lodge and dam – but all are the results of evolutionary pressure and not of conscious thought. Perhaps the first object that should be distinguished as of conscious design was a hand-axe, but certainly, by about 20 000 years ago, there was an impressive example in the bow, a device which has just that elegance of principle that has been mentioned – even though its inventors could not fully appreciate it.

Until perhaps two centuries ago, design by man proceeded only slowly, in a fashion scarcely less evolutionary in its character than design by nature. Major steps occurred only very rarely, perhaps every few hundred years, then every few generations, because there was no consciousness of the possibility, let alone any expectation, of deliberate rapid progress. Myths such as those of Prometheus and Cadmus, attributing invention to supernatural sources, often in despite of a jealous heaven, suggest a general attitude which might be expected to be hostile to any such view. Even as late as 1725, the satirist Swift in his account of Gulliver's voyage to Laputa is indiscriminately hard on those seeking to introduce new methods based on the application of science.

However, just as evolution in nature is slow but sure, and arrives at organisms of extreme refinement, so evolutionary design by man in the last 20 millennia or so has produced examples of great elegance. Such examples played an essential part in establishing the infrastructure of craftsmanship and traditional detail design on which conscious progress would one day rely. In particular, there must have been great interest in the detail and manufacture of weapons, because of both their crucial importance to survival of the bearer and also that special love given to those possessions which the owner may see as some expression or extension of his person, and which may be seen nowadays lavished on clothes and cars. This must have encouraged the best craftsmen to make small judicious experiments, of just the sort required for evolution.

Compared with weapons, the evolutionary pressure on hand-tool design would be less but, nevertheless, splendid examples were produced. Even then, when a general conscious pressure to improve developed, it found plenty of scope left.

Consider, for example, the scythe (Fig. 1.2). It was already widely spread in the Roman Empire and it is still extensively used. It is beautifully adapted

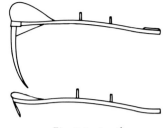

Fig. 1.2. Scythe.

to the human user, so that man and tool combine to form an efficient grass- or grain-moving whole. It is from this high degree of adaptation that most of the apparent asymmetries and oddnesses of the scythe derive – the out-of-plane angles of the blade and the handles, for example. The swing of the arms combined with the swing of the body, transferring the weight from one foot to the other, shares the work among the muscles and conserves energy, while the cradle ensures that the cut crop falls in a neat swathe ready for gathering and binding into sheaves:* all this is beautiful evolutionary design, but it took at least 2000 years. Later came composite scythe blades, made by forge-welding a narrow strip of steel between the edges of two wider strips of iron to give a cutting edge of hardened steel backed by softer, but less brittle (and less costly) iron.

1.4 Architecture

A very important field in which man began to develop his talent for design at an early stage was architecture. The design of large buildings required the use of important supporting techniques like drawing and modelling, and their construction required refinements of organisation, measurement and control. Above all, architecture developed the understanding of spatial relationships which is essential in most areas of design. Details like foundations, weatherproofing and rain disposal, and erection, all presented instructive problems of function, materials and construction.

However, in much architecture the functional aspects are very secondary to aesthetic ones, and, moreover, rather readily met (or indeed, neglected altogether, as in some badly-designed buildings which have, none the less, won awards). Another defect of architecture as a training ground for functional design was that often the economic constraint, so powerful throughout nature and engineering, was virtually absent, the patrons caring more for glory than the public good (which was the public's loss then, but is sometimes our gain now).

* Lisle, E. (1757). *Observations of Husbandry*, vol. 1, p. 235. London.

Thus in architecture we should look for examples of functional design at large undertakings whose function was not to provide glory, but some service or amenity, or else which posed serious technical difficultes, like roofing large spaces. The latter were represented in the past most strikingly by the mediaeval cathedrals,* and the former by the water supply systems of the Roman Empire.

1.5 The aqueducts of Rome

To support Rome in her great days of empire, when her population and her commercial activity were at their peak, a great system of aqueducts was built to bring water into the city from the surrounding hills. Their total length was about 350 km, and they brought in perhaps a million tons of water, over 200 million gallons, every day.

Because pipes were expensive and unreliable the water flowed in channels which had to maintain a very slight downward gradient (1 in 1000 to 3000) regardless of the ups and downs of the land. Through most of their length they ran underground, but about 50 km were above ground, sometimes raised high on arches or tiers of arches.

Other great aqueducts were built to supply major cities of the Empire: the Pont du Gard (Fig. 1.3) is a famous example, with its three tiers of arches. To build such systems required design skills of a relatively high order. The terrain had to be studied to find an economical route over suitable ground, neither excessively long nor requiring too much excavation, nor too many or high arches, which needed competent, if rudimentary, surveying. The sizes of the arches or tiers of arches had to be chosen to give a safe and economical design and, above all, the tremendous undertaking of building these great structures had to be organised and controlled. The Roman aqueducts were great feats of engineering, not to be equalled for many years, and a surprising number still stand today.

1.6 The violin

The violin is one of the most astonishing achievements of evolutionary design. A taut string, when bowed, produced only a faint sound: to make a stringed instrument, its vibrations must be transmitted to a sound box of large area which will in turn propagate sound waves into the air. In the transfer process, the volume and quality of the sound are strongly affected by the characteristics of the sounding box, which in turn depend in a very complicated fashion on the material of which it is made, its shape and its local variations of thickness, and the exact position of the bridge, which

---

* Morris, R. (1979). *Cathedrals and Abbeys of England and Wales*. London: J. M. Dent.

Fig. 1.3. Pont du Gard aqueduct.

transmits the vibrations into the top face or belly of the sound box, and the sound post, which connects the belly and the back (Fig. 1.4). Modern research has only begun to unravel the complex function of these parts and the relationship between the musical quality of the instrument and the shape and elasticity of the wooden parts of which it is built. For example, one important aspect appears to hinge on the particular frequencies at which the back will vibrate in two particular shapes ('modes') and computers have been used to try to determine where and by how much to reduce the thickness of the wood to adjust these frequencies to the values found in instru-

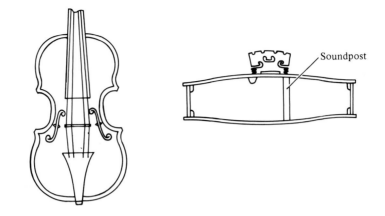

Fig. 1.4. Violin.

ments of proven excellence. Nevertheless, the violin first appears in a painting in 1530 in a fairly advanced form, without any closely similar antecedents we know of, and reaches a peak of refinement with the designs of Stradivarius, less than 200 years later.*

One explanation of the rapid development of the violin is the short cycle between conception and evaluation in use, much shorter than in the case of a ship, say. In the same way, in natural evolution, the length of a generation will affect the speed of change which is possible in an animal or a plant.

1.7 Naval architecture

For many centuries, even into the eighteenth century, the most advanced work of man was the sailing ship. Unlike buildings, ships are difficult to design in their functional aspects, and the difference between good and bad designs is a matter of life and death. Moreover, the functions are many, and, what is characteristic of much design by man and all design in nature, they interact heavily with one another. It is often the strong interactions between one function and another that make design difficult: in the case of a man-of-war, for instance, not only was it necessary for the hull to be easily driven by the sails, so that it would have the speed to overhaul its enemies or escape from them, but it had also to be able to withstand storms. Both these requirements conflicted with the problem of mounting the ordnance so as not to affect the strength of the ship as a structure, or make it top-heavy and liable to capsize, or unable to use its lowest guns in a rough sea.

The sailing ships of Western Europe afford a good opportunity to study the progress of evolutionary design. It is interesting to see how slow this

* Boyden, D. D. (1965). *The History of Violin Playing.* Oxford University Press.

appears by today's standards, in spite of strong pressures towards improvement that must have existed. An example is the introduction of knees made of iron. Knees are brackets used to join major timbers of a ship at a large angle, say, where a deck-beam meets a rib (see Fig. 1.5). If knees have to be made from a single piece of wood, as they did in the past, it is clear that they will only have much strength if the grain of the timber flows round the angle, i.e., if they are cut from a piece of a tree which naturally has the general form of a knee. In Britain during the Napoleonic wars, the supply of such pieces was not equal to the demand, and so iron knees were introduced as a substitute. However, a proposal of the kind had been made 140 years earlier, to Pepys, and even acted upon in the case of one ship (which unfortunately was burnt within a year). It seems clear that iron knees would have been a great improvement even in 1670, but their general use might have been delayed even longer but for a shortage of timber.

Another example of the relative sluggishness of evolutionary design is given by the history of ships' steering. Before about AD 1200, this was achieved by means of special large oars or paddles, lashed to one or both sides of the ship at the after end. Because of their position, they must have been very exposed to damage, difficult to handle and not very effective. They were superseded by the stern-mounted rudder around 1200, but until about 1700 this was arranged in the fashion of Fig. 1.6(a), where the tiller, a horizontal bar fixed to the top of the rudder post, extended into the ship through a slot in the planking; large amounts of water must have been shipped through this slot, which necessarily had to be much wider than the tiller to allow the rudder to turn. Later, the rudder post was carried further up and through a round hole (which it filled) to a tiller located wholly within the ship, as in Fig. 1.6(b).

For many years, also, the tiller was moved by a device called a whip-staff that no one with a practical understanding of design would propose (Fig. 1.6(c)). An upright lever was pivoted at deck level, with the lower end engaging with the tiller, and the helmsman steered by moving the upper end to

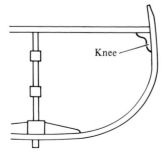

Fig. 1.5. Knee in sailing ship.

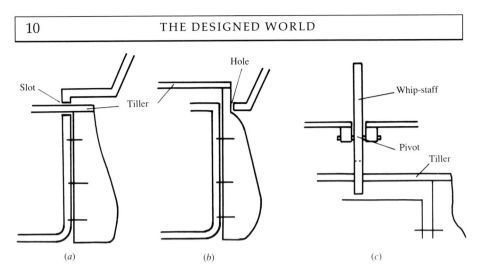

Fig. 1.6. Rudder arrangements. (a) Pre-1700. (b) Post-1700. (c) Whip-staff.

one or other side. So poor was this device that in heavy weather tackles were attached to the tiller with men pulling on them to supplement the helmsman's strength. This abomination was replaced by the familiar helmsman's wheel about the beginning of the eighteenth century. The tackles acting on the tiller were a curious anticipation of later steam steering gear, where steam cylinders acted on either side of the tiller under the control of a wheel, an early servo system (see Chapter 6).

1.7.1   The eighteenth-century man-of-war

In the eighteenth century the idea of deliberate and rapid progress in design was becoming more widely accepted, so that to some extent the ships of the time represent the peak achievement of the evolutionary period, and a very remarkable one. The hull was designed by a combination of rule of thumb and physical insight gained over many years of experience, through the medium of a drawing technique of great refinement. These drawings were considerable works of art (see Fig. 1.7). The various sets of lines were sketched in by eye, defining in effect several different hulls, since the sets would not agree. The disagreements would be removed gradually by redrawing the lines which seemed less satisfactory in each area of the hull to suit those with which the designer was better pleased, until all the sets fitted the same three-dimensional form. Drawings could also be made from various angles and with the hull rolled over, which helped to give a more complete impression of the shape. An expert studying these lines would be able to guess the likely speed, how close to the wind she would lie, how she would rise to a big wave in a stiff breeze, and so on.

The sails and rigging were very elaborate – a ship of the line would have perhaps 1400 pulley blocks – and incorporated the small developments and refinements of centuries. A very full inventory of stores was carried, includ-

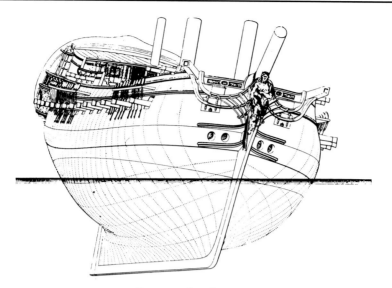

Fig. 1.7. Ship drawing.

ing spare spars, sails, ropes and blocks, with provision for all kinds of eventuality that might arise far from home, and the records are full of accounts of the resourcefulness of their crews in effecting major repairs.

It is interesting how near the naval ship had come to a living organism in its independence of the land and its free existence at sea, its mobility, its means of offence and defence and self-repair, and its provision for finding its way. Victor Hugo, rather later, wrote of such a ship:

> A man of war is composed at once of the heaviest and lightest things, because it has to do at the same time with the three forms of matter, solid, liquid and gas, and it must battle with all three. There are eleven claws of iron to seize the granite of the sea-bed, and more wings and more antennae than a mosquito to catch the wind in the storm-clouds. Its breath comes from its hundred and twenty cannon as if from enormous bugles, and replies defiantly to the thunder. The ocean seeks to lead it astray in the terrifying uniformity of its waves, but the vessel has its soul, its compass, which counsels it and always shows it the north. At night its lanterns complement the stars. Thus, against the wind it has rope and canvas, against the water, wood, against the rocks, iron, copper and lead, against darkness, light, against the immensity, a needle.*

Altogether, these ships were a remarkable achievement, and a society which could produce them was ready for the great blossoming of material progress based on innovative engineering design, which was just beginning.

---

* Hugo, V. (1862). *Les Misérables,* part 2, book 2.

1.8 Design and drawing

Design and drawing are very close – indeed, the word 'dessin' means 'drawing' in French. Most designers, most of the time, will think in images rather than words, and many great designers have also been great draughtsmen – Leonardo da Vinci is an outstanding case. An engineering designer is usually competent, or better, at free-hand perspective sketching, which he will often use in parallel with, or as a preliminary to, his formal projected technical drawings. In design, while an idea may be conveyed in words, it is often only when lines appear on paper that understanding develops and progress is made.

Many different kinds of drawing are used by the designer, of which the chief is the orthographic scale projection, familiar to most people in the form of the architect's plan and elevations: a plan is a view looking straight down on the subject of the drawing, and an elevation is a view from the front, back or side. Often the view is drawn as if a cut had been made in the subject parallel to the plane of the drawing and the part in front of the cut removed. Such views, like those in Fig. 1.8, are called sections: an architect's plan of a house is usually a horizontal section through the level of the windows of a floor, and looking down.

Next to orthographic projections, free-hand sketches are used, and may be more important in the early stages. While they are a fluent medium for the development of ideas, however, they are not good tools for precise thinking like the scale drawing.

There are in addition a number of very important, highly specialised uses of drawings, mostly unknown to the layman. Ship's lines are a relatively

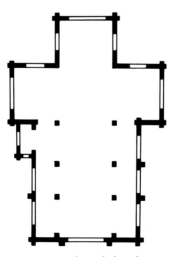

Fig. 1.8. Plan of church.

simple case, consisting essentially of a number of superimposed outline sections like a contour map. But there are drawings used in the design of mechanisms and of turbomachinery which represent motion in various ways, and others which represent forces, voltages, magnetic fields or other intangible things. All these specialised drawings could be replaced by calculations (and are usually checked by them), but often the calculation does not provide the all-important physical insights that the drawing does.

In recent years, the development of computer graphics has given the designer a powerful and elegant new tool. He can now 'draw' with a computer, which shows him what he has drawn, performs calculations upon it to order, turns it round so it can be seen from any angle, and so on. The computer enables the designer to work faster, it eliminates much drudgery and leaves him free to design, and it reduces the risk of error. At any stage it can produce a drawing of what has been done so far, in a fraction of the time taken to produce it by hand. In some cases it can go straight from the drawing to tapes which can be fed to a computer-controlled machine-tool which will make the parts required.

Edison said that genius is 1% inspiration and 99% perspiration, and he might well have said the same of designing. With the computer at one's elbow, perhaps the inspiration may grow to 5%, or even more. Above all, the computer helps the designer without taking away the immediacy of the image on the drawing board, but rather improving upon it, with a dynamic image which will grow or contract, change its angle and do other things at a command. Components may be moved about, or duplicated, in the time it takes to read the words.

In 1935, the great designer of sailing boats, Uffa Fox, wrote a fantasy, a dream, he called it, in which he was summoned to Heaven to design a sailing ship to beat the French liner Normandie across the Atlantic on her maiden voyage. He was given a board to work at, on which by pressing buttons the preferred sets of lines would be retained and all the others would be adjusted to fit – in technical terms, the lines would 'fair' themselves – or a particular line could be moved, and all the others would change to match. Nowadays, we have boards on earth which will do as much, and a great deal more. (The end of the dream was that the heavenly ship beat the Normandie, but because it was invisible, it was not reported in the papers.)*

Today we have computers that can do all that Uffa Fox dreamed of in 1935 and more, and they may do even more in the future. Figure 1.9 shows a display from a design tool called SIMAD (System for Improving Mechanical Assembly Design) which we developed at Lancaster with a grant from the Science and Engineering Research Council. The designer has outlined a design for a simple axisymmetric assembly, such as is common in pumps,

* Fox, U. (1935). *Uffa Fox's Second Book*, pp. 355–67. London: Peter Davies.

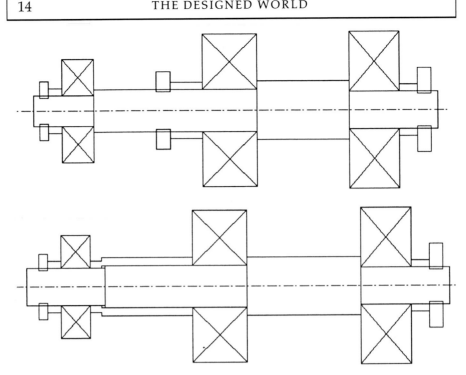

*Fig. 1.9. SIMAD (System for Improving Mechanical Assembly Design).
The designer has sketched in the assembly shown in outline at the top:
the system 'thinks' that it might be altered to reduce the cost, while still
working equally well. It shows its 'idea' (below), which saves a nut and
a thread-cutting operation on the shaft, for the designer to consider.*

will save a part and simplify assembly. In the display it is asking the
designer to approve, and to choose between two different ways of making
the change. In no way does the computer understand the assembly, and so
in the last resort it cannot itself decide, but it can very rapidly review possibilities in the light of the engineering knowledge which has been put into
it, and follow through the consequences, at least as far as the rules it 'knows'
go. With more rules and more development, tools like SIMAD will be of
great help to designers in the future, enabling them to look at far more
alternatives in the same time, and helping to evaluate them.

Other tools of the same general kind are being developed, in a sense not
so advanced as SIMAD, but much more comprehensive and able to deal
with a whole system. Some of these are able to simulate the behaviour of
the complete system and tell the designer how it will behave (or at least,
how calculation says it will behave) in a very short time. As the systems
we build become more complex (and more like living creatures), we shall
become more dependent on such aids to the designer if our products are to
be satisfactory.

1.9 Design and science

The relationship between design and science is often misunderstood. Science is the study of the natural world; it is concerned with understanding what is. Engineering design is concerned with creating new things; it makes extensive use of science, but it is a quite different activity. Usually, engineering design makes use of rather old science – for example, even in fields as advanced as the Apollo moon-flight programme, the vast majority of the mechanics used will be Newtonian, not relativistic – but occasionally the designer is creating in fields the scientist has yet to analyse. In the past it was not so; the engineer was usually working without the benefit of science. It is only the support of science that has made possible the quickened pace and great achievements of engineering today, and in most fields nowadays the designer must have a solid background of scientific knowledge.

1.10 Elegance in design

One characteristic of functional design is elegance. Most people find a buttercup beautiful, and many would say that the locomotive was at least pleasant to look at. However, the buttercup has an essential elegance, much more fundamental than its mere appearance. It is an elegant solution to a difficult problem in functional design; it has leaves to gather sunlight, oxygen and carbon dioxide from the air, and roots to extract water and minerals from the soil and hold it fast in the ground. Its stems support the leaves and flowers and transmit materials and signals (in the form of special substances). In its cells it makes and distributes many substances. It grows, it repairs damage to itself and it flowers and produces seed. It does all this in a fiercely competitive world with an extreme economy of living material, and its beautiful outward form is a reflection of its economical design.

The buttercup is a splendid piece of engineering, much more advanced and refined than the locomotive. But even so, the locomotive is an elegant design, economical in its use of energy and material, with its balanced mechanisms and well-proportioned parts, full of ingenious detail and thoughtful refinements, and the overall coherence and unity that results so often from a single purpose intelligently pursued. It has beauty for the educated eye – and because of its simple action the education need only be slight – and that beauty comes nearly all from its functional design, and very little from conscious aesthetic intention.

1.10.1 The 'Trojan' engine

Before moving on to design in nature, let us look at one brilliant example of design by man, chosen in an attempt to convey the elegant simplicity of which this branch of intellectual activity is capable, an example in which mere appearance contributes nothing.

In a two-stroke internal combustion engine, every in-stroke (see Fig. 1.10(a)) compresses a charge of air and petrol vapour which is then fired, so that the following out-stroke is a working stroke, at the end of which the exhaust gases must be displaced from the cylinder and replaced by a fresh charge of air and petrol vapour. This replacement of the exhaust gases by fresh mixture is often contrived by using the crank case to compress the next charge, and arranging for two holes, or 'ports', in the cylinder wall to be uncovered by the piston near the end of the out-stroke (see Fig. 1.10(a)). One of these ports, the exhaust port, allows the burnt gas to escape, while the other, the inlet port, lets in the precompressed fresh charge. The replacement process is dependent on the fluid dynamics and rather chancy, because the two charges are not separated. Ideally, it would be better if on the out- or working stroke the exhaust port opened first, to allow much of the burnt gas to escape and its pressure to drop before the inlet port opened. Conversely, on the in- or compression stroke, the exhaust port ought to close before the inlet port, so that none of the fresh charge escapes. In short, the exhaust port should be first to open and first to close, but this is impossible to achieve with the conventional arrangement of Fig 1.10(a).

The Trojan two-stroke engine was designed in 1910. It had four horizontally-placed cylinders in two pairs. Figure 1.10(b) shows a section of the

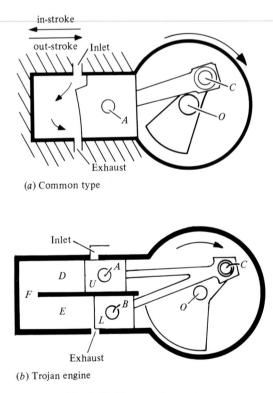

(a) Common type

(b) Trojan engine

Fig. 1.10. Two-stroke engines.

engine in which D and E are one such pair of cylinders: the two pistons U and L are connected by a single V-shaped connecting-rod ACB to the crank pin C, so that they go up and down roughly in step. The two cylinders are connected together at F to form a single combustion chamber, and they work almost as a single cylinder, having one inlet port and one exit port as shown. As seen in the figure, the crankshaft O rotates clockwise.

The figure shows the engine towards the end of the working stroke, with both pistons moving towards the right. Because of the inclination of the V-shaped connecting-rod, the lower piston is ahead and has already started to open the exhaust port, so that the burnt gases are beginning to escape. The inlet port in the upper cylinder has not yet opened. As the crank pin C crosses the centre line, both ports are open. On the return stroke, the inclination of the connecting-rod is reversed, so that the lower piston L is still ahead and closes the exhaust port before the upper piston U closes the inlet. Thus by locating the inlet and exhaust ports in two separate cylinders with the pistons slightly out-of-phase, the ideal phasing of ports has been obtained. Furthermore, the space through which the charges flow, DFE, forms a U-tube and helps to maintain separation.

One point remains to be explained. The distance between the axes of the cylinders D and E is fixed, but the inclination of the line AB varies as the crank turns. It follows that the distance AB varies. Now it would be possible to allow for this by making AC and BC two separate connecting-rods, but that would make the construction more complicated. Instead, since the variation in AB was fairly small, the single V-shaped connecting-rod ACB was arranged to bend sufficiently to accommodate the changes.

The Trojan engine was very simple, with only seven moving parts; because of its curious porting arrangements, it performed well over a wide speed range and it became a great favourite with many users.* It would almost certainly not prove competitive if revived today, because the world of engines has moved on in the meantime, but it has all those elements which make a design intellectually pleasing, apparent fortuity – at first sight there seems no sense in it, aptness – the same arrangement provides at once for phased ports and a division between the burnt and the fresh charges, and extreme simplicity, of form, if not of function, together with a piquant touch in the bending connecting-rod.

The peculiar excellence of the Trojan arrangement will naturally appeal most readily to the experienced designer, who knows how rarely an idea works out as neatly as this, but perhaps the lay reader too will recognise the ingenuity of this design.

Many other two-stroke engines have been designed since the Trojan, which have used the same principle but in a single cylinder and without the phased port opening. In them, a domed piston head has been used to obtain

* Setright, L. J. K. (1975). *Some Unusual Engines.* London: Institution of Mechanical Engineers.

something approaching the same U-shaped flow path. Just recently, there is renewed interest in this simple kind of engine, but with the improvement of our understanding of gas flow and the use of a computer, a more refined shape of piston can be designed. If the work is successful, the new engines may be much better than the Trojan in performance and yet much simpler in construction. This illustrates a common trend in engineering design, towards greater refinement and reduced complexity, made possible by greater understanding and ingenuity.

1.11 Evolution

Before Darwin, the design of living organisms was thought to be the direct work of God. Darwin and Wallace (1859) postulated that species evolved by a process of natural selection by survival of the fittest. Among the small variations between individuals of a species, some would favour survival, and on average the individuals possessed of these favourable variations would live to pass them on to more offspring. Thus the species would accumulate more and more of any favourable deviation, and, conversely, unfavourable variations would be bred out. In this way, the species would gradually adapt itself over many generations.

Three principal objections were raised to these ideas, all of which have since been removed. Two of these were that the earth was not old enough for there to have been time for the evolutionary processes, and that the intermediate forms, that according to the theory must have existed in the past, could not be found among the known fossils. The first objection fell when it was recognised that the earth was much older than had been assumed, and the second lost its force as many more fossils were found and many of the gaps in the record were filled.

The third objection was based on the contention that, in spite of natural selection, the favourable variations would not be strengthened but lost in interbreeding with individuals not possessing them. This 'blending inheritance' would keep the species nearly constant. Darwin himself doubted that his theory could be a complete account because of this effect. The work which was to end this last objection was published only a year or so after Darwin's major work, the *Origin of Species*, but its significance was not recognised until 1900. This was Mendel's work on inheritance among peas, which showed that it was not 'blending' but 'particulate' – that is to say, genetic differences were passed on in packets or 'genes' of irreducible size, and were not diluted by breeding but merely variously distributed among the members of later generations.

Since 1900, the genetic mechanism has been revealed more and more, culminating in the remarkable achievements of modern molecular biology. There can now be little doubt that natural selection is the principal mechanism of evolution. In this book that is the view that will be taken, although most of the time it will not affect the argument.

1.12 Design in nature

Any observer must be convinced of the excellence of design in nature. Unfortunately, so much of it is beyond our understanding at present that we are usually in the position of believing that the design is excellent, without being able to explain it fully. For instance, a cursory examination might suggest that the human knee is a simple pivot or hinge joint, like a door hinge. It is much more complex than that, however, and the partial explanation of its nature in Chapter 4 is only a beginning. Now, while it is fairly easy to establish that the knee is not a simple hinge, and just how it differs from a hinge, and even why some of the differences are advantageous, it is not at all easy to be sure that all its peculiarities are improvements over a plainer design. However, while there are cases where natural design does appear to be at fault, it is much more probable that it is our understanding that is lacking when anything seems to us unaccountable or wrong.

Design in nature is hedged in by limitations of the severest kind, chief amongst which are those associated with her method of production, i.e., reproduction and growth, which severely limits the kinds of materials that can be used, the sorts of mechanism possible (since the organism must be one connected whole) and the structures which can be developed: it may be speculated that it also sets limits of a much more subtle kind (see Chapter 9).

On the other hand, complexity in itself does not appear to be expensive in nature, and her prototype testing is conducted on such a lavish scale that every refinement can be tried. An engineer always has to balance manufacturing cost against performance, and so the components he designs have shapes which are easy to make, rather than the best for the task. An artificial replacement for a knee joint is always likely to be much simpler and more symmetrical than the organic parts it replaces, partly to make it easy to produce, and partly because of lack of understanding of the design problem.

In design by man, simplicity is generally to be accounted a virtue; for example, there is a strong aesthetic appeal in the Trojan engine because relatively complex requirements in terms of the phasing of port opening and closing have been met with few and simple parts. Nearly always, however, it is possible to solve a problem more completely and precisely by means of a more complicated design. The engineer, for whom complication means cost, a longer development time and greater problems in maintenance and repair, must strive to be simple and accept added complexity only reluctantly and when thoroughly convinced that the gains will exceed the losses. In nature, on the other hand, if such a constraint on complexity exists at all, it can do so only at a very high level.

Evolution has proceeded, broadly speaking, from simple life-forms to more complex ones, the higher displacing the lower. Particularly in the animal kingdom, the largest creatures have been the most advanced. Nevertheless, simpler designs are very numerous and have not been displaced, very often simply because they are small and at that scale the more advanced forms are not practical. On land, all large creatures have backbones and most have warm blood. In the sea, the invertebrates reach much larger sizes, paricularly the great squids, and even the largest fish, the sharks, are of a rather primitive kind. The biggest creatures in the sea, however, are the whales, which are among the most advanced designs in nature.

In the world of plants, the situation is much more clear-cut. On the land, the majority of plants belong to the highest kinds, the trees, the grasses and other flowering plants. In the sea, on the other hand, tiny single-celled plants drifting near the surface, the phytoplankton, are the predominant vegetation and the beginning of most of the food chains.

For all the difficulty of understanding the details of design in nature, we understand fairly well the most important aspects and the way they have determined the success of different types of plant and animal and their distribution over land and sea. Mysteries remain, notably the relatively sudden disappearance of the great reptiles, the dinosaurs, some 60 or 70 million years ago, perhaps as a result of a world-wide change to a colder climate. But most developments in evolution we understand, and particularly a sequence of major steps forward in design that might be called the great inventions of nature. Just as the invention of devices for converting chemical energy into mechanical work, the steam engine and later the internal combustion engines, gave momentum to the Industrial Revolution and made possible rapid transport, the developments of electrical engineering and flight, so there have been great innovations in living designs which have led on to remarkable changes which might not at first be seen to be related to them.

1.13 The great inventions of nature

Among these great innovations in design, the crucial inventions of nature, the earliest have left no trace of their development in the fossil record. The organisation of living material in a cell with a cell wall and a nucleus, the transmission of the blueprint of its design and its means of self-construction and the very important device of sexual reproduction, all developed in minute organisms which have left little evidence. The very remarkable and important invention of chlorophyll which is present in all green plants and enables them to convert the energy of sunlight into chemical energy belongs to the same early stages in the development of life. Before the advent of chlorophyll, life was probably in the position of industry today, dependent on chemical energy available in the surroundings.

Another big step forward was the development from single-celled organisms of multi-celled ones, without which most of the later innovations would have been impossible. It is not clear to the author why single-celled organisms could not have continued to increase in complexity and size; perhaps the reason lies in fundamental limitations of the way the genetic code works. It is more convenient in engineering and other human activities, when a system is sufficiently complex, to divide it into modules or sections or regiments or whatever, and perhaps there is some equivalent effect in nature.

Much later came the development in animals of tough outer casings which served the purposes of both skeleton and armour. This invention, on which are based insects of all kinds and the crustaceans, like shrimps, crabs and lobsters, is very economical of vital material and weight and has proved immensely successful, but it has one crucial limitation which apparently nature has been unable to overcome. These casings, or exoskeletons, cannot grow with their owners, and must be moulted with each increase in size.

Another important invention was to provide the way ahead; the internal skeleton, typified by the backbone, the distinguishing feature of that great class of animals, including the fish, the amphibians, the reptiles, the birds and the mammals, which we call 'vertebrates'. The earliest fishes, the first important group to have backbones, were often heavily armoured, but their supporting structure was an internal skeleton. The functions of protection and structural support were separated, just as Watt separated the cylinder and the condenser in his design of the steam engine, and this proved to be a great advance. Later most fishes lost their armour, because it was more of an encumbrance than a protection, just as much later the fighting man in the field shed most of his armour.

An important innovation nature made in the fish was its covering of scales, the remains of its armour; the scale demonstrates an important characteristic of design in nature, which is that generally every new thing must develop from some old thing. Thus eventually the scale was transmogrified into one of the inventions of nature which is most pleasing and admirable to the human designer, that elegant, subtle and astonishing structure, the feather, which will be discussed at length in a later chapter. In a similar way, the relatively undemanding design of bones in the fins of some fish evolved into the beautiful skeletal mechanisms in the limbs of horses, tigers and men.

Another important invention was the ability to keep the body temperature constant. It is possible that the great dinosaurs had developed this power, perhaps rather imperfectly. It shows another tendency in design, both living and human, that of controlling, rather than being controlled by circumstances.

Perhaps the most remarkable invention in plant design after chlorophyll was the use of specially adapted flowers to enlist the services of animals, usually insects, in the process of sexual reproduction. Flowers are the sexual organs of the higher land plants, and many of them rely on wind to carry the male pollen to the female stigma, a wasteful and chancy process. But by providing insects with food, other flowers achieve cross-fertilization by their agency with much less waste, particularly when, as is more often the case, the insects visit only one kind of flower at one time.

These are just a very few of the more striking and important of the many inventions of nature. Many others are discussed in this book, along with human designs using similar principles.

1.14   A limitation of natural evolution as a means of design

One of the inventions of nature was flight, first achieved in the insects. Now it might be speculated that birds evolved from insects, but such was not the case. The bird, or even the pterodactyl, requires certain advances to which the insect never progressed, particularly the internal skeleton. It also needs better systems for breathing air and distributing oxygen than the insects ever achieved.

The insects were already committed to a different and very successful design philosophy, and one which could not respond by moving in the required directions through an intermediate series of successful forms. And this is a great limitation of evolution – it cannot produce a radically new design except via a long succession of small changes, and every one of these intermediate designs must itself be an improvement that can survive to give rise to the next. No such route from the insect to the bird was possible. Although insects were flying through the great forests that were to become today's coal measures, some 300 million years ago, it was from the land reptiles that the birds would develop almost 200 million years later. This was because the invention of flight was not essentially so important as those other inventions. Similarly, man achieved flight at the beginning of this century by the use of the internal combustion engine, whereas he has only just succeeded, a lifetime later, in flying by the power of his own muscles. The effective bar to man achieving flight earlier was the lack of a sufficiently powerful engine, rather than any difficulty about the flying itself. Indeed, problems centred around energy and power, which is just the rate of making energy available, are central in much design, as will be seen in the next chapter.

2

Energy

2.1 Energy

It is often helpful to think of the designer as working in three different kinds of medium – materials, energy and information. Materials are the most readily appreciated, but it is convenient to study energy first because it introduces ideas which are essential to the consideration of materials: information is left until Chapter 6.

An aircraft can fly the Atlantic because it can store a great deal of energy in chemical form in fuel and convert that energy reasonably efficiently into other forms as required. An aircraft or a bird can take off and fly because of its ability to convert or release energy at a high rate – the rate of conversion or release of energy is called power.

The ability of a bird or a mammal to survive in cold weather depends on its ability to keep down its loss of heat, which is a form of energy, and the protection that a vehicle offers its occupants in a collision depends largely on the capacity of its structure to absorb the destructive energy of motion (kinetic energy). All life depends on energy: plant life draws its energy from the rays of the sun, and animals in turn obtain energy by eating plants or other animals. At another extreme, a clothes-peg or a nut-and-bolt depend for their functioning on their power of storing a little energy, as we shall see.

2.2 Mechanics

To understand the design of living things or machines, it is necessary to grasp certain key ideas in the science of mechanics, of which one is energy, and the others are force, work and power. The first part of this chapter is largely devoted to explaining some of these concepts, which are well worth grasping for their use in fields other than design and in books other than this one.

2.2.1 Forces

If we push against a stationary object, then we say we exert a *force* on it. We can be more precise, and state the size or magnitude of the force in some unit – in this book the *newton*, the unit of force in the SI

(International System) will be used. Notice also that a force has a direction – the direction in which we are pushing.

If a table is supporting an apple, then the apple exerts a downward force on the table, the *weight* of the apple, in fact. The word 'weight' is used in two different senses, which is a little confusing: one is the *mass* or amount of matter – if we go to buy two pounds of potatoes or a kilogram of sugar, it is a mass we are referring to and in the past the Inspector of Weights and Measures was employed to help see that we received the right *mass* of potatoes or sugar. The other is the force exerted on a mass by gravity, and it so happens that the weight (in this force sense) of an average apple is about one newton, which is rather a happy accident if you remember the tale of the young Isaac Newton being started on the train of thought which led to the concept of gravity by an apple falling on his head. To avoid confusion, engineers and scientists use the word 'weight' only in the force sense, and use the word 'mass' where a quantity of matter is meant.

Another important idea is that of a *reaction* – if the apple pushes downwards on the table with a force of one newton, then the tables pushes back with an *equal but opposite* upward force of one newton, called a *reaction*.

If an object is not accelerating, i.e., it is not changing velocity, it is said to be in *equilibrium*, and for it to be in equilibrium, all the forces on it must balance. For example, the apple on the table is not changing its velocity, which is zero, and it is subject to two forces – its own weight, the force exerted on it by the attraction of the earth's mass, and the reaction from the table. The magnitudes of these forces are equal, but the gravity force acts downwards and the reaction upwards, so that they exactly balance one another out (see Fig. 2.1). If we take away the table so that there is no reaction, the gravity force on the apple is no longer balanced and this unbalanced force causes it to accelerate downwards – to fall, in fact.

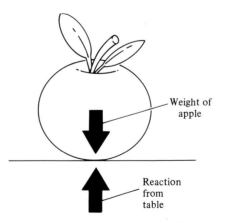

Weight of apple

Reaction from table

Fig. 2.1. A force and its reaction in equilibrium.

This balancing out of forces holds in all cases where there is no acceleration, even in cases where motion at a steady speed is taking place. Figure
2.2 shows a man pushing a laden sledge with a horizontal force P. The level
ice resists any motion of the sledge by a horizontal force F, the frictional
resistance. If the sledge is in equilibrium, i.e., unless it is accelerating, P and
F must always be exactly equal in magnitude and opposite in direction, so
they balance one another. In practice, when the sledge is moving, F will be
nearly the same whatever the velocity. Let us say, then, that F for the sledge
in motion is 120 newtons, or 120 N, as it is written for short. Then if the
man pushes with a force P of 80 N, there will be no motion, and the force
F will be 80 N to balance P – in other words, while the sledge is stationary,
only such frictional resistance as is necessary to balance P is called into play.

When the sledge is in steady motion, the resistance F and the push P are
still equal, and since F is then 120 N, so is P. But, of course, the man could
push harder, so that P was 130 N, say, and then, since F would remain at
120 N, there would no longer be equilibrium, and the sledge would accelerate under the effect of the unbalanced 10 N.

2.2.2 Work and energy

In the engineering sense, work is done when the point of application of a
force moves in the direction of that force. This definition may sound
obscure, but an example will clarify the concept. If the man pushes the
sledge along, he does work on it, because he applies the force P and produces
movement in the direction of that force. The amount of work is the product
of the force P and the distance moved in the direction of P. However, if he
does not push hard enough to move the sledge, then he does no work, even
though he keeps it up for hours and exhausts himself in the process.

To do work requires the expenditure of *energy*, for energy is simply the
capacity to do work. It follows that work and energy are measured in the
same units, since one unit of energy is simply the capacity to do one unit
of work.

In the International System of units work, being force times distance, is
measured in newtons times metres, or newton metres, written Nm for
short, and these are also the units for energy. However, work and energy
are such important concepts and are used so frequently that the newton
metre is given a special name, the joule, after Joule, who established the

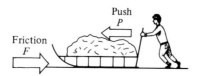

Fig. 2.2. Force and reaction with motion.

relationship between mechanical work and heat. Thus if the man pushes the
sledge 10 kilometres (10 000 metres) and the force P is 120 N, then the
work he does is

120 N×10 000 m or 1.2 million joules (1.2 MJ).

Notice that it is convenient to use the abbreviation J for joules, and in this
case to combine it with M for mega, a megajoule, MJ, being a million joules.
   Now energy cannot be conjured from nowhere. If we require to perform
a task involving, say, 1.2 MJ of work, then we must first assure ourselves
of the necessary 1.2 MJ of energy to perform it. In the case of a man pushing
a sledge, the energy comes from substances in his body, fats and carbohydrates, extracted from his food. In the case of an aircraft, the work it does
overcoming the resistance of the air to its flight is provided by energy stored
in fuel which is chemically very similar to the fats nature uses as energy
stores. Indeed, the fuel comes from the fossil fats of living organisms which
died hundreds of millions of years ago, and which in turn derived the energy
they stored from the sun's radiation.

2.3   The forms of energy

Energy, the capacity to do work, is possessed by many bodies and can take
many different forms. There is kinetic energy, which is possessed by a mass
in motion such as a bullet in flight, or a spinning flywheel, or a running
man. A man running at 15 miles per hour, or about 7 metres per second,
has kinetic energy of about 1500 J, which is about the same as that of a rifle
bullet. There is thermal energy, or heat: to bring to the boil a kettle of
water (about one litre) requires the addition of about 200 times as much
energy as must be added to the runner to bring him to 7 m/s or to the rifle
bullet to bring it to 600 m/s. There is chemical energy, which can be even
more concentrated. One litre of petrol or one kilogram of pure fat stores
enough chemical energy to bring over 100 kettles to the boil, or to set
20 000 rifle bullets or mile runners in motion, about 40 million joules
(40 MJ). To be precise, these large quantities of energy are contained in 1 kg
of fat, together with sufficient oxygen to burn it in completely, a total of
about 4.4 kg, but usually the designer does not need to worry about carrying
the oxygen in his design because it is available in the air. There is nuclear
energy, which is much more concentrated still and is the source of the sun's
radiation upon which all life depends. One kilogram of hydrogen undergoing a fusion reaction (in which two atoms of hydrogen combine to form
one atom of helium) liberates about 24 million million joules, or almost a
million times as much as burning fat.
   However, there are some relatively very dilute forms of energy which
are of great importance in design. Clock springs and elastic such as is used

to drive model aircraft store *strain* energy, so-called because the material used is strained – bent in the spring, stretched in the elastic strands. Clothes-pegs, paper clips, elastic bands, press-stud fasteners and other fixing devices, even bolts and nuts, work because of their ability to store strain energy.

Some clocks are driven by weights which fall, turning a drum: clearly, a raised weight stores energy. In a hydroelectric scheme, water trapped behind a high dam stores energy in this form, which is called 'potential energy'. The water is used to drive a turbine which in turn drives an electrical generator (Fig. 2.3).

With one exception, any form of energy can be transformed into any other form, ideally, on a unit per unit basis. For example, in the hydroelectric station, the potential energy of the water is changed into kinetic energy in the turbine nozzles and finally into electrical energy. The one exception is heat, which can only be partly transformed, as will be seen later.

2.4 Conservation of energy

In all the cases we shall consider in this book, energy is conserved, that is to say, in all the transformations of energy from one form to another, the total quantity of energy remains unchanged. This is called 'the law of conservation of energy', and also 'the first law of thermodynamics'. Suppose that just over a kilogram of water, having a weight of 10 N, leaves the lake and falls through a total height of 100 m via the turbine. The water gives up potential energy equal to its lost capacity to do work, which is simply

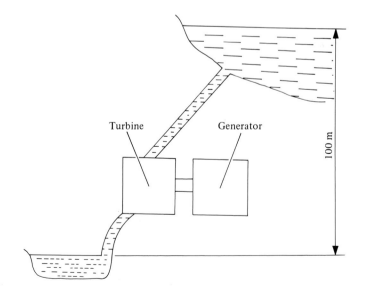

Fig. 2.3. *Hydroelectric scheme.*

its weight times the distance fallen, that is, 10 N×100 m=1000 Nm or
1000 J. Now this 'lost' energy cannot really be lost, by the law just enunci-
ated. About 850 J will be delivered by the generator as electrical energy,
while the other 150 J appears as heat, in the water itself and in the electrical
windings and armature. The 150 J of heat, although not 'lost' in the absolute
sense, are lost for all practical purposes, since only a small proportion can
ever be converted back to any other form. Of the 150 J, 100 might be energy
converted into the heat form in the turbine, which would produce a rise of
two- or three-hundredths of a degree Centigrade in the temperature of the
water as it passed through, and would be difficult to detect, let alone recover.
The other 50 would produce a much larger rise in temperature of the air
being passed through the electrical generator to keep it cool, and it would
be possible to use this warm air to drive a steam engine; this possibility will
be examined later.

2.4.1   The pendulum

Consider a pendulum swinging to and fro. When it comes to the end of a
swing (position A, Fig. 2.4) it slows down and stops momentarily before
starting to swing back. At that instant it has no velocity, and hence no
kinetic energy. In the middle of the swing (position B, Fig. 2.4) it is moving
fairly fast, and so has kinetic energy. Where has this energy come from?
The answer is, from the potential energy lost when the centre of the bob
(which is assumed to represent virtually all the mass of the pendulum) falls
from A to B. As the swing continues from B to C, the kinetic energy is
converted to potential energy again, until at C it has all gone, only to
reappear as the bob swings back again towards B. Thus by the time position

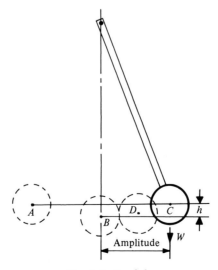

Fig. 2.4. Pendulum.

A is reached again, the little bit of potential energy represented by the difference in height *h* of *A* and *B* times the weight of the bob has been converted to kinetic energy and back again *twice*. At a position such as *D*, part of this little bit of energy is in potential form and part is kinetic, but with a frictionless pendulum the sum of the two is the same in all positions.

Suppose we took a film of the pendulum. It would be impossible to tell whether the film was being run forwards or backwards, i.e., which frame came before or after another; the film would look the same whichever way it was run. A phenomenon, like the swinging of a pendulum, which provides no means of telling the *direction of time* on a film of it, is called *reversible*.

However, real pendulums have a little friction, from the air and at their point of suspension, and the swings gradually reduce, so that with a long enough piece of film it would be possible to tell which way it should run. The friction, which is an *irreversible* process, indicates the direction of time.

2.4.2 A bouncing ball

Suppose we drop a rubber ball on a hard surface; if it is a suitable kind of rubber (called 'low-hysteresis' by the engineer) it will bounce back almost to the height from which it was dropped – the bounce is nearly reversible. In one bounce there are four transformations of energy. Firstly, the ball falls, converting potential energy into kinetic energy. On hitting the ground, the ball stops and the kinetic energy disappears, being converted very rapidly into strain energy in the material of the ball, which is for an instant stationary but squashed. The squashed ball immediately springs back to its original shape, converting its strain energy back into kinetic energy. The ball rebounds, travelling almost as fast as when it hit the ground. Finally, the gravitational force on the rising ball slows it down until all its kinetic energy has been turned to potential energy, and it stops momentarily before starting to fall again.

In each of these four transformations a small fraction of the energy is 'lost' – or more exactly, is dissipated into some other form. For example, when the ball is falling, because of air-resistance, some of the potential energy is transformed, not into kinetic energy of the ball, but into heat. When the ball is squashed on impact, a little of the kinetic energy becomes heat, and so on. In general, 'lost' energy usually turns up as heat.

Before leaving the ball and the pendulum, note that the frequency of the pendulum, that is, the rate of swinging, is constant whether the swings are small or large, but the bouncing of the ball becomes much quicker as the height of bounce decreases.

2.4.3 The convertibility of heat

In most transformations of energy, part becomes heat. Only part of the heat can be transformed back into other forms, so such transformations are not completely reversible. The convertibility of heat depends on the *thermo-*

dynamic temperature at which it is. For the purposes of this book, thermo-dynamic temperature (or absolute temperature, which is practically the same thing) is just:

centigrade temperature+273 degrees.

For example, the thermodynamic temperature corresponding to 27 °C is

27+273=300 K

where K stands for kelvins, which are the same size as degrees centigrade but start from a different zero, the absolute or thermodynamic zero.

If T is the thermodynamic temperature at which some heat energy is available, and T_0 is the thermodynamic temperature of the surroundings, then T_0/T is the fraction of the heat which cannot be converted to any other form of energy. This fact can be used to examine the practicality of con-verting some of the heat developed in the generator of the hydroelectric scheme into further electrical energy by means of a steam engine or a steam turbine. Air (say) would be circulated through the generator to cool it, and the warm air would then be used to heat the boiler, providing steam to a steam turbine (a steam turbine would be more efficient and more easily coupled to the generator than a reciprocating steam engine with cylinders and pistons). The surroundings of the hydroelectric station might be at 15 °C, say, so that T_0 would be 15+273=288 K, and we would probably not want to run the generator hotter than about 65 °C or thermodynamic temperature 338 K, so the fraction of heat absolutely irrecoverable would be

$$\frac{T_0}{T} = \frac{\text{ambient thermodynamic temperature}}{\text{supply thermodynamic temperature}}$$

$$= \frac{288}{338} = 0.85,$$

leaving only 0.15 recoverable. Remembering that only 50 of every 100 J extracted from the water became heat in the generator, the maximum recov-erable would be 0.15×50 or 7.5 J in every 1000. In practice the yield would be even smaller, less than one-half as much. One reason is that the temper-ature in the boiler would have to be rather lower than the temperature of the cooling air, and the condenser would have to be a little warmer than the surroundings, reducing the convertible fraction of energy even further.

If the fraction of heat generated in the first place had been larger, and if the temperature at which it was available had been higher, then the scheme of adding a steam turbine would have been much more attractive. Both these criteria are met by the exhaust heat of a diesel engine, and 'steam' turbines with heat supplied by the exhaust have been used with diesel engines to increase the power output for a given fuel consumption: the 'steam' is not steam, however, but another substance, of the kind used in refrigerators and aerosols.

2.5   The Second Law of Thermodynamics

Suppose we have some heat available at a thermodynamic temperature T of 600 K and the thermodynamic temperature T_0 of the surroundings is 300 K. Then the unconvertible fraction is

$$\frac{300}{600} = \frac{1}{2}.$$

It may occur to the reader to ask why we should not convert half the heat to some other form, and then half the remainder, and so on, which would clearly make nonsense of the idea of limited convertibility. The answer is that to convert the convertible fraction, we have to deliver up the unconvertible residue to the surroundings, or reject it, as the usual term is. It is then at T_0, so it is altogether unconvertible.

Imagine a red-hot horseshoe left cooling in a blacksmith's forge. To begin with, it is very hot, and the heat in it could be partially converted to other forms of energy. However, when it has grown cold, that possibility of conversion to other forms has disappeared from the universe for ever. The heat energy itself is still there in the surroundings but in the absence of any temperature difference it cannot be changed to any other form. There has been no loss in energy, but there has been a loss of convertible or *available* energy.

In every irreversible process (and all processes in the real world above the atomic scale are irreversible to some extent) some available energy is lost from the universe. Eventually, so far as we can tell, all the available energy will be gone, all energy will be heat and the temperature of the universe will be the same everywhere. There will be no life, no motion even above the atomic scale – the universe will have run down.* Fortunately, this 'heat-death of the universe' is in the unimaginably remote future, so we can reasonably postpone worrying about it for a few thousand million years, and concentrate on the earth's, or rather man's, own energy crisis, which is really an available energy crisis.

The limited convertibility of heat is one of a number of aspects of what is generally called the Second Law of Thermodynamics (of which C. P. Snow said no man could be called educated who did not understand it). This law may be stated in the delightfully simple form

heat cannot of itself flow from a colder to a hotter body

and leads to many important but often remote-seeming consequences.

* Thomson, S. P. (1910). *The Life of William Thomson, Baron Kelvin of Largs*, vol. 1, p. 290. London: Macmillan.

A bucket of water from the sea contains sufficient heat energy to drive a supertanker about two metres, if it could only be converted: the Second Law explains why it cannot, and why it is impossible to make what the engineer calls a 'perpetual motion machine of the second kind', a machine able to do work continuously at the expense only of heat drawn from its surroundings. If it were not so, we should have no need of fuel and no energy crisis, plants would not need sunlight and animals would not need most of the food they eat. A supertanker would drive itself by extracting heat energy from the sea, and a tree would grow by cooling the air around it.

Living organisms and machines need *available* energy. Inanimate nature squanders available energy with great profligacy. The radiation from the sun has a temperature of several thousand degrees, so that on earth over 90% of it is available energy, convertible to other forms. It is common to describe such energy as 'high-grade'. However, most of it is immediately reduced to ambient temperature and the available energy is lost, like that of the cooling horseshoe. A very little is converted to forms other than heat, in the kinetic energy of the winds, the kinetic and potential energy of waves, and the potential energy of the rain that supplies the hydroelectric reservoir. Some is intercepted by plants, which use a small proportion of it to synthesise energy-storing foods, thus saving some of the available energy. In striking contrast to inanimate nature, both living organisms and human engineering are characterised by design for the reduction of irreversibilities and the husbanding of available energy. On the whole, man is rather more successful than living organisms, which are handicapped by their inability to use high temperatures and have to use very subtle and elaborate chains of chemical reactions instead.

2.6 Power

Power is simply the rate of doing work, i.e., the amount of work done in unit time. Thus, if the man pushing the sledge with a force P of 120 N is moving at 1.5 m/s the power he is providing is

$$120 \, N \times 1.5 \, m/s = 180 \, N \, m/s = 180 \, J/s$$

Since the work done is force × distance, the work done per unit time is force × distance moved per unit time. But distance moved per unit time is velocity, so that

power = force × velocity.

The unit of power is the joule per second, which has a special and familiar name: it is called the watt, after James Watt, who, besides improving the steam engine in many ways, chiefly by his invention of the separate condenser, first put the ideas of work and power on a satisfactory basis and originated that earlier unit of power, the horsepower. The symbol for watt is W, so we write the power delivered by the man pushing the sledge at 180 W. A kilowatt is 1000 W and is written kW. One horsepower is 746 W. Many modern power stations develop about 2000 MW, or 2 million kW.

These concepts of force, work, energy and power may take some time to grasp, but are relatively simple and of vital importance to an understanding of design. For example, it is largely considerations of energy and power that dictate the form of all things that fly, run, walk or travel on or under the water.

2.6.1 Energy, power and flight: birds

In winged flight, the forward motion of the wings through the air provides a lift force L to support the animal or aircraft. In steady, level flight the lift L exactly balances the weight W as shown in Fig. 2.5, just as the weight of the apple is balanced by the reaction of the table or the weight of the sledge is balanced by the reaction from the ice. The forward motion which generates the lift is resisted by the air, which exerts a drag force D (corresponding to the friction force F in the case of the sledge) on the aircraft or bird, which is overcome by a forward thrust T produced by jets, propellers or wing-flapping as the case may be. In level flight without acceleration,

$$T = D \text{ and } L = W$$

Now because the forward motion generates both the lift and the drag, the two are related. Since lift is necessary but drag is undesirable, the

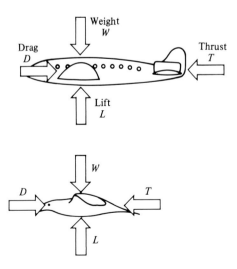

Fig. 2.5. Balance (equilibrium) of forces on a bird or aircraft in level flight. L=W, T=D.

designer will try to increase the one while reducing the other, and his success in this respect may be measured by the ratio

$$\frac{\text{lift}}{\text{drag}} (L/D \text{ ratio})$$

which he achieves.

The greater the lift/drag ratio, the more successful the designer has been and the less power will be required for flight.

The range of lift/drag ratios in birds is not quite as good as that achieved by aircraft – say, 6–24 as against 10–40. This is partly explained by the more severe constraints on the design of birds which may have to fly among and land upon branches and build nests as well as finding food and laying eggs – extra requirements which would tax an aircraft designer exceedingly. However, there is a wide range of moderately aerodynamic birds and commercial airliners which have comparable lift/drag ratios of around 15. Let us look at the power requirements and range of flight of such typical birds and aircraft.

First let us consider a bird of mass 0.3 kg (rather more than a magpie and less than a partridge) and hence a weight W of about 10×0.3 N, or 3 N. The lift L in steady, level flight will be 3 N, and the drag will be about

$$\frac{L}{15} = \frac{3\,\text{N}}{15} = 0.2\,\text{N}.$$

If the speed of flight is 15 m/s, then the power required is

work per unit time=force×velocity
$$=0.2\,\text{N} \times 15\,\text{m/s} = 3\,\text{N m/s or 3 W}$$

(that is, about one two-hundredth of the power taken by an electric iron).

Let us suppose this bird flies 3000 km, which is about the farthest any bird is known to migrate without food, i.e., without refuelling. The work done is the drag times the distance, and since 1000 km is a million metres, this is

0.2 N×3 million metres=0.6 million joules or 0.6 MJ.

Now this implies that the bird needs to store 0.6 MJ of energy to be able to undertake such a flight, even if it can convert that energy completely into useful work. In practice, the bird's fuel is fat, which stores about 40 MJ per kg, but it can only convert about 7.5% of that, or, say, 3 MJ per kg, into useful work. It follows that the fat stored for the flight needs to be about

$$\frac{0.6}{3}\text{kg or 0.2 kg, roughly}.$$

Now in this rough calculation we have assumed a constant weight, but in fact the bird will decrease in mass and so in weight, and to have an *average* mass of 0.3 kg it would start at about 0.4 kg and arrive at about 0.2 kg. Thus the bird would start with a body weight about half-fat, and arrive with very little fat. (Some readers will recognise that it is not the arithmetic mean of the initial and final mass which counts, but the exact solution involves integration and natural logarithms, and makes hardly any difference anyway: in the related case of rocket flight, the ratio of initial to 'all-burnt' mass is much greater and so is the difference between the arithmetic and logarithmic means.)*

Since the drag depends only on the weight and the aerodynamic design of the bird, and since work is the product of force and distance, the *speed* of flight does not greatly affect the fuel required for a given range. However, it does affect the power required which is equal to drag times speed, and in our example was 3 W at 15 m/s. Now it requires about 1 kg of bird muscle to produce about 50 W of useful flying power, so 3 W requires about 3/50 kg=0.06 kg, and of our bird's net weight of 0.2 kg about 30% needs to be flying muscle, leaving 70% for skeleton, other muscles, head, digestive system, lungs, blood system, feet and feathers. All this shows what experience would suggest, that flight is a very difficult thing, just within the limits of design with living materials and muscles. Indeed, an amateurish design of bird which was not of the utmost economy in bone structure, muscle application, aerodynamic form, internal aerodynamic design of the air passages and many other aspects, would not be able to take off.

Because of scale effects, flight is less difficult for very small creatures, which is why although birds must always be designed primarily as flying machines, with most of their body mass devoted to the one purpose, in insects flight often appears to be almost an optional extra, with wings added to a beetle, say, which would look a satisfactory entity without them. One manifestation of the difficulty of the power requirements for flight is that true sustained hovering, which requires about twice as much power per unit mass as forward flight, is limited in the animal kingdom to creatures with masses of about 20 g or less.

2.6.2   Energy, power and flight: aircraft

Aircraft have one severe disadvantage compared to birds, that of large size, but they have three great advantages; firstly, the relatively small range of functions demanded, secondly, the much higher power/mass ratio of aero-engines compared with muscles and, thirdly, the more complete conversion of stored chemical energy into useful work which they can achieve. It is the second of these advantages which explains why helicopters which are capable of true hovering can be designed with masses up to several million times that of the largest hovering bird.

---

* Pennycuick, C. (1972). *Animal Flight*. London: Arnold.

Consider a large passenger aircraft, with a lift/drag ratio of 15 and an all-up mass of 150 metric tons, or 150 000 kg. Its weight (and hence the lift) will be about 150 000×10 N=1.5 million N (1.5 MN). Then the drag will be 1/15 of its weight or 100 000 N (100 kN), so that to fly 3000 km (about 1600 nautical miles) will require energy equal to drag×distance

$$= 100\,kN \times 3000\,km = 300\,000\,MJ.$$

Now an aeroengine, such as a modern turbofan, can convert about one-third of the stored energy in its fuel to useful work, or about 1/3×40 MJ per kg, so the mass required is roughly

$$\frac{300\,000\,MJ}{\frac{1}{3} \times 40\,MJ\,/\,kg} = 22.5\,tons,$$

or about 15% of the all-up mass. In practice, a large reserve must be provided for safety.

The power required is force×velocity, or if the latter is 250 m/s (about 490 nautical miles per hour)

$$100\,kN \times 250\,m\,/\,s = 25,000\,kW.$$

Since aeroengines may have a power/mass ratio *at cruise* of 6 kW/kg, the engine mass would be about 4 tons, about 3% of all-up mass.

2.6.3 Efficiency

It was remarked that bird flying-muscle and wings can convert only about 7.5% of the chemical energy of fat into useful propulsive work. In engineering terms, we say the efficiency is 7.5% or 0.075.

The *efficiency* of an energy conversion device is the ratio of the useful energy out to the energy in. Indeed, more generally,

$$\text{efficiency} = \frac{useful \text{ output}}{\text{input}}$$

Because of the conservation of energy, the *total* energy output of any device must be equal to the total input. Thus in the case of muscle, it is the mechanical work, the force×distance effect it exerts in shortening, which is the useful output, and this is about 0.2 of the chemical energy input, i.e., the efficiency of muscle is about 0.2, or 20%: the rest mainly becomes heat, which is why exercise warms us up. We write,

$$\text{muscle efficiency} = \frac{\text{muscle work}}{\text{chemical energy input}}$$

Further losses occur in converting the shortening work of the muscle into propulsive work, which is the propulsive force T×distance flown, i.e.,

$$\text{propulsive efficiency} = \frac{\text{useful propulsive work}}{\text{muscle work}}$$

Then, overall efficiency $= \dfrac{\text{useful propulsive work}}{\text{chemical energy input}}$

$$= \frac{\text{useful propulsive work}}{\text{muscle work}}$$

$$\times \frac{\text{muscle work}}{\text{chemical energy input}}$$

$$= \text{propulsive efficiency} \times \text{muscle efficiency}$$

If the propulsive efficiency is 40%, or 0.4, then the overall efficiency is 0.4×0.2=0.08, or 8%. Flapping wings appear to have poor propulsive efficiency compared with propellers or jets, and muscles are less efficient than piston engines or gas turbines.

2.7   Fish and ships: scale effects

Fish, ships and submarines are supported by their buoyancy, and so although they are subject to drag opposing their motion, this drag is generally not related to a lift force and so not directly related to their mass. While the drag on a bird or an aircraft is largely independent of its speed, however, the drag on a ship or a fish rises very rapidly as it goes faster. As we saw, the energy required to fly a given distance is roughly the same whether flight is slow or fast, but the case is quite difference with swimming or sailing.

The drag on a swimming or floating body is roughly proportional to the square of its speed, e.g., if a fish goes twice as fast, the drag it has to overcome goes up 2×2=4 times. Moreover, since power=force×speed, the power goes up 4×2=8 times, i.e., as the *cube* of the speed, whereas the power required for flight goes up directly as the speed.

Another important difference between swimming and flying lies in the variation of power requirements with size. This is the first example we have met in this book of *scale effects*, a very important category of constraints on design.

Suppose we make a fish twice as big in every one of its linear dimensions – an exact scale model, if you like, except that the model is bigger than the original. We make the big fish just twice as long, twice as thick through, with just twice the distance from the centre of its eye to the corner of its jaw, and so on, so that every length whatsoever is exactly twice that on the little fish. Now the surface area of the big fish will be exactly 2×2= 4 times that of the little fish. Suppose we consider a fish-scale on the side of the little fish which is conveniently square and 1 cm by 1 cm, so that its area is 1 cm². On the big fish this scale will be 2 cm by 2 cm, so that its area will be 4 cm², or 4 times as large, as will be every area we care to choose. It can be seen that the total surface area of the big fish will be just 4 times as large as that of the little fish.

We can study volume change in the same way. To every piece of the little fish there corresponds a piece of the big fish which is just twice as long, twice as deep and twice as thick, and so has eight times the volume.

If we do not restrict the scale factor to the value of 2, we can say,

> area is proportional to the square of the linear scale factor
> volume is proportional to the cube of the linear scale factor

or, more simply,

> area is proportional to length squared
> volume is proportional to length cubed.

These ideas about scaling are very important throughout the designed world, as the example of the two fish will illustrate.

The drag on a fish is chiefly due to fluid friction, which is roughly proportional to its surface area, so that if the big fish is swimming at the same speed as the little fish, it requires four times as much power to do so. But as it has eight times the volume of the little fish, and so eight times as much muscle, it might be expected to have eight times as much power, i.e., twice as much as it needs to swim at the same speed. Now as we have just seen that power is proportional to the speed cubed for the same size body, this means that the big fish can swim $2^{1/3} \approx 1.26$ times as fast as the little one. Thus little fish, which are pursued and eaten by larger fish, are severely handicapped by the scale factor: other things being equal, a predatory fish which is eight times as long as its prey may be able to swim $8^{1/3} = 2$ times as fast, which leaves dodging as the only hope of escape for the prey. More thorough analyses give a bigger advantage still to the large fish.

The case of ships is much more complicated, partly because they are subject to another form of drag, not directly related to area, because of the waves they make. Also, it is not speed itself which is of interest, but the economic conveying of cargo. The costs of transport by sea fall into two groups, which we may call A-costs, which are reduced by increasing speed, and B-costs, which are increased by increasing speed. A-costs include most of the cost of the ship itself and the wages of the crew: a faster ship means more cargo carried per year, so the A-costs per ton kilometre carried are inversely proportional to speed. B-costs include the cost of the engines and the fuel: since power is proportional to speed cubed, both the capital cost of the engines and the cost of fuel burnt per year are proportional to speed cubed. However, as the ton kilometres carried per year are proportional to the speed, the net effect is that B-costs per ton kilometre are proportional to speed squared.

Thus as we increase the designed speed of our ship, A-costs fall and B-costs rise. Very fast ships are uneconomical because of their high B-costs: very slow ships are uneconomical because of their high A-costs (see Fig.

2.6). Somewhere in between is an economical design speed, which occurs when the A-costs are just twice the B-costs, at the speed corresponding to T in Fig. 2.6. (Those readers who are familiar with the differential calculus will know how to prove this by writing the total cost as

$$C = \frac{a}{x} + bx^2$$

where a and b are constants, x is the speed and the first and second terms of C represent the A- and B-costs, respectively: by putting dC/dx=0, the minimum of C is found).

It is possible to study the effect of size on the unit costs of ships* using the same arguments as in the study of size in fish, and it will be found that larger ships are more economical basically because they require relatively less power (there are, of course, other reasons, such as more cargo for the same number of crew).

2.8   Peak power requirements

In the designs of nature and man, problems arise over peak power requirements. Naval ships and fish may need to expend several times their cruising power in high-speed dashes, and as this cannot be done efficiently with one set of engines or muscles, two are commonly provided, cruise and main engines in the naval vessel, red and white muscles in the fish. When cruising, the main engines in the ship and the white muscles in the fish are inactive.

Birds also use the red and white muscle system to cope with the high-power demand at take-off. Red muscle owes its colour to the presence of a

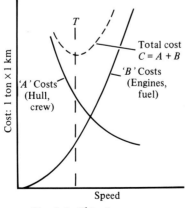

Fig. 2.6. Shipping costs.

* Rolt, L. T. C. (1970). *Victorian Engineering*, pp. 85–6. London: Penguin.

protein called myoglobin which takes part in the process by which fat is combined with oxygen to liberate energy, in the form of about 20–25% useful work and 80–75% heat. Thus red muscle is adapted to use the concentrated energy store, fat, which requires a corresponding supply of oxygen, via the bloodstream and from the respiratory system. White muscle uses a much less concentrated store, glycogen, which has the advantage that energy can be derived from it in the absence of oxygen. White muscle is capable of high-power outputs for a short time, after which its supply of energy has to be renewed by relatively slow chemical processes and the transport of materials through the tissues. Red muscle can use fat within itself, and so is suited to prolonged effort.

Thus chicken breasts are predominantly white meat, and the legs dark meat, because this bird is descended from species which were on their feet all day, and flew only in short, swift bursts. Rabbit meat is pale, because the animal feeds near its burrow for which it springs at any sign of danger, but hare meat is dark, because the hare must be able to run for a considerable time to escape a pursuer.

2.8.1   Oxygen debt

The chemistry of work production in muscles is very complicated and not fully understood, but is based broadly on processes using oxygen. However, some of the stages can proceed without oxygen but produce only a fraction of the energy of the whole chain of reactions. The materials that can be used in this 'anaerobic', or oxygenless, production of work are limited and soon exhausted, after which the rate of working of the muscular system is reduced to that set by the oxygen supply. Thus an athlete may be capable of producing a short burst of energy at up to about 1 kW but that soon fades as his anaerobic fuel supply is exhausted – within a minute the level drops below 700 W, and in two minutes the burst is virtually over, leaving only a steady rate of about 400 W based on oxygen-using processes. The burst cannot be repeated until the system has had time to renew its anaerobic fuel reserve, the exhaustion of which is called an 'oxygen debt', since extra oxygen is needed to restore these materials. In the design of an animal, the nature and amount of muscle will determine the power that can be put out in this anaerobic burst, but the steady power requires that not only the muscles, but also the lungs, heart and blood system, be of a size to support it.

Since most design in nature is aimed at minimising mass for a given performance, the incurring of an oxygen debt at maximum power levels is a rational principle, enabling economies to be made in the capacity of the respiratory and circulatory systems. In engineering design, similar situations and solutions arise less often, but with increasing frequency as we advance.

2.8.2 Peak power in aircraft

Several ways of providing high-power for take-off have been used in aircraft. Because the designers have succeeded in producing jet engines of great flexibility in performance and low weight which are capable of maintaining high efficiencies at part load, the common solution in civil aircraft is to fit powerful engines and throttle back in normal flight.

Rocket-assisted take-off (RATO) is a closer parallel to the case of birds. Rockets are very light for the power they provide, but burn propellants very fast, so that the combination of small jet engines and rockets may show a weight saving compared with larger jet engines if the rockets are used only for a short time. Rockets do not require an outside source of oxygen, and so could be regarded as providing an 'anaerobic burst'.

It is also possible to increase the power of jet engines by burning extra fuel in the jet leaving the turbine, which steps up the thrust ('after-burning'). This process is inefficient, but is a way of providing bursts of additional power for short periods without much increase in weight.

2.9 Energy conservation and storage

2.9.1 Energy conservation in insect flight

Figure 2.7 shows three stages of the downstroke of the wings of a flying bird. In positions (a) and (c), the beginning and end of the stroke, the wing has no downward velocity, but in mid-stroke (b), the wing is moving downwards quite fast. From (a) to (b), the breast muscles are both acting on the air and accelerating the mass of the wing downwards, and the work done in accelerating the wing appears as kinetic energy, energy of motion, of the feathers, bones and flesh. From (b) to (c) the flying muscles are still acting to produce a downwards force on the air, but the downward velocity, and hence the kinetic energy, of the wings is decreasing and helping to provide the aerodynamic force. Thus the flying muscles have two fluctuating powers to provide – one to do work on the air and one to accelerate and decelerate the wing. This second power, the inertial power, as it has been called, would disappear if the wing had no mass: it is a nuisance because although ideally the power is alternately positive in the first part of the stroke and negative in the second part, it is impossible to recover the negative or 'wing-back-to-

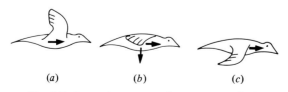

(a)   (b)   (c)

Fig. 2.7. Loss of energy in flapping-wing flight.

bird' power by the use of muscles alone. However, since the positive and negative inertial works are the same, a system which could recover all the negative work would mean that no net inertial power had to be provided. This insects have, as we shall see.

Inertial power is the reason it is desirable to keep down the mass of tennis and squash rackets or, a closer parallel to the bird, oars. Every stroke in tennis or rowing requires the expenditure of energy to accelerate the racket or oar, energy which ideally could be recovered later when deceleration occurs at the end of the stroke.

Unfortunately, muscles are not able to recover energy in this way: worse, they have to expand energy to have work done on them. If you lift a weight, your muscles shorten and pull your bones to different positions, doing work in the process. If now you slowly lower the weight again, it does work on your muscles but you feel the effort required to restrain the motion nearly as much as you felt the effort of lifting – subjectively, you feel you are still doing work though in fact the weight is doing work on you.

This subjective impression is justified – your muscles are still 'working', to the extent that they are still using up chemical energy. Unfortunately, muscles have to use energy all the time to produce force, whether they are shortening (when they are doing work), extending under load (when work is being done on them) or remaining the same length (when there is no work, positive or negative). That is why it is tiring to stand, or to keep pushing at a sledge that will not move.

Thus in flight by wing-flapping, the muscles have to supply the positive part of the inertial power, but cannot recover, or may even have to use more fuel because of, the negative part. Insects overcome this waste by using springs to recover the kinetic energy of the wings at the end of the stroke, and to accelerate them again at the beginning of the next. Figure 2.8 is a diagram to explain the arrangement, where the wing is hinged to the body at P and restrained by springs above and below. In Fig 2.8(a), at the beginning of the downstroke, the top spring is compressed and the bottom one stretched and both are accelerating the wing downwards. By mid-

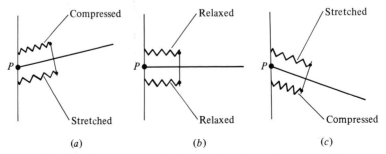

Fig. 2.8. Energy storage in insect wing (very diagrammatic).

stroke (Fig. 2.8(*b*)) the springs are both relaxed and the wing velocity is at a maximum. As the stroke proceeds, the wing stretches the top spring and compresses the bottom one, returning its kinetic energy for re-use on the following upstroke, so that it has stopped by the time it reaches position (*c*). The actual arangement of the springs is quite different, but the action is the same.*

Thus in the insect wing, the muscles do not have to supply inertial power (except for a very small amount to make up the losses in the springs, and an initial impulse to start the flapping going) but only the flying power proper. A useful way of describing this is to say that the muscles 'see' a wing of zero mass. This use of the verb 'see' is common in engineering design, particularly in the electronic world, and is useful and expressive: it means here that as far as the muscles could tell by what happens to them, the wing has no mass. (This is only true if flapping occurs at one particular speed.)

Another way of looking at this is that if we put the insect in a vacuum and bent and released the wing, it would vibrate at a certain frequency, called the 'natural' frequency, like the prong of a tuning fork or a ruler held in a vice at one end. If air were then admitted, the vibration would quickly stop because of the air forces on the wing surface – it is virtually only air forces that the muscles have to overcome in flight, provided flapping takes place at that same natural frequency.

2.9.2 Apparent mass of a vibrating system

Consider again the pendulum of Fig. 2.4, and imagine yourself trying to swing it by holding the bob in your hand and moving it to and fro: in engineering language you are 'forcing' it to vibrate. At the natural frequency of the pendulum it will go quite happily by itself and so there need be no force between hand and bob – your hand feels (or 'sees', in the terminology just introduced) no mass. If you increase or decrease the frequency, then you will have to exert force on the bob, which will increase the further from the natural frequency you go.

Thus part of the penalty the insect pays for its energy-saving wing-springs is that the system works well only over a limited range of flapping speed.

2.9.3 Simple vibratory systems

The pendulum, the tuning fork, the insect wing, are all simple vibratory systems, in which an amount of energy alternates between two forms – kinetic and strain or potential energy. When the wing, say, is momentarily

---

* Weis-Fogh T. (1961). Power in flapping flight. In *The Cell and the Organism*, ed. Ramsay, J. A. & Wigglesworth, V. B. Cambridge University Press.

stationary at the top of its stroke, all this energy is in the strain form, but immediately it begins to be converted into kinetic energy, and so on. The alternations take place at a fixed frequency, the natural frequency.

The energy forms in a simple vibratory system do not have to be kinetic and strain or potential. They can be two forms of electrical energy, as in the receiving circuits of a radio or television set, which respond much more to the signal they are tuned to (i.e., to those of the frequency at which the adjustable natural frequency of the circuits has been set) than to any other. The less irreversibility there is in such a tuned circuit (in engineering jargon, the higher its Q) the more selective it will be.

2.9.4 Energy conservation in locomotion – the switch-back railway

Consider the power requirements of an underground railway with stations at short distances apart. A train leaving station A expends energy mainly in accelerating itself – that is, it is converting electrical energy into kinetic energy. Before it even reaches top speed, however, it is necessary to switch off the current and apply the brakes so as to stop in station B.

One way of saving energy would be to build the railway as in Fig. 2.9: as soon as the train pulled out of station A it would begin to gather speed downhill, and it would slow down automatically as it approached station B. If there were no friction or air-resistance, the motor could be switched off once the train started out of the station, but in practice it would then stop at about P. Nevertheless, smaller motors could be used and much less electricity would be consumed. For trains running at a top speed of 100 km/h (about 62 m.p.h.) the dip d between stations would need to be about 40 m.

2.9.5 Regenerative braking

With electric traction there is another way of achieving something like the economy of the switch-back railway without its manifest difficulties: instead of using brakes to slow the train when coming into stations, the electric motors are made to act as generators which convert the kinetic energy of the train back into electrical energy, slowing it down in the process. With many trains in the system, those which are accelerating will thus be drawing some of the energy they need from other trains which are slowing down, only the difference being supplied from outside generating stations. However, because of the inefficiencies of the electric motors, both those which

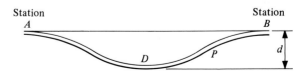

Fig. 2.9. Switch-back underground railway.

are acting as motors and those which are acting as generators-cum-brakes, and because of the power needed to overcome air-resistance, the reduction in outside power required will not be very large. Moreover, the motors are no smaller with the regenerative scheme, and the control system is much more complicated, so regenerative braking is not usually used.

There is one advantage that regenerative braking can offer underground railways that is often overlooked. Practically all the electrical energy supplied to the railway is eventually converted to heat. When such a railway is first opened, the walls are surrounded by cool earth which soaks up this heat, becoming hotter in the process. As the process continues, the heat soaks further out into the earth and slows down the rate of soakage at the tunnel wall, the temperature of which rises faster, and the atmosphere in the railway becomes unpleasantly hot. Regenerative braking reduces this problem by removing part of the heat flow, specifically, that from cooling brakes.

2.9.6 Peak power requirements in cars: the energy-storing car

Peak power requirements in cars are set by the needs of overtaking, which requires rapid acceleration, and hill-climbing, when, besides the normal power needed to drive the car along, extra is needed to push it up the hill. The ratio between the peak power and the normal steady requirement may be very high, perhaps as much as 10:1.

Unfortunately, car engines are very inefficient at low loads, so that the need to provide high-power involves poor efficiency at low-power – which means most of the time in normal driving. Also, since cars in towns are always braking at road junctions and accelerating away from them again, it would be useful to apply the regenerative braking idea to them.

A way of doing this is to use some energy store – let us leave the question of what kind of store for a while – combined with a small engine. Then when accelerating, for overtaking, say, the power of the engine would be supplemented with energy drawn from the store. Braking would be achieved by drawing energy from the motion of the car, which would therefore slow down, and storing it in the store.

Figure 2.10 is a simplified diagram of such a scheme, using a flywheel as an energy store. Between the wheels and the flywheel is what is called a 'continuously variable transmission', or CVT for short. A CVT is like the gear-box on an ordinary car, except that instead of, say, four forward gears, bottom, second, third and top, it has an infinite number and can move smoothly or continuously through them. To fix ideas, suppose the flywheel can run between 100 and 25 times as fast as the wheels, and that when the ratio, flywheel speed/wheel speed, is 25:1 the kinetic energy in the flywheel is equal to that in the car. Imagine this car running at 100 km/h with the flywheel running 25 times as fast as the wheels and the kinetic energies of

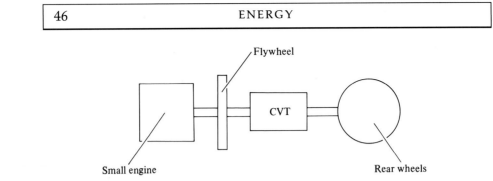

Fig. 2.10. Energy storage system in car.

car and flywheel equal to 0.5 MJ each. Now imagine that the speed ratio between flywheel and wheels is gradually changed to 50:1, so that whatever speed the car is going at, the flywheel is now running at *relatively* twice the speed. Since kinetic energy is proportional to speed squared, the fly-wheel now has $2 \times 2 = 4$ times as much as the car. Suppose that the road is level and resistance is neglected and the engine is idling, so that the system consisting of the car and the flywheel has lost no energy in the meantime, and that the kinetic energy of the car is now E, so that that of the flywheel is $4E$. Then

$$4E + E = \text{total kinetic energy}$$
$$= \text{original kinetic energy}$$
$$= 0.5 \text{ MJ (flywheel)} + 0.5 \text{ MJ (car)}$$

Then E is 1/5 MJ, i.e., changing the CVT ratio from 25:1 to 50:1 has reduced the car's kinetic energy to 2/5, and so its speed from 100 km/h to $\sqrt{2/5} \times 100$ km/h=63 km/h. If the CVT ratio is further changed to 100:1, then the car's share of the original 1 MJ falls to 1/17, or about 0.06 MJ, and its speed falls to 34 km/h. The remaining 0.94 MJ is now all in the flywheel, and by *slowly* moving the ratio back to 25:1, the extra 0.44 MJ could be returned to the car, which would accelerate up to 100 km/h again. The change of ratio must be fairly slow or the required acceleration of the car will be more than the grip of the tyres can achieve, and they will skid. With such a car, when stopping at traffic signals, most of the energy norm-ally wasted in braking would be stored in the flywheel, and used again to accelerate away when the lights changed.

The system has to be much more complicated than this – for example, we can never stop the car without disconnecting the flywheel – but the advantages are considerable. Firstly, a much smaller engine can be used, just enough to drive the car steadily at motorway speed, since overtaking requirements are met by drawing on the energy of the flywheel. This engine would be working much harder relative to its size, and such an engine is more efficient and has a cleaner exhaust than a larger engine which is underloaded most of the time. Thus fuel consumption would be reduced, not only by the smaller energy requirement achieved by saving most of the waste of braking, but also by the more efficient production of that energy.

Three energy stores have been suggested for such systems – flywheels, electrical accumulators and hydraulic accumulators, which are really springs in which a highly-compressed gas forms the spring material. All of them are rather heavy, the flywheel carries a risk of explosion, the electrical accumulator has a disadvantage that we shall see and the hydraulic accumulator is relatively heavy per unit of energy stored.

2.9.7 The quickness of fleas

The electrical accumulator is unsuitable as an energy store for a car, because although it can store sufficient energy it cannot release it quickly enough – it is not *powerful* enough.

A small jumping creature like a flea is handicapped by a scale effect which may be explained as follows: if a man and a flea were to jump the same height, they would have to leave the ground with the same velocity, and so the average velocity relative to the ground during jumping would have to be the same. But because the man's legs are, say, 1000 times as long as the flea's, his muscles would be working for 1000 times as long – say one-quarter of a second instead of the one four-thousandth of a second available to the flea. Now the muscles of the flea simply cannot release energy as quickly as that – they are not *powerful* enough – and its performance would be very poor if it were not for the specialised design of its jumping legs. A flea jumps by releasing energy stored in springs: these springs are loaded relatively slowly by means of muscles, and then released suddenly. The loading takes about 50 milliseconds (thousands of a second) and release about 1 millisecond. One of the defects of cold-blooded creatures is that the speed of muscular processes depends on their temperature, so that it is recorded that cold fleas may take as long as two seconds to load their jumping machinery. The release mechanism depends on a common device in design, the 'over top-dead-centre' principle which may also be seen, for instance, in an energy-saving detail in some of the leg-joints of a horse. These bend slightly backwards when the animal is standing, so that the weight on them tends to fold them the wrong way (see Fig. 2.11): this means that the muscles which work the joints can relax and do not have to expend energy all the time to keep the horse standing.

2.9.8 The quickness of bows

The flea, then, since it cannot obtain the necessary quickness of energy-release from muscle, converts muscular work to strain energy stored in a spring, which can be released very quickly. That is what the archer does: he draws the bow, transforming chemical energy in his muscles, which can only be released slowly, into strain energy in the bow, which is simply a

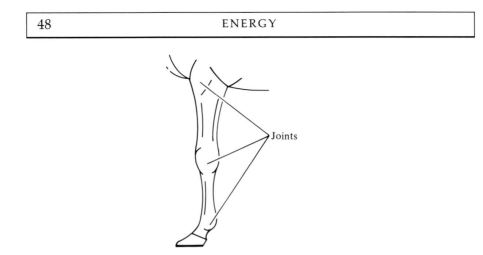

Fig. 2.11. Locking joints in horse's foreleg.

leaf spring. This strain energy can be released very quickly, in a much shorter time than the muscles of the archer could deliver it directly.

Even if the archer's hand could do work at the same rate as the bow, he could not throw an arrow at the velocity at which the bow delivers it, because the work would appear as kinetic energy of the arrow *and* his hand and part of his arm. As this total mass is much greater than that of the arrow and the associated part of the string, the velocity would be correspondingly less. This effect is less with a spear, of course, because the spear is much heavier relative to a hand than the arrow.

The bow is a very elegant design, not only in its detail, but also in the basic concept, which ensures that nearly all the strain energy is transformed to kinetic energy in the arrow. To see why this is so, consider the imaginary catapult of Fig. 2.12, in which a stave of yew AB, similar in form to half a very long bow, is firmly fixed in the ground at A. A string BC has a loop at C which is slipped over the flight end of the arrow, which is pulled back and released. When this badly-designed weapon is fired the end B of the stave and all the string move with the arrow and at the velocity of the arrow, which has to share its kinetic energy with them. Parts of the stave below B travel less fast, but still take some of the energy.

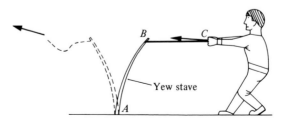

Fig. 2.12. An inefficient 'bow'.

Now consider the bow in Fig. 2.13. Immediately after release, in position (a), the tips B of the bow are travelling at only a fraction of the speed of the arrow, and their effective mass as far as sharing in the kinetic energy goes is reduced as the square of that fraction, since kinetic energy is proportional to speed squared. This is already an improvement over Fig. 2.12, but there is more to come. Towards the end of discharge, in position (b), the fraction, velocity of B/velocity of C, becomes rapidly smaller as the angle BCB of the string becomes flatter. The share of the kinetic energy taken by the bow becomes less, and the kinetic energy is fed back from the bow to the arrow. The mechanics are exactly the same as those of the flywheel energy store in the accelerating car, with the bow as the flywheel and the arrow as the car. As the 'gear ratio' between the two changes, so that the bow moves more slowly relative to the arrow, kinetic energy is fed from bow to arrow. Because of this property of the system, its *kinematics*, as such relationships between motions are called, most of the work put in by the archer eventually becomes kinetic energy of the arrow. In other words, the efficiency is high, where here

$$\text{efficiency} = \frac{\text{useful output}}{\text{input}}$$

$$= \frac{\text{kinetic energy of arrow}}{\text{work done by archer}}$$

2.9.9 Missile throwing without energy storage

Let us carry out a design exercise using the principles which can be found in the examples of the flea and the bow, the purely academic exercise of trying to design a missile-throwing device that does not employ energy storage. You can postulate if you like some island tribesmen not possessed of wood or other springy material suitable for bows and having to defend themselves against invaders.

Two relevant conditions for the design of such a weapon have emerged (both of which apply to bows as well). Firstly, the velocity of parts of the

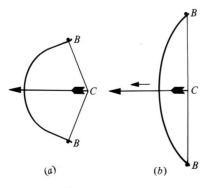

Fig. 2.13. Bow.

system other than the missile itself must be kept small. Secondly, the time for energy-input by the thrower must be long, so that as much energy can be transferred as possible.

Since there must be some mass acting on the missile and moving with it, that mass should be as small as possible (like the string in the case of the bow) and certainly less than the mass of a hand and forearm. Also, the missile end of this driving element, as we may call it, must be travelling faster than the hand end – this will be called the kinematic principle, and is the embodiment of the first of our conditions. This kinematic principle will also help with the second condition, for if the missile end of the driving element goes faster than the hand end, it will also go further, and if it goes further at the same sort of velocity, it will take longer and allow more time for the muscles to do work. Also, from the argument of the spear versus the arrow, the heavier missile is more promising to start with.

For a primitive tribe, the only likely forms for the driving element are a stick or a string. It remains only to sketch these two elements in conditions which give the required kinematics (a high value of the ratio of missile velocity to hand velocity, which is called simply the *velocity ratio* in mechanics). In Fig. 2.14 this is done for a spear, using as a driving element a stick – this device was used by many primitive cultures and is called a spear-thrower. With a thrower length of rather less than a metre, the distance through which the spear travels during acceleration is roughly doubled compared with a weapon held directly in the hand.

The other likely driving element, a string, has the virtue of low mass, but cannot be used except to pull – it has to be kept in tension. This can be done by whirling the missile in a circle, as in Fig. 2.15. Notice that the string and hence the force P in it are not radial, but directly partly inward and also partly along the path of the projectile, as shown. The force P is just the same in its effect as the two forces R and S (shown in the figure) acting together. R and S are called the *components* of P, and it is the component S which is along the direction of motion which is doing work on the missile and increasing its kinetic energy. The component R does no work, since there is no radial movement.

Fig. 2.14. Spear-thrower.

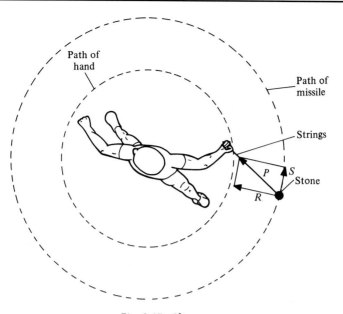

Fig. 2.15. Slinger.

Because the missile is travelling in a larger circle than the hand, it must be moving faster, so the velocity ratio is high, and the distance travelled by the hand is long: in consequence the time available to do work is long, and both conditions essential to a good design have been met. This weapon, the sling, was important among various nations in the past (e.g., David and Goliath). It was usually made with two strings attached to a cloth or leather cup. The missile, usually a stone, was put in the cup and released at the right moment by letting go of one string.

2.10 Systematic invention and design

We have just re-invented two important ancient devices by a systematic approach. It is easy with hindsight, you may say, but nevertheless, the process is almost mechanical once the two essential conditions and the two likely driving elements have been recognised: by increasing the abstractness of our treatment, we are led inexorably to good solutions. Few design problems can be treated quite so rigorously, but most can be made much easier by such processes, which are examined in Chapter 10.

2.11 Energy and design

In the short space of this chapter, it has only been possible to outline important concepts and touch on a very few of the many aspects of design in the living and man-made world which they illuminate. The energetics of flight have been discussed, but not those of running, which display the problems of loss of kinetic energy of limbs more critically and elaborately than does flapping flight, nor those of space flight, the most demanding of all upon the resources of the designer.

Scale effects, introduced briefly in connection with fish, ships and, later, fleas, occur all the time in nature and engineering, though by no means always in connection with energy.

The idea of regenerative braking, the saving and re-use of the kinetic energy of a vehicle when it slows or stops, is an example of the elegant economies often to be found in design, as are the spring-mounted wings of insects. The use of an energy store in combination with an engine in order to provide a more economical power supply for the propulsion of a car than an engine alone again shows a characteristic of design, the combination of components to form a whole which is well-matched to a particular task.

Many important aspects of energy in design have not been mentioned in this chapter; some of these are discussed later.

2.12   Specific energy and specific power

It is interesting to have some comparison of the capacity of fuels and energy-storage devices, which is best done on a basis of energy stored per unit mass. A few interesting cases are given in Table 2.1. Specific power, that is, power per unit mass, is also given for a few sources in Table 2.2.

Table 2.1. *Typical values of stored specific energy*

	Specific energy, kJ/kg
Water behind proposed Severn Tidal Barrage	0.04
In clockwork motor spring	0.04
Stretched sinew, as used in ancient catapults (estimated)	0.1
Flea's jumping energy storage (resilin)	1.5
Stretched sinew, economically used	10
Flywheel, rim speed 250 m/s	25
*Hot water in kettle (100 °C)	39
Electrical accumulator	80
*Hot water at 230 °C (as in steam catapult, Chapter 10)	230
Gunpowder	2.8×10^3
Nitrocellulose (gun-cotton)	4.4×10^3
Liquid rocket fuel (liquid oxygen + kerosene)	9×10^3
Hydrocarbons (petrol, oil, etc.)	40×10^3
Fusion reaction	24×10^9

* Values for *available* energy only, ambient temperature 20 °C.

Table 2.2. *Some typical values of specific power*

	kW/kg
Muscle	0.05
Lead/acid accumulator	0.18
Electric motor	0.25
Aircraft piston engine	1.5
Hydraulic motor	2.5
Aircraft turbine engine	6.0

3

Materials

3.1 Materials, design and craftsmanship

The appreciation of materials, their special properties, their fitness for certain purposes, the deep satisfaction given by their cunning and sympathetic use in articles combining function and ornament, is an important part of human culture. Scythes and violins, walnut and mahogany furniture, tweed cloth and silver cutlery, glass and bronze, all show the subtle alliance of craftsmanship and the nature of the raw material in the creation of artefacts combining beauty and use. All this was achieved with very little in the way of science, before we enjoyed anything of the understanding of materials we have now, and it is one of the fundamental failures of imagination of our age that it does not recognise that this tradition has not perished, but has flourished and transcended itself in some modern engineering products. For example, a record-player pick-up cartridge may marry a tiny precisely-shaped diamond, a fine strip of bronze three times as tough as anything known 100 years ago, delicate coils of wire as fine as a spider's web, a powerful little magnet made of rare metals or oxides, whose existence was unsuspected a century ago, and intricate and perfectly-fitting parts of strong plastics and metals.

Moreover, all this was offered, not to kings or bankers, but to any citizen of the developed world. For perhaps half a day's pay he could buy this triumph of craftsmanship and ingenuity, beyond anything made by Faberge. Now the vinyl disc has been displaced by the compact disc, with an optical system in place of a mechanical one. The materials are not so interesting, but still all-important, chosen now for optoelectronic rather than mechanical properties, and the value in terms of design and craftsmanship is perhaps even more remarkable.

3.2 Factors affecting choice of material

The many different materials of the pick-up cartridge are chosen for a variety of reasons. Some are chosen for special physical properties possessed by very few substances, as in the case of the magnet. The stylus is of diamond because it is the hardest material we have, and the wire is of copper because it is almost the best conductor of electricity and is, moreover, ductile and

54

sufficiently strong. The plastics are chosen out of a very wide range of similar materials on a large number of considerations taken together, only some of which have to do with function. They are picked for strength, stiffness, appearance and stability of shape with time, but also for ease of manufacture in the right forms with the right accuracy and a good surface finish.

Production considerations are of the greatest importance in design in nature and by man. So strong is the interaction between the three aspects of material, method of manufacture and form that the designer must take them all into consideration at once; form must be determined in conjunction with method of manufacture, and only occasionally is it possible to substitute a different material in a design without having to make radical changes for reasons either of function or manufacture.

Consider a garden chair. It may be made of wood, or plastic, or aluminium alloy, or steel, or cast iron or of some combination of these. Some parts may be of yet other materials, such as canvas. The production method will also be important; for example, the aluminium alloy may be in wrought form, or cast; the wood may be plain solid wood, or laminated, or plywood or even wickerwork. In all these materials a great variety of garden chairs have been designed, and in almost every case the form is dictated to a large extent by the material.

Now consider the choice of material for a human thigh-bone, a femur. It has to be strong and it also has to be stiff. Many materials used by man would be much better than bone from both these viewpoints, but living substance cannot make them: moreover, it is difficult to see how a steel or glass-fibre bone could be arranged to grow with the owner. Bone consists mostly of small crystals of a mineral substance, a complex of calcium phosphate and calcium hydroxide, called bone salt, embedded in a matrix of strong protein fibres: roughly speaking, the crystals provide stiffness and strength, and the protein provides strength and prevents brittleness. Bone is only about one-third as stiff and half as strong for the same weight as, say, aluminium alloy or alloy steel, though that is an astonishing performance given the unpromising nature of the basic constituents, but it can grow with the body it forms part of, while remaining functional the whole time.

Three factors govern the choice of materials in design

(a) properties
(b) production (reproduction and growth in nature)
(c) price.

Even in nature, price is important: a spider's web catches food, but unless it catches more than enough to replace the materials used up in its construction its biological cost will be too high. Note that the biological cost is often reduced by the spider recycling some of the web material, and also that the strength of the material must be taken into account in assessing the cost – if it is stronger, less will be needed. Similarly, high-tensile steel reinforcement bars used to strengthen concrete are more expensive per ton than ordinary steel, but because less are required to do a given task, they are often cheaper in use.

The very important question of production will be dealt with fully in Chapter 9, while this chapter is concerned mostly with properties.

3.2.1 Tensile strength

Of all the properties of material which are important to designers, the one which comes most readily to mind is strength, and of all the kinds of strength, tensile strength, such as is exhibited by a rope or wire in tension, is most familiar.

Suppose we have a wire of cross-sectional area one square millimetre (which corresponds to a diameter of about 1.14 mm) and it is subjected to a pull of 200 N, perhaps because it is being used to suspend an object of about 20 kg mass from the ceiling (see Fig. 3.1). We say that the *tensile stress* in the wire is 200 newtons per square millimetre, or 200 N/mm² for short. Stress is defined as force per unit area, or

$$\text{stress} = \frac{\text{force}}{\text{area}}.$$

1.14 mm
diameter

20 kg

200 N

Fig. 3.1. Wire in tension.

If the wire is of ordinary mild steel and we increase the load in it, it will break when the force is about 450 N, or when the stress is 450 N/mm². This breaking stress is called the *ultimate tensile strength* (UTS for short).

If we have a wire of twice the cross-sectional area, we would expect it to sustain twice the load before breaking, and indeed, it will, i.e., it will fail at the same *stress*. In practice, designers try to limit the stresses imposed upon materials to a *working stress* which is only a fraction of that which will break them.

3.2.2 Ductility

Under stresses of up to about 250 N/mm², the mild steel wire will stretch very slightly and *elastically* – that is to say, if released it returns to its original length. At higher stresses it stretches by proportionately much larger amounts (Fig. 3.2), and these larger extensions remain when the stress is removed: they are permanent. Such permanent extension is called *plastic deformation*, and materials that show it are called *ductile*.

If you bend a piece of wire very slightly in your fingers, it will spring back when you release it. But if you bend it much more and release it, it will spring back only a little of the way. This is because the material has deformed plastically, being permanently stretched on the outside of the bend and permanently compressed on the inside. (There is always the same small amount of spring-back, however, after permanently bending a wire, and this has to be allowed for by engineers designing tools to bend wire or strip into useful shapes.)

Ductility is very valuable in a material, both as an aid to production and in use. As a simple example of the former, consider those toys and other

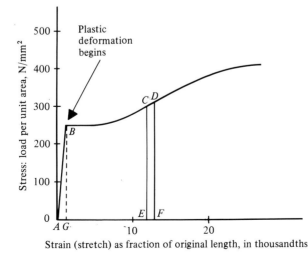

Fig. 3.2. *The stretching of mild steel.*

articles made of thin sheet steel ('tin') which are held together by tabs which are pushed through mating slots and bent over or twisted to prevent them pulling out again – it is only the ductility of the material that makes such a construction possible. Nails or bolts may be made out of ductile wire by 'cold-heading' – squashing one end of a piece lengthwise to thicken it and spread it out into a head. Saucepans can be pressed out of a flat sheet of steel rather as flan-cases are pressed out of a sheet of dough (dough is a ductile material).

When a steel wire is stretched elastically, energy is stored in it *reversibly*, i.e., it will give back the energy, as in a clock spring. When it is stretched beyond its elastic limit, the process is irreversible, and only the little bit of elastic energy can be recovered. The plastic deformation can be very large, and so the energy absorbed is much greater than in the tiny amount of elastic deformation, perhaps 500 times as much.

Mild steel, deformed plastically, can absorb about 10 000 J/kg. If a car crashed at 28 m.p.h., its kinetic energy of about 62 500 J could thus be absorbed by a mere 6 kg of steel. The difficulty is that even when a sheet steel structure is badly smashed, only a very small proportion of the steel need be severely deformed. This may be seen by examining a deformed steel can, such as that in Fig. 3.3, where severe deformation is limited to a few narrow bends and most of the metal is relatively unaffected. These strips may be less than a millimetre in effective width and may constitute less than 2% of the material in the can, which is made of rather thicker stuff relative to its size than a car is. To overcome this problem, the engineering designers incorporate structures (like the rims of the can) which cannot be buckled without deforming a relatively large fraction of the steel, as in the example shown in Fig. 3.4.

3.2.3 Brittleness

Most inorganic materials are brittle, not ductile. When stretched, they deform elastically, but when overloaded, instead of undergoing large plastic deformations, they simply snap. Stone, glass, ice and cold toffee are brittle. Consequently, the energy they absorb before breaking is very much smaller

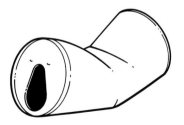

Fig. 3.3. Deformed can.

Fig. 3.4. Deformation in compression of a tube in sheet steel.

than for a ductile material: a brittle car, for instance, would shatter in an accident and so afford its occupants no protection.

Figure 3.5 compares the stress–strain diagrams for a brittle and a ductile material. That for the ductile material (A) is the same as Fig. 3.2, a steep straight line corresponding to elastic stretching and then the long flattish curve which represents plastic stretching, and eventual breaking. The brittle material (B) goes up the same straight elastic line to begin with, but continues straight until it breaks, with no plastic stretching to speak of. The brittle material is twice as strong, that is, it needs twice the stress to break it, but the energy absorbed in breaking is given by the shaded area under each curve, and this is much greater for the ductile material (usually more so than the figure suggests).

Under the circumstances shown in the inset of Fig. 3.5, where the load P is caused by dropping the weight till it hits the stop, it is the energy-absorbing capacity of the link which determines whether it will survive; if it can absorb the kinetic energy of the weight when it hits the stop, it will not break, and here a ductile material is much superior, even though it may be less strong under steady loading.

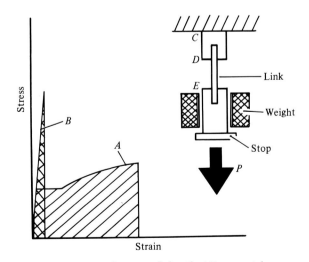

Fig. 3.5. Brittle (B) and ductile (A) material.

It is not only when there is an obvious requirement for energy absorption that brittle materials are inferior to ductile ones. Sometimes the 'loading' which has to be withstood is that of a forced deformation. For example, the windscreen of a car may be shattered by relatively small deformations of the frame surrounding it; in coal mines, the arches supporting the roofs of the roadways are slowly deformed by pressures from the rock, which moves slowly but irresistibly. They must bend, but not break, so that they continue to prevent rock-falls. For such a purpose, a brittle material would be useless. Above all, in brittle materials a very local overstress will cause a crack which spreads and leads to complete failure where a ductile material would merely give at that point and spread the load to less heavily-taxed material elsewhere.

The design of many animals and plants, particularly the higher ones, calls for rigid materials, for bones, for tree trunks, etc. Living organisms, however, cannot produce parts of strong ductile metals, and have to use specially-structured materials in order to achieve rigidity combined with the power to absorb shocks without breaking, essential in bones, for example.

A brittle material breaks cleanly across, absorbing little energy because in the product (force×distance) the distance factor is very small. To make a material of brittle components that will absorb energy in breaking, it should be designed so that it will not snap cleanly but rather tear apart, with many small breaks and many small surfaces gliding along each other, as may be seen in the splintered surfaces of broken wood. Thus glass-fibre-reinforced plastic (GRP for short) is a composite made of two materials which are basically brittle, but is not itself brittle. This happy result is possible because of the structure of the material, which consists of very fine filaments of glass embedded in a matrix of plastic.

Wood is another composite material, with very pronounced *anisotropy* – that is to say, with different properties in one direction to another. Along the grain, it is many times stronger than across the grain.

3.2.4 Compressive and shear strength

Tensile strength is the resistance to pulling apart lengthwise of a material: compressive strength is its resistance to crushing lengthwise, as in Fig. 3.6(*a*). Ductile substances are ductile also in compression, and a short cylinder crushed axially will compress elastically to begin with, and then yield and deform into a barrel shape as in Fig. 3.6(*b*). Wood compressed along the grain fails in the characteristic fashion shown in Fig. 3.6(*c*), with the individual fibres buckling in a narrow zone at 45° to the grain, the reason for which will be studied later. Concrete and brittle metals like cast iron or hardened steel also fail on planes at 45° to the axis of thrust, either like wood or, characteristically in the case of concrete, so as to leave pyramids

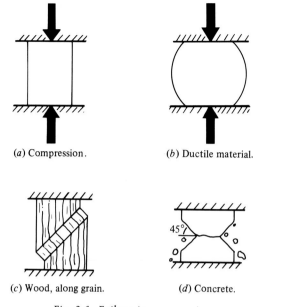

(a) Compression. (b) Ductile material.

(c) Wood, along grain. (d) Concrete.

Fig. 3.6. Failure in compression.

or cones of material facing each other, as in Fig. 3.6(d), where the fragments originally between the two pyramids are shown as having fallen away.

Fundamentally, materials do not fail in compression, but in a way (or mode) called shear. We cannot annihilate the material by squashing it – it has to go somewhere, and the only way it can go is sideways. Figure 3.7(a) shows a block loaded in compression. It is made of very strong material, in two pieces glued together with weak ductile glue along the faces AB. What will happen as the load increases? Clearly, the faces AB will slide over one another, shearing the glue and shortening the block, as in Fig. 3.7(b).

Figure 3.7(c) shows another such block, but with three joint faces at increasing angles to the transverse direction, AB, CD and EF. On which will failure occur? As the angle of inclination increases, so the tendency to slide will increase (more strictly, the component of force along the plane increases), but, on the other hand, the area of the plane, and hence of glue resisting sliding, will increase. The ratio, force along plane/area of plane, i.e., the shear stress in the glue, is a maximum at an angle of 45°, where it is equal to just half the direct stress, i.e., the axial force divided by the cross-sectional area.

Clearly, shear stresses also exist in tension, for shear failure will occur in one of our glued blocks if we stretch it just as if we compress it (Fig. 3.7(d)): again, the maximum shear stress occurs on planes at 45° and is just half the direct (tensile) stress. Indeed, when the wire began to give, or yielded, at

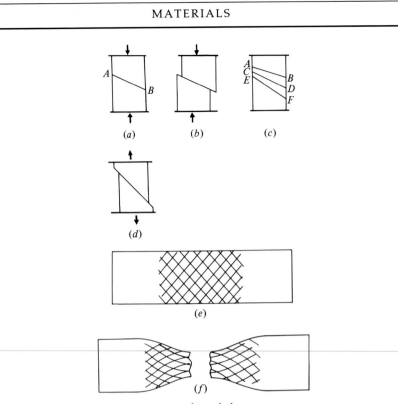

Fig. 3.7. Shear failure.

point *B* of Fig. 3.2, it was, strictly speaking, due not to the direct stress of 260 N/mm², but the shear stress of half that, or 130 N/mm², for a true tensile failure would tear across the material of the wire.

To see the difference between these two basic mechanisms of failure, shear and tension, make a bar of plasticine, say, 15 cm×3 cm×1 cm. Plasticine is a ductile material much like mild steel in its plastic behaviour, but a good deal easier for most of us to break. On one of the 15 cm×3 cm faces lightly score a grid of lines at 45° to make a net of squares (Fig. 3.7(*e*)). Now stretch the bar slowly until it breaks, observing carefully what happens. Usually the bar necks down locally, the squares in the necked region becoming diamond-shaped, and eventually the neck breaks across in a tensile failure. But the necking and the extension of the squares into diamonds is due to shearing deformation (Fig. 3.7(*f*)).

3.2.5 Specific strength

We can express the strength of materials in terms of their ultimate tensile or compressive strengths, and some typical values for important materials are given later. However, we are often interested in the mass of material that is needed for a given task, so that some measure of strength per unit mass is needed. A suitable measure for a member used in tension is thus:

$$\frac{\text{ultimate tensile strength}}{\text{mass per unit volume}} = \text{specific tensile strength}$$

which turns out to be in some respects a similar concept to specific energy. If we have a certain mass of material M with a certain volume v and a strength f, then

$$\text{specific strength} = \frac{f}{M/v} = \frac{fv}{M}$$

and fv is in the same kind of units (in the jargon, has the same *dimensions*) as energy. For stress is a force per unit area, and volume is area times length, and force/area×area×length gives force×length, which has the dimensions of energy.

Now mild steel breaks at about 460 N/mm² and has a density of about 7800 kg/m³, and hence the specific strength can be calculated to be about 60 kJ/kg. In practice, for safety, we could only use it at a stress of, say, 150 N/mm², which corresponds to a specific strength of about 20 kJ/kg. This means that we can use 1 kg of mild steel, in a suitable form, to support a load of 20 kN over a length of 1 m, or 2 kN over a length of 10 m, and so on.

Another way of looking at specific tensile strength is to note that it is related to the length of material which can hang vertically without breaking under its own weight. You can easily show for yourself, using the fact that a mass of 1 kg has a weight of about 10 N, that the maximum free-hanging length of a substance is about 1 km for every 10 kJ/kg of specific tensile strength, so that for mild steel it is about 6 km. Figure 3.8 compares these lengths for some man-made and natural materials.

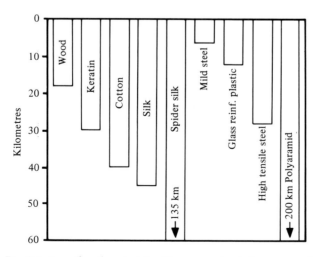

Fig. 3.8. Lengths of materials able to sustain their own weight.

3.2.6 Stiffness

An important property of materials is *stiffness*, the resistance they offer to being bent or deflected, as against *strength*, the ultimate resistance to breaking or collapse. If bones were made entirely of their protein constituents, they would be nearly as strong but only about a tenth as stiff: the chief function of the mineral or bone salt content is to add stiffness, because a rubbery skeleton would not be very satisfactory.*

Imagine a wooden rod, the diameter of a pencil, and a few centimetres long, stuck in a hole in the floor so it stands up vertically like a column; a man can balance on it and it will bear his whole weight, even if he is a very heavy man. But take a rod of the same diameter but as tall as a man, and fix it upright in the same way, and it will not support the weight of an apple, let alone a man!

Now if we simply looked at the stress in the rod as we did the stress in the wire, and divided the weight of the man by the area of the cross-section of the rod, i.e., the area of a circle equal to it in diameter, we should get the same answer whatever the length of the rod, and this would give us a true picture if we were suspending the man by the rod, so that the rod was being stretched or in *tension*. For a rod in tension, the length makes no difference to the strength.

However, when we squash a long rod lengthwise, in *compression*, as in the case just described, the length is critical. Short columns, whose length is only a few diameters, do have a strength independent of their length, and will collapse roughly as described earlier, by shearing. But long rods collapse at very much lower loads, the weight of an apple rather than the weight of a man, long before the limiting shear stress is reached, by *bowing out sideways*, as in Fig. 3.9. In resisting this bowing out, or buckling as it is called, the property which counts is stiffness, and mere strength is of no avail. As an extreme example, a rope may be very strong in tension, but is quite useless in compression, when it collapses under its own weight without breaking.

Now the problem of buckling is associated with slenderness. If we took 100 of our pencil-thick man-high wooden rods and glued them together in a bundle, they would support, not just 100 times the weight, but nearer 10 000 times as much. If we took 1000 of our long thin rods and made a bundle, they would carry the weight of 1000 men, and it would be the *strength* of the wood that was the determining factor, not its stiffness, because now our column-bundle would have a diameter about one-eighth of its length, and would no longer be 'slender', but 'stocky'.

* For an account of progress in imitating bone for medical purposes, see Bonfield, W. *et al.* (1983). Hydroxapatite-reinforced polyethylene composites for bone replacement, *Biomaterials and Biomechanics*, 1983, pp. 421–6.

Fig. 3.9. Buckling of a thin member in compression (strut).

Most structures which exhibit functional design, skeletons, bridges, looms, lorries, grasses, have slender components, and so stiffness is an important property in them.

Notice that if in glueing up our many rods we made the bundles hollow, they would be much stockier, and so much less prone to buckling (although the glue would need to be better).

3.2.7 Vibration and stiffness

Resistance to buckling is not the only reason for needing stiffness. Instruments and machine-tools need stiffness if they are to measure accurately and make accurate parts. A tape measure made of a stretchy substance like elastic would not be much use: we prefer steel, which may be pulled taut without stretching much (although surveyors using steel tapes measure the force they apply to them and the sag in them and make corrections – that is, when they are not using more modern electronic equipment). A lathe is built very much more solidly than is necessary to prevent it breaking, so that the cutting forces do not push the tool out of the work and make it cut oversize.

There is another reason for making machine-tools stiff, which is to avoid vibration. In the case of the lathe, if the tool is forced outwards, the depth of the cut, and hence the cutting force will change (Fig. 3.10). Under certain circumstances, this can lead to violent vibrations, in which the tool moves rapidly in and out relative to the work, leaving a wavy surface and sometimes breaking the tool or damaging the machine.

The phenomenon, called 'chatter' because of the noise it makes, is an example of a *self-excited vibration*. Vibrations are alternating motions of

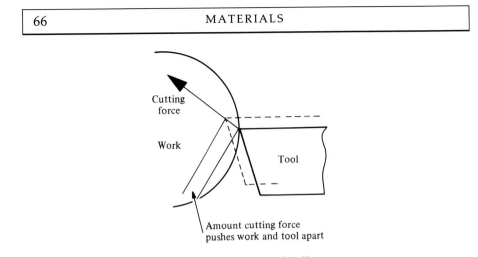

Cutting
force

Work

Tool

Amount cutting force
pushes work and tool apart

Fig. 3.10. Importance of stiffness.

more or less constant frequency, like those of a struck piano-string, a bowed violin string or a car-body when it is stationary with the engine running. The car-body vibrates at the same frequency as the engine is turning – we say that it is being *forced* to vibrate by the engine at the engine's own speed, and so this is a *forced vibration*. The piano-string is set vibrating by a blow, and then left alone: this is a *free* vibration, and it gradually dies away because of *damping* – a gradual loss of energy, partly in the form of radiated sound energy, so that a full description of the motion is a *damped free vibration*. The frequency of a free vibration is called the *natural frequency* (damping of a small amount makes only a small difference to the frequency of a free vibration). In the case of the pendulum in Chapter 2, the free vibrations it made were gradually reduced in amplitude, or size, by air-resistance and friction in the pivot.

When a vibration is forced at a frequency very close to, or at, the natural frequency, then it can become very violent, especially if the damping is low: this phenomenon is called *resonance*. Marching soldiers break step on a bridge, in case their regular tread should coincide with a natural frequency of the bridge, set up a resonant vibration and break it. As we saw with the pendulum, very little force is needed to keep it vibrating at its natural frequency, just enough to overcome damping.

A familiar tale of resonant vibration is that of the soprano shattering the wineglass. If we strike the glass gently it rings, vibrating at its natural frequency. If the singer can hit and hold the right note, the glass will resonate, absorbing some of the sound energy falling on it. The distance that a point on the glass is vibrating, called the amplitude of vibration, will increase as more energy is absorbed, until the rate of gain of energy absorption from the voice is equal to the rate of loss of energy through damping. Such a balance is reached because the rate of energy input is proportional to the amplitude, but the rate of energy loss is proportional to amplitude squared (see Fig. 3.11, where A is the point at which the two energy rates, or powers, balance). There will be an amplitude at which the glass will break, because the bending stresses in the glass are too high. If this amplitude is less than that corresponding to A, the glass will break.

A self-excited vibration takes place at the natural frequency, instead of the frequency dictated by some external source, such as the speed of the car engine or the pitch of the singer's voice. Consider a child on a swing receiving a push from another every time he starts to swing down and away: the child/swing system has a natural frequency, and it is this that dictates the frequency of the pushes, because the friend watches the motion and pushes at the same time in each vibration. A bowed violin string is in self-excited vibration of a different kind. The friction of the bow drags the string aside until eventually the string slips on the hairs of the bow and flies back. It is able to do this because the horsehairs of the bow have little teeth, and once they lose their grip they wll not catch the string again until the string stops, which it does after half a vibration, just as the pendulum stops momentarily half a vibration after it is released. The hairs then catch the string again, and the process is repeated, hundreds of times in every second. The string vibrates and the sound is produced at its natural frequency, because it is this natural frequency that determines how long it is between the hair teeth losing their grip and restoring it again.

A striking example of a self-excited vibration was the collapse of the Tacoma Narrows bridge in 1940: under the influence of a steady wind blowing across it, the deck of this 853 m span suspension bridge began to 'gallop' up and down and twist until it failed completely after about 5 hours. At the time, not enough was known about self-excited vibrations of that kind.

The Tacoma bridge would not have failed if it had been made of much stiffer material, or of a much stiffer design or had had much higher damping. There is no scope for the first change in design, partly because the stiffness of a suspension bridge derives largely from its weight, and partly

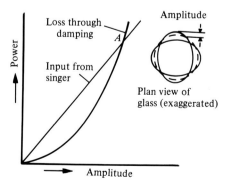

Fig. 3.11. Energy input to vibrating glass.

because stiffer materials are not yet available which are practical alternatives for bridge construction: only stiffer designs and higher damping are possibilities, and both are used now. Similar self-excited vibrations, called *flutter*, occur in wings and other parts of aircraft if they are insufficiently stiff.

3.2.8   Stiffness of materials

Steel is about 20 times as stiff as wood, but as it is also about 15 times as heavy, it is only a little stiffer weight for weight. Steel is about three times as stiff as aluminium alloys, but it is about three times as heavy as well. Thus a lot of important materials have about the same stiffness/weight ratio.

However, some of the new materials available in fibre form have much higher ratios of stiffness to density. Boron is about twice as stiff as steel, but less than one-third as dense, so its stiffness/weight ratio is more than six times as great. Carbon is about the same stiffness as boron, but it is less dense, so that its stiffness/weight ratio is more than eight times that of steel. In practice, these very stiff (and strong) materials are used in conjunction with plastics in a *composite* form, so that the practical properties are less outstanding, but still very high indeed.

Some important materials, however, fall well below the steel–wood–aluminium stiffness/weight ratio – bone (about one-third), concrete (about one-fifth) and most plastics are examples. They are, nevertheless, very useful.

It is not always obvious when stiffness is necessary. In high-performance vehicles, stiffness is important, even though flexibility is needed in the suspension. The reason is that the motion allowed by the suspension must be a very controlled one: for example, it is important that when deflected upward relative to the body, a rear-wheel of a car is held rigidly facing forward, or, if it does turn slightly about the vertical or 'steer', is still held rigidly in that steered position at an angle which has been calculated to be desirable.

This need accounts for the use in racing cars of materials like carbon-fibre reinforced polymer (CFRP). It is important to keep mass down and stiffness up, and so the highest specific stiffness is needed and the structure must use it to the best advantage, which means using a monocoque, a hollow shell form (from the French 'coque', a shell, also used for the hull of a ship or the shell of a nut). A monocoque structure has nearly all its material in an outer skin which is made as unbroken as possible: it uses material particularly well against twisting between front and rear axles, and stiffness against such twisting is particularly desirable. Space frames are not quite as efficient because the material is only stressed in one direction.

3.2.9   Flexibility

Stiffness is not always a virtue, and in nature most components are flexible and only a few are stiff. Man in his designs also uses many materials for their flexibility, and may well be tending to move towards practice in the living world in this respect. A century ago, there were few very flexible materials available to engineers, and those there were of organic origin: now there are a whole range of synthetic rubbers and flexible plastics, with and without fibre reinforcements.

Flexible materials mostly depend upon chain molecules, and man's understanding of, and ability to make and work with, such substances is very recent. However, the flexibility of materials like steel, tiny as it is, can be put to great use, as will be seen in the chapter on mechanisms.

3.3 Molecular structure

The properties of materials are determined by their molecular structure, a subject which is beyond the scope of this book. Briefly, the forms of most importance in solid materials are the crystal, in which the atoms are arranged in a regular three-dimensional lattice, and the polymer, formed of cross-linked chains of molcules. Perfect crystals are very strong, but real crystals, unless grown under special conditions, contain defects called dislocations, which are rather like the effects of a dropped stitch in knitting. Under applied stress, metal crystal dislocations move rather easily through the structure, causing a small slip in the plane in which they move. It is the accumulation of movements of this kind which causes yield in ductile metals.

Most materials under stress develop cracks, which may propagate or extend through the material. It is rather like tearing cloth – once you can start a little tear, it may propagate rather easily right across a sheet. The condition for this to happen depends on energy – if the crack or tear is to extend unit distance, the energy to break the material must come from the tearing agency – your hands perhaps – which exert a force F, say, and would move apart a little if the crack extended, x, say. Then if Fx is greater than the energy needed to break the material, the crack or tear will propagate.

You may have noticed that when tearing cloth, say old garments into cleaning rags, the tear stops at a seam or other interruption of the regular weave of the fabric. Seams are stronger than the rest of the material, but other interruptions will have the same effect – for instance, a slit at right-angles to the tear. Crack propagation may be prevented by introducing deliberate discontinuities to act like seams or slits in cloth, but this device belongs to microscopic, rather than molecular, levels.

Dislocations, however, may be stopped from moving by discontinuities on the molecular scale, produced by alloying elements, and these raise the stress at which yield begins.

3.3.1 Polymers

Chain molecules form the basis of most living structural materials and also of plastics and synthetic rubbers. Instead of forming a lattice, the molecules are in long chains or strings, which may lie in all directions or may lie mainly in one direction like the fibres in carded wool. Each of these molecules may be thought of as having little 'arms' of various kinds sticking out along their length, which can take hold of certain of the 'arms' on other molecules, a process which is called 'cross-linking'. If all the molecules lay parallel and all the possible 'arms' were linked, a regular lattice would be formed, a crystal, in fact, and sometimes this condition is approached. An interesting example is the way in which silk is made by silkworms or web by spiders. Inside the caterpillar, silk is a liquid, consisting of long chain molecules moving about randomly relative to each other. This liquid is expelled through a narrow tube in the spinneret, which causes the chain molecules to align themselves with the direction of the flow, and hence with each other, bringing the little arms alongside one another. So much cross-linking occurs under these very favourable conditions that the liquid becomes a strong solid filament. The same method is used by man for making certain synthetic fibres.

Most of nature's building materials are of cross-linked chain molecules – cellulose and chitin, the chief structural substances in trees and insects, respectively, which are both carbohydrates, the protein keratin, the basis of birds' feathers and human hair and nails, among other things, and so on.

3.4 Mechanical structure

Both the individual crystals in metals and the changes between areas with many and few cross-links in a natural substance like collagen are visible through a microscope. Frequently, however, materials are also mechanically, rather than chemically, structured, as already mentioned in the case of bone. An example is wood, which consists of long tubes of cellulose bound together with substances called *lignins*, which also contribute to the strength.

Composites are materials having more than one component, such as bone, wood and GRP. Metal alloys commonly have more than one component – for example, Damascus and Japanese sword blades contained two principal *phases*, one with a higher carbon content, very hard and giving the cutting edge which was fabled to be able to cut through a silk handkerchief as it floated to the ground, the other softer but giving toughness and the capacity to absorb the shock of impact on enemy armour without breaking. These two phases occurred in layers, a product of that same remarkable production method of repeated successive folding and flattening which is used also to make flaky pastry. These layers produced characteristic banded patterns on the blade. But metal alloys are not usually regarded as composites in cases where there is no deliberate mechanical control over the shapes in which the phases occur.

Composites like bone and GRP are made largely of brittle substances like bone salt and glass, and yet have good properties of impact resistance provided that certain principles are observed in their design. For example, in GRP the strength lies mostly in the glass-fibres, and if these are too short, the full strength of which the composite is capable will not be developed: under tension, the material may fail by the fibres pulling out of the matrix, while remaining unbroken themselves. Also, the materials must be chemically compatible: the glass used in GRP cannot be used to reinforce concrete because concrete attacks it chemically, and a special glass must be used instead. The materials must bond well together, they must not expand too differently with rise in temperature, and so on. As with marriage, the possible difficulties are so many that it is surprising to find so many succesful alliances. One of the oldest and most happy is that between steel and concrete, still the most important man-made composite and full of bonuses like the protection of the steel against corrosion and fire.

3.4.1 Structuring of composites

A great potential advantage of composites lies in the possibility of varying their properties locally or directionally to suit the design. For example, wood is a composite which is also strongly structured in that most of the fibres lie along the grain, giving very great stength, but for loads acting in one direction only. In a thigh-bone, the structure runs along the shank, but is varied in orientation in the ends to take joint loads and muscle and ligament attachments.

A very neat example of a composite designed for a special purpose is given by a helicopter drive-shaft connecting the tail-rotor with the main rotor head. This was made in the form of a thin-walled tube of fibre-reinforced plastic, the reinforcement being in 10 layers, each with all the fibres laid in one direction. Because of the importance of avoiding whirling, a vibration in which the shaft moves like a skipping rope, most of the fibres, six layers of them, were laid along the length of the shaft (see Fig. 3.12). The inside and outside layers (1 and 10 in the figure) were laid circumferentially, and the second and third layers helically at 45°, one layer to a left-hand, one to a right-hand thread. The circumferential layers prevented any tendency of the very thin-walled shaft to buckle like a collapsed drinking straw, and for that purpose they needed to be on the inside and outside. The two helical layers were to resist torsion (twisting), and as this was a relatively light task, they were given the position least competed for by the two other purposes, resisting whirling and buckling. Thus there were closely-argued reasons for the nature of each layer and the order in which it was placed. In this case, the fibres were of boron.

It is unusual for the designs of man to have this degree of local structuring of the material, but the designs of nature are frequently much more

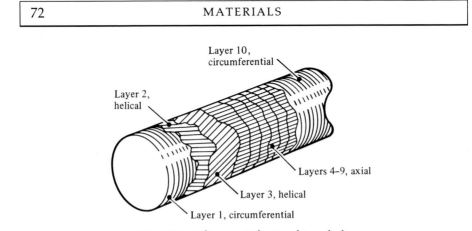

Fig. 3.12. Helicopter tail-rotor drive-shaft.

elaborate. As a production method, the growth of living organisms has many disadvantages, as we shall see, but among its advantages is the possibility of a subtle matching of the material structure to the requirements of function, to a degree which makes the helicopter tail-rotor drive-shaft seem unrefined.

Very often, design by man has, consciously or unconsciously, followed patterns already followed by nature, but in the case of structuring material it is not at all clear which way we are going. Indeed, it may be that in the pre-industrial era, when of necessity, more use was made of natural materials which were themselves structured and 'grainy', there was a more general appreciation of the needs and possibilities of composite construction. But the great difficulty of engineering with natural materials, both inorganic and organic, is their lack of consistency, and it was natural that the increasing availability of metals, with their stability, strength and reliability, should have established a climate of thought in which these attributes should have been felt to be the prerogative of homogeneous materials. A good example was the displacement of wood by iron and steel in ship construction, where the positive affection felt for wood was replaced, once familiarity with the new materials was gained, by the greatest trust being placed in the metals, with their freedom from splitting, shrinkage, shipworm, nail-sickness and all the other ills that affect wood.

However, in the last few decades pressures have arisen in engineering to go to composite constructions for critical functions because they can outperform metal. Sometimes this has been where very stiff, light structures have been needed in high-speed machines like looms, where carbon-fibres offer much higher stiffness for the same mass. In other cases higher specific strengths have been possible with composite materials. Moreover, the fibres and matrixes of these materials are synthetic, and therefore much more reliable in properties and in supply than natural materials. However, although the component materials now have consistent properties, it remains difficult to ensure consistency in the production process.

For example, GRP may be made by laying pieces of glass cloth in a mould and impregnating them with a liquid resin by hand using a brush, a process which is not easily controlled and may lead to variable proportions of glass and resin in the finished product, variable overlaps at the edges of pieces of cloth, etc. On the other hand, good control is possible with cloth impregnated with resin by machine, cut to exact shapes and laid up accurately in heated dies which are pressed together to form the finished product. Some GRP is made by winding filaments on a former by machine, and this process is capable of extreme precision, as well as producing a material structure very exactly suited to its function.

We may be on the brink of an era in which man-made composites will occupy as important a position in our material civilisation as natural ones did in the past.

3.4.2 The composite bow

A remarkable example of the use by man of several natural materials in combination, in such a way as to obtain a much better result than could be had with one alone, is the Turkish composite bow. This was a reflex bow, which is to say that with the string removed it took a form like that in Fig. 3.13(a); when strung it took the traditional 'Cupid's bow' form of Fig. 3.13(b). The cross-section is shown in Fig. 3.13(c), in which it will be seen that the bow is built up of three layers. The centre layer is of wood; the back, that is, the part of the bow away from the archer, which is stretched in stringing and stretched even more when the bow is drawn, is of sinew, which is strong and elastic in tension; the front of the bow, which is compressed by stringing and compressed even more by drawing, is of horn which is strong and springy when used in compression. A surprising detail is that the sinew is not, as might be expected, in long lengths, but in pieces only about 50 mm long glued to the back of the bow.

The last of these bows were made about 200 years ago, but their excellence may be judged by the following. Prior to 1910, the record for the distance an arrow had been fired by an archer was 340 m, achieved with a long-bow of osage-orange wood that required a force of over 700 N to draw

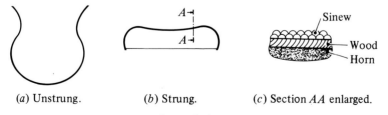

(a) Unstrung.　　　(b) Strung.　　　(c) Section AA enlarged.

Fig. 3.13. The Turkish composite bow.

it. At an archery contest at Le Touquet in 1910, Ingo Simon, using an old Turkish composite bow requiring a force of 440 N to draw it, fired an arrow 434 m (in a letter to P. Klopsteg, Simon wrote that the force needed to draw his bow was only 290 N).*

Similar bows, but of less refinement, were made by many races in the past, for example, the Eskimo and the North American Indians, using the materials that they had available. In this century a North American Indian showed how his people used to make bows using a glue made from boiled salmon skins, whereas the Turks made their glue from sinew. The basic material of such glues, whether made from fish skin or sinew, is collagen, one of the remarkable structural proteins found in natural design.

The composite bow shared the efficiency of the long-bow but was better matched to the muscles of the archer, as will be shown in Section 6.6.4, because of its reflex form. However, without the superior capacity of sinew for storing energy compared with wood (and, indeed, with almost every other material), it would scarcely have been possible to make a practical reflex bow, yet another example of the critical role of materials in design.

3.4.3 The comparative strength of composites

One of the great attractions of composite materials which has been noted is the ability to use some of the strong and stiff fibres which have been developed in recent years, and so to obtain lightness and strength not otherwise possible. For instance, glass-reinforced plastic of an ordinary kind gives a strength of about 40% of that of mild steel, but at about 20% of the weight. Carbon-fibre of a high-strength type can be as strong as steel, although its density is about one one-quarter as much, so that its specific strength can be four times as great. More recently still, some aramid (polyaromatic amide) fibres have strengths of almost 3000 N/mm^2, 50% greater even than that of very hard drawn steel wire, and less than one-fifth the density.

It must be remembered, however, that fibres provide strength in one direction only and must also be embedded in a plastic matrix which approaches the same mass but does not usually contribute to the strength. Above all, the problem of making joints between separate parts is more difficult than with homogeneous materials like metal or plastic.

3.4.4 Wood

Wood is a remarkable example of a structured composite, the basic constituents being cellulose, a fibrous polymer, and lignin, a kind of resin which acts as the matrix sticking the cellulose together. The wood grows in the form of elongated cells, mainly all lying parallel to one another, along the

---

* Klopsteg, P. E. (1947). *The Turkish Bow*. Evanston: P. E. Klopsteg.

grain. The cellulose forms the walls of these long thin cells, which form tubes up which sap rises. In the parts towards the centre the cells are dead, and their only remaining function is structural.

Now the fracture of both cellulose and lignin is brittle and absorbs little energy, so that special means are needed to provide the toughness that is such a familiar characteristic of wood. Toughness depends on the energy absorbed during fracture, and this is large in wood in spite of the brittle nature of the constituents, because the area which has to be broken is very large. This is achieved by the special geometrical arangement of the cellulose fibres. These run along the walls of the tubular cells in a helical fashion, as shown in Fig. 3.14. The helices have a long pitch (distance in which one turn is completed), so that the angle between the fibres and the axis of the cell is not great.

When wood is stretched, the helices tend to straighten and pull away from the walls, into the centres of the cells, and this is what happens if the tension is great enough. To do so, the helices have to break the lignin sticking them together, and the area over which they are stuck is very large. Breaking such a large area of lignin, in spite of its brittle nature, requires a large amount of energy, so that this subtle structuring of wood overcomes the deficiencies of its constituents and endows it with the toughness it needs to fulfil its functions.

3.5 Fluid materials

Not all the materials used by designers are solids: many are liquids or gases. Some of the important purposes to which they are put are the transport of

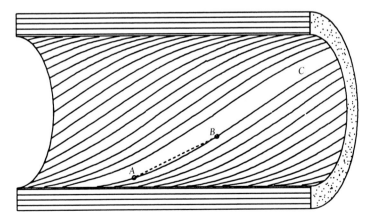

Fig. 3.14. Structure of wood. ABC *represents a typical helical fibre. Under tension it tends to pull away inwards from the wall, because the direct distance between any two points* A, B *is less than the distance along the helix.*

energy, materials and signals. For example, blood is used to convey energy in the form of heat and materials of many kinds, and it may also carry signals in chemical form, such as the hormone adrenalin which readies the whole body for extreme exertion. Liquids, or more rarely gases, are used to lubricate joints and bearings, and gases, or more rarely liquids, are used to act as springs and store energy, as in the hydraulic accumulator mentioned as a possible energy store for the small-engined car, where oil transmits energy and a compressed gas stores it. Gases are used as the working fluids in heat engines and refrigerators: it is interesting that in his prime movers for converting the chemical energy of fuels into mechanical work (internal combustion engines and steam power plants) man uses an energy conversion process (burning) in the gaseous phase, whereas nature in her prime movers (muscles) uses a solid phase converter.

One other use of fluids, not often recognised, is that of supporting loads, as is done, for example, by the air in vehicle tyres or the body fluids in soft-bodied creatures like worms and squids.

The most important fluids by far in both living organisms and engineering are air and water, although the water often has additives. In engineering these may be used to inhibit corrosion, to prevent freezing and to provide some lubricating properties: in that very important water-based fluid blood, even if we discount the solid components which it may be regarded as merely transporting and consider only the liquid plasma, the additions and their functions are both very numerous.

Engineering designers, while using air or water for preference, with oil as their next choice, often use other fluids where special properties are required. For example, the rotors of the big electrical generators in power stations nowadays usually have streams of hydrogen passing through them to cool them: this very light gas that forms dangerously explosive mixtures with air and is difficult to keep sealed in, is used because of its excellent cooling properties. Even though it presents many difficulties in use, it increases the efficiency of the generators by better cooling than can be provided by the air that was once used. Water would be a better coolant than any gas, although it presents even greater problems, and it is likely to replace hydrogen in the future. Liquid sodium metal has been used for cooling the valves of engines, nuclear reactors and turbine blades because of its ability to transfer heat very rapidly.

Lubricating oils are chosen for their viscosity, but it is desirable also that their viscosity should not fall too rapidly with temperature, as otherwise they will be either too thick when cold, or too thin when hot. In addition, they must not deteriorate too rapidly when hot, they must not foam excessively when churned by moving parts, they must not attack other materials in the system, like rubber seals or protective paints, and so on. Many oils have substances added to them so that if lubrication breaks down locally and overheating takes place, they attack the metal very superficially producing a skin which resists seizure, the local welding together of the moving parts.

3.6 'Smart' materials

For most of history man has had to be content with the materials which were to hand – stone, wood, plant and animal fibres, leather, bone, and so on – and those developed by largely empirical techniques. It is difficult for us today to grasp, for instance, what determination and vision it must have taken to learn to make iron, for example. It needs a good hearth and a great deal of fuel burnt at a high rate for a long time, and it is a wonder it was done so early. But with the scientific knowledge we have today, within rather narrow limits, we can produce materials with the properties we want. A good example is the development of materials for gas turbine blades, which have to withstand high stresses at high temperatures, where they are not far off melting.

Another example is photochromic materials, which are used in sunglasses which darken to admit less light when the level of radiation increases, or glass which reduces the loss of heat from buildings by radiation through windows. Such materials are sometimes called 'smart'. Some react to an outside influence in a desired way, like the photochromic ones, while others have unchanging properties, like the window glass.

Some packaging may indicate its age and hence that of its contents by changing colour, while other kinds may be formulated to biodegrade once they have fulfilled their task. Materials may be designed to indicate when they have been overstressed, as in climbing ropes, or subjected to too high a temperature.

Another group of materials which are finding many uses nowadays are some alloys with 'shape memory'. Such materials can be manufactured in one shape and then formed into another, by pressing for instance: when heated above a certain temperature, called the transition temperature, they will try strongly to revert to the shape they were before pressing, and this behaviour can be used to effect some motion, for instance to open a sprinkler valve.

Another group of specially-designed materials are the zeolites, materials with complex molecular lattices which can be used to separate substances from mixtures. If one of two mixed substances can enter the interstices of the lattice of a zeolite and the other cannot, then we have a means of separating them. A great variety of zeolites can be made, and their composition may be determined to suit the task for which they are needed. Other substances can also be used in this way, as with the membranes which are used in reverse osmosis, to remove salt from water by pumping it through them, leaving the salt behind.

3.7 Energy, materials and information

The designer builds in materials and his designs must handle three resources, energy, materials and information (in which it is convenient to include instructions or orders). There is a sense in which information itself must always be embodied in energy or material – pulses in telephone wires or nerves, a packet of light waves from a traffic signal or in an optical fibre, ink marks on paper, regions of magnetisation on a tape or the deoxyribonucleic acid (DNA) that is the blueprint for making a living organism. Nevertheless, it is frequently convenient and useful to regard information as an entity in its own right, as will be done later in this book.

The last two chapters have dealt with the nature and qualities of energy and materials, the media in which organisms and artefacts are designed, and have gone further to pursue a few illustrations of the influence of these resources on design, such as the way the peak power needs of birds and ships are met and the structure of bone and composite bows.

The next three chapters deal with three important manifestations of design – mechanisms, structures and systems – which are the most conspicuous in any study of living organisms or the works of man.

4

Mechanism

4.1 Mechanism

The engineer and nature work with energy, materials and information. Three of the characteristics of the designs they produce are mechanism, structure and systems, common features of organisms which will be discussed in the next three chapters.

Mechanisms are familiar enough, in clocks, sewing machines, locks and catches, switches, the human hand, and so on. They can be considered as assemblages of simple elements, such as pivots and hinges, sliders, gears, and so on. Their functions generally are to transmit and modify forces and movements: the object of a door catch is to move a bolt under the action of a spring so that the door will be held closed, and to convert a turn of the handle into the retraction of that bolt when it is required to open the door. A door lock has a rather different function because it requires a very special kind of movement to withdraw the bolt, which depends on a knowledge of the code (in the case of a combination lock) or possession of a key which has the required movements coded into it. Historically, locks represent the human designer's first traffic in the third of his media, information, about 6000 years ago. They also exhibit, in a simple form, a crucial design principle, that of *stereospecificity*, that is, of actions like unlocking a door which depend on the interaction of components of very special shape. Stereospecificity is crucial to the highly-selective chemical processes in which the design of living organisms is founded.*

4.2 Pivots

This survey of mechanisms will begin with that very simple component, the pivot or hinge, like the hinges of a door or the bearings of a wheel, an elbow or the hinge between the two half-shells of the kind of shellfish called a 'bivalve'.

The wheel bearing is different from the other three examples because it allows unlimited rotation; it can go on turning in the same direction indefinitely, complete rotation after complete rotation. A wheel cannot be

---

* Ferry, G. (1985). The locksmiths of the nervous system. *New Scientist*, 6 May, p. 18.

connected to a body by a pipe, or a blood vessel or a nerve-fibre, because
any such link would be broken after, at most, a few complete turns. This is
one of the reasons the giant propeller-driven bird of Chapter 6 remains a
mere fantasy.*

If living mechanisms cannot achieve unlimited rotation in a pivot, at least,
not above the molecular scale, they have advantages in the production of
pivots not shared by man until very recently, and then, only to a very
limited extent. The hinge of a door, say, cannot be made complete at one
go: the pin has to be put through the two halves and then hammered over
or fixed with a nut afterwards (see Fig. 4.1). In nature, however, hinges
grow in such a way that they are assembled from the beginning: the jaw-
bone cannot be removed from a badger's skull without breaking bone, and
the shells of some shellfish cannot be separated without damage. One part
has grown inextricably interlinked with the other, rather as in those chains
of many links carved from a single piece of wood. Just recently man has
learnt how to cast two pieces hinged together, but the technique is very
new and useful only in rare cases.

4.2.1 A butterfly valve

Figure 4.2(*a*) shows a simple butterfly valve, consisting of a circular disc
pivoting on a diameter *AA* and serving to close off a pipe or tube. Figure
4.2(*b*) shows diagrammatically the view down *AA* with the valve closed; the
open position is shown by a dotted line.

Now such a simple device would seem incapable of much refinement. It
is found in the carburettors of cars in virtually this simple form, and there

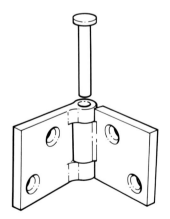

Fig. 4.1. Hinge.

---

* At the time of writing, the only joints in living organisms capable of unlimited rotation were thought to occur in
the flagellae of certain bacteria, but now there is evidence that a protozoan, familiarly called 'Rubberneckia', demonstrates the same power. *The Collecting Net*, vol. 4, No. 4, July 1986, pp. 11–12. The Marine Biological Laboratory,
Woods Hole, Mass., U.S.A.

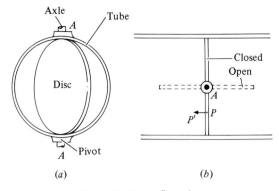

Fig. 4.2. Butterfly valve.

it works very satisfactorily, because not much is asked of it. But regarded as a tap or valve is usually regarded, as a way of cutting off flow completely, it is not very good since it tends to leak. To see why this is so, consider an even simpler valve, the common bath plug.

Many bath plugs have the form of a slice of a cone (Fig. 4.3(a)), so that they push into the plug-hole until they fit tightly. A cylindrical bath plug (Fig. 4.3(b)) would not be nearly such a good design. The desirable feature of the conical form is that as it enters the hole the gap between the two decreases steadily all the way round. This progressive narrowing of the gap until contact and a seal is achieved will be called 'closure'. The cylindrical bath plug and the simple butterfly valve do not provide closure. For many purposes, butterfly valves which do provide closure are desirable. Let us study the problem of designing such a valve.

First, we need one very simple idea that is central to much of the design of mechanisms. Consider the movement of the point P in Fig 4.2(b) when the disc is given a very small clockwise turn: it moves in the direction of P', at right-angles to AP. We say that when when a body turns about a pivot A, any point in that body moves at right-angles to the line joining it to A.

Armed with this simple but powerful idea, look at the version of Fig. 4.4(a) which, like all the cases we shall consider, closes by turning clockwise or right-handed. Here there is closure at the points X, where the motion is

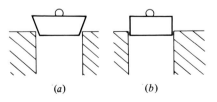

Fig. 4.3. The closure of bath plugs.

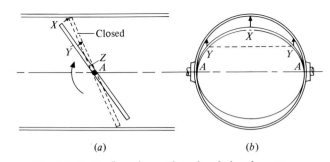

Fig. 4.4. Butterfly valve with inclined closed position.

in the direction shown by the arrows, at right-angles to AX. This motion is partly along and partly towards the tube wall, and it is the 'partly towards' element or *component* that gives closure. What happens at a point such as Y on the edge of the disc? To see this, consider the end view looking down the tube (Fig. 4.4(b)). The motion at Y has a component away from the axis of the tube, as shown by the arrows, which is moving the chord YY of the disc towards the side of the tube, where the width of the tube is less, so there is closure. At points like Z, however, the closure effect is very small. Moreover, there is bound to be trouble at A, the pivot itself. Ideally, the disc should have an unbroken sealing face all the way round, like the bath plug, and the seal should not be broken by a pivot.

In Fig. 4.4(a), consider the areas in which there is closure on a valve rotating clockwise about A. These are shown shaded in Fig. 4.5: at the edges of the shaded zones the closure vanishes, so that near to them it is poor, i.e., at Z, V or W, whereas in the interior of the shaded zone, at X or Y, the closure is good. What we require is a design of disc, such as the trial shape shown: it should miss A by a wide margin, and have good closure everywhere. Our trial shape clearly meets the criterion of keeping clear of the pivot A; how is it on closure?

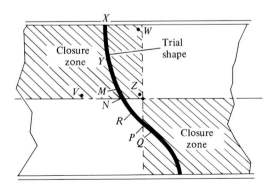

Fig. 4.5. Closure zone: central pivot.

The answer, alas, is that it is unacceptable. The closure is good, except from M to Q. Near M and Q it is poor, but between N and P the surfaces which should be moving *towards* their sealing on the tube wall are actually moving *away* during the last part of the rotation. Indeed, if the valve sealed, it would not open, because for it to open the part of the disc at R would have to move into a narrower part of the tube.

The fact is, the arrangement of the shaded areas does not allow us to design a disc which meets the criteria. The shape of the areas can be changed by changing the form of the tube and moving the pivot A off-centre, and *both* changes are needed to solve the problem.

Now in nature sealing faces of the most subtle shapes can be made and made to fit each other beautifully, as where the lips of the shells of molluscs close on one another. But the manufacturing methods of engineering can only produce simple, easily-made shapes if they are to be both accurate in form and cheap. Prominent among these shapes is the cylinder, the original form of the butterfly valve, after which the next best, nearly as easy to make, is the cone. A cylinder will not do in this case, so the next shape to try is a cone.

Figure 4.6 shows the closure zones for an off-centre pivot A in a conical tube (only a short portion, E to F, need be conical – the rest can be cylindrical) – together with the zone of closure. This zone is bounded by two curved boundaries B, one of which goes through A. Now there is a great broad shaded area going past A, so we can easily draw a shape of disc which misses A by a wide margin and remains well inside the shaded zone, where closure is good. Indeed, we now have an embarrassment of riches, for we have to choose among the many shapes which would meet the criteria.

It is a general principle of design, when there is a choice, not to make an arbitrary decision but to use the choice to advantage. We could choose the shape to make the closure uniform, so that as the valve closed, the distance between the sealing surfaces was the same all the way round, i.e., when the

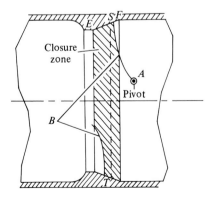

Fig. 4.6. Closure zone: offset pivot in conical tube.

gap was one millimetre, it would be one millimetre all the way round, when it was half a millimetre, it would be half a millimetre all round, and so on.

Alternatively, we could use the choice to make the shape easy to manufacture. The sealing surface on the disc is a narrow strip which can be turned on a lathe. Imagine the valve shut and the disc rotating about the axis CC (Fig. 4.7), the axis of the cone. If the disc is fixed to the face plate of a lathe and rotated about CC in this way, and a cutting tool traces out a path corresponding to the wall of the tube, it will cut the right shape. But, unless the sealing surface lies in a strip in a plane at right-angles to CC the cutting will be interrupted or intermittent, which is not good. For example, with the disc form shown in Fig. 4.7, the tool will not be cutting a full circle. When it is at G, for example, it will be cutting only that portion of the disc near H, and round the rest of the circle it will not be in contact with the disc and not cutting metal. Now this interrupted cutting must be done slowly, there is a danger of breaking the tool, and the accuracy and finish of the work will be poor. It would therefore be helpful if the disc were flat and, when the valve was closed, perpendicular to the axis of the tube.

It so happens there is only one shape of disc which meets the criterion of uniform closure, and that is the shape represented by ST in Fig. 4.6: thus the ideal disc as regards closure happens also to be ideal for manufacture (which is not a common occurrence in design).

The mathematics necessary to find the shaded areas of Fig. 4.6 is part of elementary co-ordinate geometry: the curves B are the two branches of a hyperbola. The mathematics involved in finding the shape that will meet the ideal closure criterion is only a little more advanced, but neither will be given here. The whole is an interesting branch of a science of great importance in design, that of *kinematics*: it deals with the relative motions of bodies. Some aspects of kinematics are very abstruse, but we shall chiefly need one simple proposition which has already been used in Fig. 4.4, when

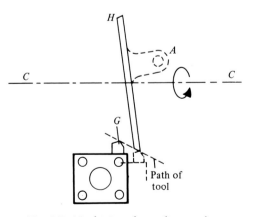

Fig. 4.7. Machining the sealing surface.

it was noted that when the disc *AX* turned about *A*, the motion of *X* was at right-angles to *AX*.

This idea, that when a body turns through a small angle about a pivot at *O*, say, the motion of point *P* is at right-angles to *OP*, recurs very often in design, as in the next two examples, the front suspension of a car and the human knee, both of which are what the kinematicist calls 'four-bar chains'. However, it can also throw a very clear light indeed on the butterfly valve problem.

Suppose the disc were replaced by a sphere, as in Fig. 4.8, fitting the conical tube round a circle at *ST*. Let *O* be the centre of the sphere and *A* any pivot point such that *OA* is at right-angles to the axis *CC*. Now if we pivot the sphere about *A*, the motion of *O* is pependicular to *OA*, that is, along *CC*. Under these circumstances, there is uniform closure (or opening) all round the circle at *ST*, just as with the bath plug. Of course, besides moving along *CC* the sphere will also rotate about its centre, but this makes no difference to the closure, because if we rotate a sphere about its centre there is no movement normal to its surface. If we now cut away all the sphere except that near the pivot and the plane *ST*, we have the solution we found before, except for the slight difference between the conical and the spherical shape, which is insignificant over the narrow sealing zone.

The argument of the last paragraph replaces, completely and more satisfactorily, all the mathematics of the original demonstration. It is simpler and better. This is often the case with functional design: the end point of a train of argument is elegant and pleasing, but of crystalline lucidity when looked at in the right way.

4.3 Car front suspensions

The front-wheel of a car turns on a short, non-rotating stub-axle. This stub-axle has to move up and down relative to the body of the car, so that when the wheel goes over a bump the full shock is not transmitted to the

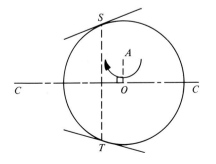

Fig. 4.8. Another approach to the closure problem.

passengers but is absorbed by the suspension. The stub-axle also has to turn about a vertical axis when steering, but for the present this second motion can be ignored. The engineer would say that, relative to the car-body, the stub-axle has two 'degrees of freedom', the up-and-down motion concerned with the spring and the turning motion involved in steering.

One method of achieving the 'up-and-down' freedom is shown in Fig. 4.9. The stub-axle is part of a wheel-carrier which is pivoted to links AB and CD, which in turn are pivoted to the car-body at A and D. A coil spring acts between the lower link AB and the body, pushing the car up relative to the wheel. It is because the links AB, CD have a form reminiscent of a chicken wishbone that the suspension is so-called.

When the wheel passes over a bump, it will rise relative to the car, moving exactly as if it were turning about the point I, fixed in the car body where the lines BA and CD produced intersect.

To see this, consider the wheel moving slightly upwards, the car-body being regarded as fixed. The point C, being constrained by the link CD, will move at right-angles to CD, that is to say, at right-angles to IC. Similarly, the point B moves at right-angles to IB, so that both points, C and B, move exactly as if the wheel-carrier were turning about a fixed point at I. In fact,

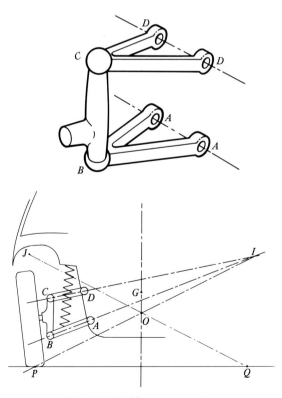

Fig. 4.9. Wishbone suspension.

for small motions only, the wheel-carrier and everything attached to it, including the point P, move exactly as if they were pivoted at I, and P moves at right-angles to IP. The point I is called the instantaneous centre for the wheel-carrier and stub-axle, because they move as if they were pivoted there. Note particularly that the centre of the contact patch of the tyre on the road, P, will move at right-angles to PI.

The wishbone suspension is an example of a 'four-bar' chain, so-called because it is composed of four links (AB, BC, CD, DA) connected in sequence to form a closed loop ABCD. Generally one link (here AD) is fixed and we are interested in the motion of the opposite link (here BC) relative to the fixed link.

Now consider the movement of the car in roll relative to a perfectly level surface. Since the motion of P relative to I is at right-angles to IP, so also if we regard P as fixed, the motion of I relative to P is at right-angles to IP. Similarly, if J and Q are the instantaneous centre and centre of contact for the opposite wheel (see Fig. 4.9), J moves at right-angles to JQ. It follows that O, the point on the centreline where IP and JQ intersect, is the instantaneous centre of the car relative to the road, and it is called the roll centre, rationally enough since it is the point about which the body will roll.

If we stand facing the side of the car and push it above the level of the roll centre, it will lean away from us, if we push it below the roll centre, it will lean towards us, and if we push it opposite the roll centre it will scarcely move at all.

Consider the implications for cornering. If the centre of gravity of the car is above the roll centre, say, at G in Fig. 4.9, then when cornering the car will tend to roll outwards. If the roll centre is above the centre of gravity, the car will tend to roll inwards. Why not put the roll centre at the centre of gravity, and avoid roll altogether?

The answer lies in the kinematics again. If the body of the car moves down in Fig. 4.9 the movement of P and Q relative to O is upwards and outwards, at right-angles to OP and OQ respectively. This outward movement will be more pronounced if O is higher, and while the flexibility of the tyre can cope with a little of it, too much will cause rapid wear and affect the grip on the road. The designer therefore puts O above the road, but not as high as G.*

There is thus a tendency of the car to roll outwards on a bend, which is limited by the stiffness of the springs. Normally, if the springs were made stiff enough to reduce roll to an acceptable amount, they would be too hard and give a rough ride, so an anti-roll rod or similar device is added, which does not affect the stiffness of the suspension when both sides go up or down together but resists differential movement between the two sides.

---

* For the geometric and other aspects of car suspensions, see Bastow, D. (1980). *Car Suspension and Handling*. London: Pentech Press.

Figure 4.10 shows an alternative to the wishbone suspension; the lower wishbone *AB* is kept, but the upper one is replaced by a telescopic member *BD*, which is rather like a bicycle pump. For clarity, *BD* has been drawn much more like a bicycle pump than it is really is. The stub-axle is fixed rigidly to the body of the 'pump' and the handle of the 'pump' is pivoted to the body at *D*.

It is not easy to see where the instantaneous centre *I* is in Fig. 4.10 but it is at the intersection of *BA* and the line through *D* at right-angles to *BD*.

This 'Macpherson strut' suspension is attractive to designers for a number of reasons. Compared with the wishbone suspenson, it takes up less valuable space in the car (D_2 in Fig. 4.10 shows roughly where the upper suspension point of the wishbone tye would come). The 'bicycle pump' arrangement can readily accommodate a shock-absorber and the spring can conveniently be fitted round it. For steering, the wheel-carrier has to be pivoted at top and bottom so it can turn, and the top pivot is readily provided by the rotation of the 'pump-body' on the 'handle'. Finally, the angular movement at *D* is small, so that a simple flexing bearing, requiring no lubrication, will suffice. These are typical of the kind of considerations which the designer must weigh in his work.

4.4   The slider-crank chain

To the kinematicist, the strut suspension is just an 'inversion' of one of the best known of all mechanisms, the piston, crank and connecting-rod arrangement seen in most reciprocating engines. Figure 4.11 shows this mechanism, with the parts labelled to correspond with the parts of Fig. 4.10. In each case, the points *B* and *D* can slide closer together or farther apart, and there are pivots at *B* and *D*. In the suspension, however, the points *A*

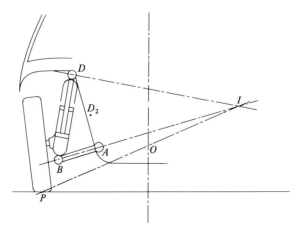

Fig. 4.10. Strut suspension.

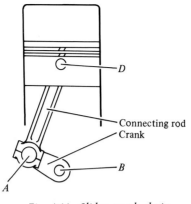

Fig. 4.11. Slider-crank chain.

and D are fixed, while in Fig. 4.11, the point B and the direction of BD are fixed. Furthermore, the slider-crank chain can be shown to be just a special case of the four-bar chain.

4.4.1   Insect flight mechanisms

A small insect flies by means of a pair of slider-crank chains. Figure 4.12(a) is a diagram of the cross-section of an insect thorax – the part to which the

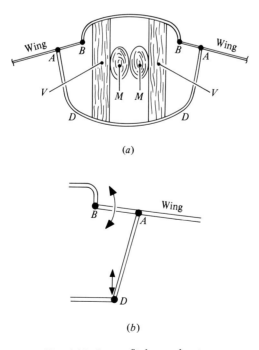

(a)

(b)

Fig. 4.12. Insect flight mechanism.

wings are attached. The tough exoskeleton, the armour and structural part of the insect, is roughly tubular in section at this point and forms the wall *DABBAD*. It is flexible near *D*, and also at *A* and *B*, so that in effect there are hinges there. When the vertical bands of muscle *V* contract, the lower wall of the thorax *DD* rises, forcing the points *A* upwards relative to the points *B* and so raising the wings. Then the muscles *V* relax, and longitudinal muscles *M* contract, shortening the thorax so that the viscera or innards of the insects are squashed outwards, forcing *DD* down and so causing a downward wingstroke. The mechanism is, in effect, a pair of slider-crank chains (Fig. 4.12(*b*)), with the parts *BA* having the role of cranks, the parts *AD* the role of connecting-rods and the points *D* corresponding to the pistons. Unlike the crank in an engine, which turns continuously, the parts *BA* swing only through a small angle and back again.

The insect flight mechanism is in reality a great deal more complicated than this account suggests, involving movements out of the plane of the paper in Fig. 4.12.* The reason for this mechanism, which is very unlike that used to operate the wings of birds and bats, or, for that matter, the legs of insects, lies in the peculiar effects of scale on flight. Small creatures can fly more easily than large ones but to do so they must flap their wings very fast. Ordinary animal muscle contracts by about 20% of its length and is capable of at most 100 cycles of operation per second.

Most insects need to beat their wings faster than this, which requires the use of a different kind of muscle which can operate much faster but only contracts by about 3%. With such small (but forceful) contractions, the flying mechanism must magnify the small movement extremely, and this can be achieved by making the distance *AB* sufficiently short. This is best done with flexible hinges, not pivots.

Let us now look at an application of four-bar chains in the human body.

4.5   The human knee

If an engineer were designing the human knee, he would probably think in terms of a simple pivot or hinge joint. But while there are such joints in nature, they are not ideally suited to the production processes and materials of living organisms, and the knee uses no pivot or pin in its construction. The connection between the bones above and below the knee is made by crossing ligaments, the cruciate ligaments, which work rather like a pivot. The diagrams in Fig. 4.13 show the knee, seen from the side, in three different states of flexure – straight, fully bent and half-way between. In each position the ligaments prevent the bones being pulled out of contact or slid sideways one relative to the other, because they are very strong and stretch

* Pringle, J. W. S. (1957). *Insect Flight.* Cambridge University Press.

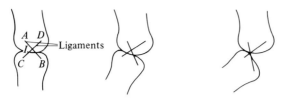

Fig. 4.13. Knee in various states of flexure.

very little. As long as they remain taut, we can regard the cruciate ligaments as rigid links, so that *ABCD* is a four-bar chain. The instantaneous centre is where *AB* and *CD* cross, that is, at *I* in Fig. 4.13, so that the only motion allowed to the lower leg relative to the upper is rotation about the point *I* (which, incidentally, moves quite a distance backward as the knee is bent).

In order to keep the ligaments taut, the ends of the bones must be specially shaped to remain in contact all the time (in reality, the bones are not in contact directly with each other, but with the cartilage which separates them, which has been omitted from the diagrams for clarity).

4.5.1   A model knee

You might like to try an exercise in knee design, which is easily done with card, scissors, pencil and drawing pins. On the middle of a piece of card about 15 cm square, sketch the shape of the head of the bones of the lower leg: you might make the head flat, as in Fig. 4.13, or slightly hollow, if you prefer, as in Fig. 4.14(*a*). Cut away the card above the top of your outline, so you have something rather like Fig. 4.14(*a*). Now take a large piece of

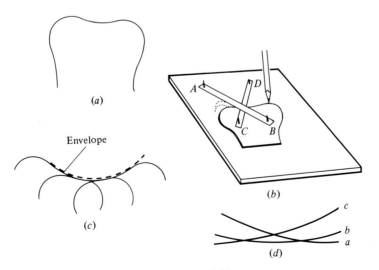

Fig. 4.14. A model knee.

card, say 25 cm square, on which you are eventually going to draw the bone of the upper leg.

Push two drawing pins through the small card from underneath, at the points representing the attachments of the ligaments to the lower leg, B and C in Fig. 4.14(b), and, similarly, put two drawing pins in the larger card to represent the attachments A and D to the upper leg. Now put the small card on the large one, and add 'ligaments' made of strips of card about 8 mm wide and of suitable length, pushing them down over the drawing pins (but minding your fingers) as in Fig. 4.14(b). You now have a working model of the action of the knee, for you can swing the lower leg about on the upper leg. However, the action is incomplete because it lacks the articulation between the bones, which demands a special shape of the end of the upper bone (femur) to work upon the lower form which you have chosen, with the attachment points and ligament lengths you have also chosen.

To find the required shape of the end of the femur, all you need do is move the lower leg by small stages through its required range of bending, at each stage drawing round the form of the upper surface of the lower bone, as is shown in Fig. 4.14(b). Figure 4.14(c) shows the kind of result you may obtain: all the curves will touch a common 'envelope' curve which is the required shape, and if you cut away the unwanted card below it, the model is complete and will show both the action of the ligaments and the articulation of the bones.

It may be that your choice of shape for the lower bones and of end points for the ligaments will be unfortunate, and the design will not be good: Fig. 4.14(d) shows one way in which this may happen. The lines marked a, b and c are three of the tracings made on the larger card. Now b is the shape of the upper face which would touch the lower face everywhere in position b, but is cut away entirely by cutting along the lines a and c corresponding to positions on either side. This defect is called 'interference'. In position b, if the ligaments were held taut by pulling on the lower leg, there would be a gap between the bones. If you put your weight on such a knee in position b, the bones would be pushed together, the ligaments would go slack, permitting sideways bending perhaps, and your leg might well collapse under you. You must try again, with different ligament attachment points, until you find a design in which there is an envelope curve such that every one of your traced lines touches it, and none of them misses in the way line b does in Fig. 4.14(d).

This exercise illustrates the great range of choices possible in even a very simple design problem, and the difficulty of deciding between them. In this case, there are four ligament attachment points and the profile of the lower bones to choose. But a real joint is more complicated: it is three-dimensional, and it only acts purely as a hinge when it is straight or nearly straight, behaving in a more complicated manner when it is bent.* The articulation of the bones is not exactly of the kind in the model. The ligaments stretch, and the shape of the joint faces is such as to change the stretch, and hence the tension, in them, and allow the more complicated three-dimensional behaviour of a real knee.

4.5.2 The behaviour of the instantaneous centre

You may find some difficulty in accepting the notion of an instantaneous centre such as I in Fig. 4.13. When the joint is bending, with the upper leg held still, the lower leg is rotating about I, that is to say, it moves as if there were a fixed pivot at I. But as you will see if you make the model, I is not fixed, but moving quite rapidly: in fact, the lower leg rotates about the *fixed* point momentarily coincident with I. You can demonstrate this very clearly, if you wish, with the aid of the model knee.

To the top of the 'lower leg' card, attach a piece of tracing paper which extends under the 'ligaments' and covers the region around I (Fig. 4.15). Replace the 'ligaments' with wider strips of card with holes cut in them which coincide in one position of the 'lower leg', as in the figure (strips about 4 cm wide with 2 cm holes will do well). Rule lines joining the drawing pin holes in the 'ligaments' AB and CD, so you can easily see where they cross. (If you can use thin transparent plastic for the ligaments, you will not need the large holes and you will have a very good model.)

Set the model so that the large holes fit over one another and make a pinhole at the intersection of AB and CD (if you use transparent plastic you will have to lift the ligaments off the drawing pins without moving anything else), rule lines AB and CD on the tracing paper by carefully placing a ruler against the drawing pins, and then push a pin through the tracing paper and 'upper leg'. Now replace the ligaments if necessary, move the lower leg slightly and watch the pinholes and the intersection of AB and CD.

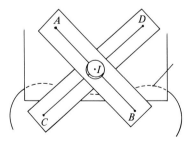

Fig. 4.15. The behaviour of the instantaneous centre.

---

* Barnett, C. H., Davies, D. V. and MacConnaill, M. A. (1961). *Synovial Joints, their Structure and Mechanics.* London: Longmans.

If you have done all this carefully, you will see that the pinholes remain very close together while the intersection of *AB* and *CD* moves off several millimetres. In the starting position we really have three '*I*'s, say, I_1, the pinhole in the upper leg, I_2, the intersection of *AB* and *CD*, and I_3, the pinhole in the tracing paper which is fixed to the lower leg. I_1 and I_3 remain close together, showing that the representation of the motion as a pivot between upper and lower legs at *I* is a very good one.

Notice that in the knee joint, where something approaching a pivot is required, the ligaments or links are crossed to bring the instantaneous centre in the middle of the mechanism. In the wishbone suspension (Fig. 4.9) where something approaching parallel motion is required, the links are nearly parallel, so that the instantaneous centre is a long way away, and the motion is mostly a translation, a bodily displacement, without much rotation. Indeed, if we make the links parallel, then *I* is pushed away to infinity, and the motion is a pure translation, without any rotation. This is what is done in a parallel ruler (see Fig. 4.16) where *I* is at infinity.

4.6   The wheel

A very convincing demonstration of the use of the concept of the instantaneous centre is given by the wheel. Imagine a wheel rolling on the road without slip, as in Fig. 4.17(*a*). There are no links in this case, but it is easy to see that the point on the wheel where it touches the road is instantaneously stationary, since the road is stationary, and so the instantaneous centre, *I*, is at that point. Then any *small* motion of the wheel is equivalent to a rotation through a small angle about *I*. The movement of *O*, the centre of the wheel, in such a motion will be perpendicular to *OI*, i.e., in the horizontal direction, which clearly it must be.

Now apply this idea to Fig. 4.17(*b*), which shows an axle carrying a pair of wheels resting on an inclined plane. A string is wound anti-clockwise round the axle and led off up the plane and parallel to it at *S*. The wheel and plane are rough enough to prevent slip: what happens when the string

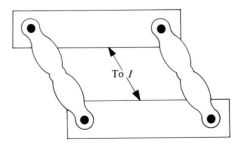

Fig. 4.16. Parallel ruler.

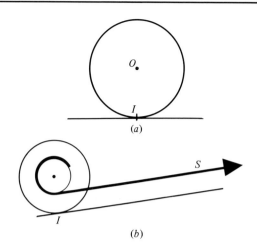

Fig. 4.17. Instantaneous centre of wheel.

is pulled just hard enough to produce motion? Now the instantaneous centre is at *I*, and the pull in the string tends to turn the wheels and axle *clockwise* about *I*. Thus the use of the idea of an instantaneous centre leads us to expect the wheels to roll clockwise, i.e., *up* the plane. Moreover, if the wheels turn clockwise they will wind string on to the axle.

Now these conclusions may seem at variance with common sense at first. That the wheels should roll up the plane, winding up string as they go, may appear unlikely or even impossible, but they do. Moreover, once the concept of the instantaneous centre and its implications have by familiarity become second nature to us, this behaviour seems the most natural thing in the world. The writer remembers very well that he was greatly surprised by this phenomenon at the age of 18 when he first saw it but, nevertheless, he finds it difficult now to see anything surprising in it. This is an example of the way in which simple but powerful concepts can improve our insight into the physical world.

If you wish to prove this result experimentally it is easily done, except that it is not always easy to prevent slip. Use a gentle slope – say, one in eight – and an axle which is small in diameter compared with the wheels: a pencil pushed through two cardboard discs works very well. Alternatively, a cotton reel of the waisted kind will do, but use a gentler slope still and lay a piece of sandpaper on it to increase the friction and prevent slip.

4.7 Degrees of freedom: conformity

Mechanisms like the four-bar chain are said to have one degree of freedom. There is only one way the knee can move, it can only increase or decrease the angle between the upper and lower leg (this is not quite true, for when the knee is bent you can also twist it a little). If we consider the front-wheel of a car, it has three degrees of freedom relative to the body of the car – it can turn on its axle, it can turn to the right or the left for steering and it can move up and down on the suspension to cope with bumps. The human lower jaw has three degrees of freedom: we can lower or raise it, opening or closing the mouth, we can thrust it forward or draw it back relative to the upper jaw or we can move the chin to left or right. The reason this is possible is that the hinges between the jaw-bone and the skull are capable, not just of flexing, but also of sliding backwards or forwards a little, as shown by the arrows in Fig. 4.18. If both slide forwards together, the chin moves forward: if the left side goes forward and the right backwards, the chin moves to the right.

There is a price to be paid for these extra degrees of freedom. If we only require a simple hinge, we can have a round pin in a round hole, and the two parts will touch each other all over a large surface (Fig. 4.19(a)). Such an arrangement is strong and not prone to wear. But if the hinge must also allow a sliding action, the pin must fit in a slot which it touches only along two lines (Fig. 4.19(b)).

There will be very high stresses along such a line, and failure or rapid wear may occur. This difficulty can be overcome by fitting the pin in a hole in a block in a slot (Fig. 4.19(c)). The pin can now be turned and slid, and all contacts are on large areas, but we have increased the number of parts. This solution was used in the axles of railway locomotives (see Fig. 4.19(d)), where the axle runs in a bearing which can slide up and down against a spring in a guide. The locomotive axle is a very close parallel to the human jaw, except the axle is capable of unlimited rotation. There is also this difference, that in the jaw only the upper side of the slot or guide exists, and this one-sided guide requires that the jaw be held against the skull by ligaments, just as the knee is held together by ligaments. Such a joint the engineer calls 'force-closed'.

There is yet one more complication in the three degrees of freedom movement of a jaw or an axle, which may be seen in Fig. 4.18(b), which shows

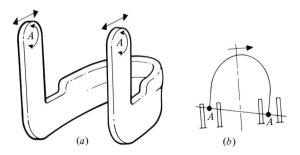

(a)   (b)

Fig. 4.18. Freedoms of the jaw.

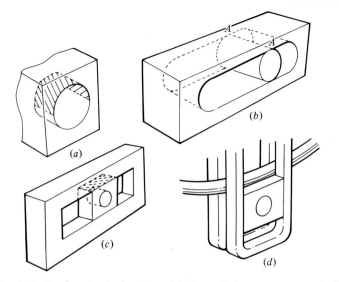

Fig. 4.19. Conformity in bearings: (a) Pin in hole: contact over shaded area. (b) Pin in slot: 'line' contact at AA. (c) Pin in hole in block in slot: contact on shaded areas. (d) Locomotive axle box (diagrammatic).

the jaw viewed from above with the chin moved to the right. The hinge axis AA is now not at right-angles to the 'guides' on the skull, and this spoils the fit of the 'bearings'. The role of the block in Fig. 4.19(c) is filled in the jaw by cartilages between the two bones; the guide has a convex surface and the jaw a slightly concave one to fit, and by these means a satisfactory effective bearing area is achieved. The freedoms of the jaw in animals are related to their diet. The jaw of a cow chewing the cud can be seen to be making large sideways movements, associated with the grinding of grass and similar matter between large, flat-topped molars (the word means grinders). The jaws of carnivores, such as cats and dogs, have a simpler and stronger bearing arrangement.

The human body has a great number of degrees of freedom. For instance, the upper arm has three degrees of freedom at the shoulder, which is a ball-joint. If you start with your right arm hanging by your side, you can raise it in a straight line either forwards or to the side, making two degrees of freedom. To assure yourself there is a third, raise your arm level with the shoulder and pointing straight ahead, with the index finger extended. Now bend the elbow till your index finger points straight up. Finally, without any further bending of the elbow and keeping the upper arm still pointing straight forward from the shoulder, rotate the forearm anticlockwise until the index finger points to the left.

Now in this last action, the upper arm rotates about its own axis, and the only joint in which movement occurs is the shoulder, thus demonstrating clearly that it has three degrees of freedom, as shown by the arrows in Fig.

4.20. The shoulder is indeed a remarkably mobile joint, and only excellent design is able to give such a variety of motions through such large angles, while retaining strength and small size. It is not surprising that it dislocates from time to time.

Proceeding down the arm, the elbow is practically a simple hinge, conferring just one more degree of freedom. Between the elbow and the wrist, however, a twist can occur, through about half a turn or 180°, giving a fifth freedom counting from the shoulder. The wrist itself can bend backwards and forwards, each finger can bend at its base and in its length and also at its base in the spreading direction, and that brings the number to 22, and so on for a few more. It is this great number of degrees of freedom which gives human beings the potential for great manipulative skill, seen perhaps at its most extraordinary in a concert pianist.

The robots that men make for use in production have many fewer degrees of freedom, perhaps a third as many, and they could not play the piano.

Lavish as nature is with complexity, because of her methods of production, degrees of freedom are not provided where they are not greatly needed, for they carry a heavy cost. If you press hard with one extended finger, on a table, say, the force you can exert is limited by the power of your muscles to prevent the finger bending. If the finger could also bend sideways, extra muscles would be needed to prevent this happening. If the space available for the two sets of muscles was only the same as for the one set we have, then the force which could be exerted would be reduced to half.

4.7.1 Nesting of degrees of freedom

Since degrees of freedom are costly, the design of an animal requires that the freedoms which are provided are so arranged as to give the best value. The common pattern among jointed limb animals, insects, crustaceans, amphibians, reptiles, birds and mammals, is to provide a ball-joint with two or three degrees of freedom at the junction of limb and body, followed by hinges further out.

The arrangement of different degrees of freedom in succession, say, from the body to the end of a limb in an animal, or from the foundations to the cutting tool or the workpiece in a machine-tool, is called their nesting order.

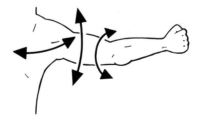

Fig. 4.20. Three degrees of freedom at shoulder.

To see why the common nesting order in jointed limbs makes sense, consider what would happen if the human shoulder were a simple hinge, say, with its axis of pivoting running from side to side, and the elbow were a ball-joint, an arrangement shown in Fig. 4.21(a).

Now if the body were kept still, the elbow could lie anywhere on the circle CC. With the elbow at any point E on the circle, the hand would be able to reach any point on a sphere F with its centre at E. By moving the elbow, the total space within reach would be the sum of all such spheres, which is a torus or doughnut shape with its middle at the shoulder (Fig. 4.21(a)). But with the real arm (see Fig. 4.21(b)) the hand can reach all points in a roughly spherical space, completely enclosing the doughnut and about 50% larger. Thus the ball-joint at the shoulder enables more points to be reached.

Another aspect of this problem is that a ball-joint is much larger and needs more muscles to work it than a simple pivot. Consequently, the arrangement in Fig. 4.21(a) would involve transferring mass from the shoulder to the elbow and the upper arm. But rapid motion requires that as little mass be moved as possible, so that limbs should be as light as possible and muscles located in the body as far as possible, a principle which has important aesthetic implications* (see Chapter 8). Figure 4.22 is an attempt to imagine the appearance of a human being with a pivot joint at the right shoulder and a ball-joint at the right elbow. The right shoulder is somewhat reduced in bulk, reflecting a provision of muscles reduced to less than one-half, since only one degree of freedom is now provided for instead of three,

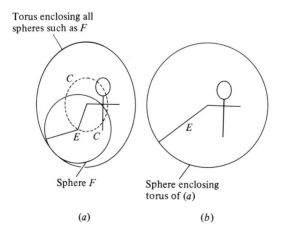

Torus enclosing all spheres such as F

C

E C

Sphere F

E

Sphere enclosing torus of (a)

(a)   (b)

Fig. 4.21. Rearranged arm joints: with three degrees of freedom at elbow, one at shoulder, only the torus shown in (a) can be reached, against the sphere of the actual arrangement in (b).

---

* This argument has recently been developed in more detail in the design of a robot capable of fast precise movement: see the ASEA pendulum robot. *Engineering Designer*, pp. 15–18, vol. 12, No. 2, March/April 1986.

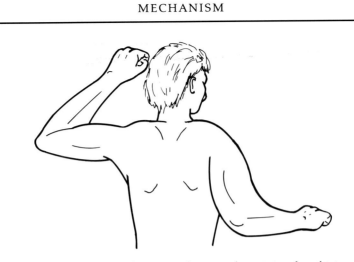

Fig. 4.22. Man with right arm with reversed nesting order of joints.

but the upper arm has to have three sets of muscles to operate the three freedoms of the ball-jointed elbow: the left arm, which is normal, has just the usual single set (biceps and triceps) needed for working the one degree of freedom of the ordinary elbow.

Clearly, this right arm, with the nesting order of universal joint and pivot reversed, appears uncouth, because we are used to our own form and regard it as natural. But it goes beyond that, for the arrangement is inherently bad, with its smaller scope for movement and its increased inertia in the limb.

As another example of nesting order, take the stub-axles carrying the wheel in the car front-wheel suspensions of Figs. 4.9 and 4.10. The stub-axle has two degrees of freedom, up-and-down and steering, and starting from the car-body the nesting order is first the up-and-down freedom, and then the steering freedom. But there is an important principle of design in vehicles, which is often given as:

Keep the unsprung weight to a minimum

(strictly, it should be mass, not weight). Now this maxim closely parallels that of keeping mass in the body rather than the limb and embodies the same general design philosophy, that mass should be kept down in moving things, and where it is unavoidable it should be kept as far as possible in the parts that move least.

The principle of keeping the unsprung mass to a minimum would lead us to the opposite nesting order, that is, steering nearest the body, up-and-down motion further away, for then the mass of the steering pivots, for example, would be sprung whereas in Figs. 4.9 and 4.10 it is unsprung. In the case of a car, the awkwardnesses of using the 'ideal' nesting order are too great, although it is used in motor-cycles.

4.8   Flexural pivots

Sometimes, where a pivot has only to turn through a small angle, it is sufficient to arrange for a part of the mechanism to be easily bent. For example, the pivot at the top of the strut in the strut suspension is of this kind (*D* in Fig. 4.10) and so too are the 'hinges' in the insect thorax (see Fig. 4.12). Most people will be familiar, too, with plastic boxes where the lid and box are moulded as one part, with a thin, bendy section where the hinge is required. However, such hinges can be used even in a material as apparently unbending as concrete.

Figure 4.23 shows part of a bridge consisting of a deck resting on columns. If the deck expands in the sun, the columns are forced apart at the top, and to accommodate this movement the columns are hinged at top and bottom. The concrete is necked down locally to a small area which is heavily loaded in compression by the weight of the deck, and it is also confined by the surrounding concrete and by heavy circumferential steel reinforcing rings. Under these conditions the concrete can successfully be bent through the small angles required.

4.9   The use of degrees of freedom to balance forces

Sometimes a degree of freedom is provided in a mechanism in order to ensure two forces are equal. For example, a horse may be harnessed to a plough (or better, to a horse hoe, which is smaller and can be pulled by one horse) in the fashion shown in Fig. 4.24(*a*). The horse pulls on two lines or traces, one on each side, which are attached to the ends of a short bar, called a swingletree.

The centre of the swingletree is shackled to the hoe, so that it can swing (or swingle, if you prefer) until the forces in the two traces are equal. For

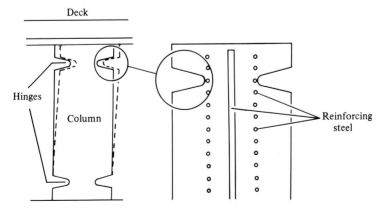

Fig. 4.23. Hinged concrete columns.

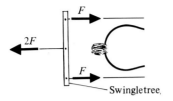

(a) Hoeing with one horse

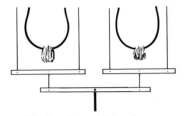

(b) Ploughing with two horses

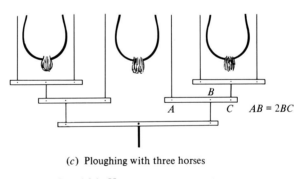

(c) Ploughing with three horses

Fig. 4.24. Harness arrangements.

ploughing with two horses, each is harnessed by traces to its own swingletree, and then the centres of each of these two swingletrees are linked to
the arms of a third, central swingletree (Fig. 4.24(b)) which balances the
combined forces exerted by one horse against those exerted by the other,
so that each of the four traces carries one-quarter of the load. An arrangement of traces and swingletrees for three horses is shown in Fig. 4.24(c),
which the reader may check for himself: note that swingletrees with unequal
arms are required.

The differential in the driving axle of a car has the same underlying
function as the swingletree. It ensures that the traction forces at the two
driven wheels are the same. This has a disadvantage if one driving wheel is
on a hard surface and capable of exerting a large traction force, and the

other in sand, where it can only exert a small traction force. Because the differential equalises the two forces, the car might as well, for all the traction it has, have both wheels in the sand. The wheel on the good surface can only generate the same force as the one in the sand: if the driver depresses the accelerator, the wheel in the sand will simply spin while the other remains stationary. In such a case, the car could extricate itself if the degree of freedom in the differential could be suppressed. Vehicles for operation in rough terrains often have a special kind of differential, called 'limited slip', which suppresses its own balancing degree of freedom if one wheel tends to turn much faster than the other: it possesses the advantage of a differential without its disadvantage, but it is expensive and cumbersome and so it is not fitted to ordinary cars.

When a vehicle negotiates a bend, the wheels on the outside of the turn travel farther than those on the inside. However, if an inside wheel and an outside wheel are both fixed solidly to the same axle, they can only turn at the same speed, and in order for a vehicle with such an axle to turn one or other wheel must slip. By making the axle in two halves connected by a differential this problem is overcome, because the extra degree of freedom permits one wheel to turn at a different speed to the other.

There are thus two ways of describing the function of a differential. One is that it makes the driving forces at the wheels equal, and the other is that it enables one wheel to turn at a different speed to the other while both continue to drive. The principles of mechanics teach us that these two functions are effectively the same. The first description involves forces – it is a *statical* property – and the second involves motions– it is a *kinematic* property. If a system has one of these properties, it will also have the other, because of the relationship between forces and motions which is encapsulated in the idea of work.

This role of the work concept may be seen most easily in the action of the swingletree in Fig. 4.24(a). Suppose the force in each trace is F, so that the force on the hoe is $2F$. If we move just one trace forward a distance x, leaving the other where it is, then the hoe moves forward half that distance, i.e., $\frac{1}{2}x$. Now the work done at the trace is the force in the trace times the distance it moves, or $F \times x$. The work done at the hoe is the force at the hoe, $2F$, times the distance the hoe moves forward, $\frac{1}{2}x$, or $2F \times \frac{1}{2}x = F \times x$, i.e., the same as the work done by the horse, which we could have deduced from the conservation of energy. In the same way, we can deduce that in the three-horse arrangement of Fig. 4.24(c), where the force in each of the six traces is F and that at the plough is $6F$, a movement of any one trace produces a sixth of that movement at the plough. Thus we can deduce a kinematic property, the relation between the movement of one trace and that of the plough, from a statical property, that each trace carries one-sixth of the load.

A four-wheel drive vehicle has a differential in the front axle, one in the rear axle and a third connecting the two axles. The effect of these three differentials is to equalise the traction forces at the four wheels, just as the effect of the three swingletrees in Fig. 4.24(2b) is to equalise the pulls in the four traces.

4.10   Locks and keys

The lock and key represent the earliest appearance in engineering of ideas of primary importance in the design of living organisms. A particular lock needs a particular key to open it – this is the idea of *specificity*, a specific key is required to open a given lock.

In the same way, numerous chemical processes take place in living organisms through the agency of enzymes, which are very specific in their operation: a specific enzyme brings about a specific chemical reaction. If we chew a piece of bread, we can soon detect a sweetness in our mouth which comes from the conversion of starch in the bread into sugar. This conversion is brought about by enzymes in the saliva.

Specificity is at the root of language. A particular succession of sounds is associated with a particular object or action. A specific word unlocks certain thoughts in the mind. Of course, the lock and key is a very small beginning, and language was already highly developed, even perhaps overdeveloped, before this first tiny step on the road to a mechanical equivalent of language was made. It is a long way from the first locks, via looms and mechanical music makers, to the first mechanical computers, and then a long way further to the electonic computer.

The way in which the first known locks worked is shown in Fig. 4.25. The door was secured by a bar sliding horizontally: the key was inserted through a slot in the door and used to slide the bar back – to the right, in the figure. A box or block fixed to the door above the bar was pierced vertically by a number of holes in which pins were free to slide. When the

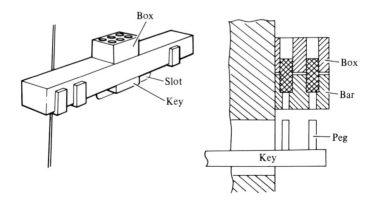

Fig. 4.25. Early form of lock.

bar was in the 'locked' position – to the left in the figure – holes in the top of it came opposite the holes in the box. The pins then dropped part way out of the box into the bar, so preventing it from sliding, as shown at the right in Fig. 4.25.

The key carried a number of pegs which matched the holes in the bar and fitted into them from below. These pegs pushed the pins up so that they were entirely in the box and no longer prevented the bar from sliding. To achieve this, the pegs had to come up through the bar just as far as the top surface and no further. If they fell short, the pins would still lock the bar: if they protruded through the top of the bar into the box, they, the pegs, would lock the bar. We shall see later how this fact can be made the basis of a more refined lock.

This early lock was exceedingly specific: only a key with a peg in the right place for every pin, and no superfluous pegs, would open it. For example, suppose the area in which the holes for pins was distributed was about 80 mm by 60 mm and the accuracy of fit and the quality of work-manship was such that each peg must be within 1 mm of the correct place to enter the hole in the bar. Then the chance of a particular peg accidentally being in the right place would be one in 40×30, or 1200, since there would be 1200 distinguishable places, spaced 2 mm apart each way, and only one would do. With only four pegs, the chances of a randomly made key fitting would be something like 100 000 million to one against.

Suppose now short pins are dropped into the holes in the bar, as in Fig. 4.26(a); these short pins are of various lengths, but all of them fall short of the top surface of the bar. Suppose also that the slot in the door which enables the key to reach the lock is made narrower (the reason for this will appear later).

Now the pins in the box will fall into the holes in the bar until they are stopped by the short pins. For the key to open the lock, each peg must be just the right length to push the short pin up until it is just flush with the surface of the bar (see Fig. 4.26(b)). If it is too short, the pin from the box will still lock bar and box together: if it is too long, the short pin will rise

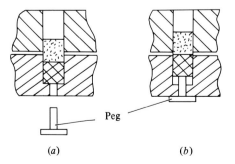

Peg

(a)   (b)

Fig. 4.26. Principle of cylinder lock.

partly into the box and lock bar and box together. Because the slot has been made narrower, it will not now admit a key with pegs long enough to come level with the top of the bar. Thus each peg must now, besides being in the right place, be just the right length to make up with the short pin the thickness of the bar.

In the 1850s Yale introduced his patent cylinder lock which had eight pins in a single line. The position of the pins was the same in all locks, but between them and the key were short pins in five different lengths. This meant that the number of different locks (and keys) possible was 5^8, or 390 625, not counting variations in the shape of the key slot.

4.11   The action of enzymes

Enzymes are complex organic chemicals manufactured by living organisms to effect various processes, such as the conversion of starch to sugar or the breaking of the protective membrane walls of bacteria. After they have performed their specific task, they remain unchanged themselves and ready to repeat the process.

Alexander Fleming, a few years before his discovery of penicillin, discovered an enzyme which he called 'lysozyme', because of its power to dissolve or 'lyse' the bacterial cell wall and so destroy it.

The molecule of lysozyme consists of a chain of amino acids, chemically linked together to form what is called a polypeptide. There are 20 different kinds of amino acid involved, in various numbers from 1 to 13 of each, to a total of 129. They are linked together in a particular order to form a chain with side chains, and four particular pairs of side chains are linked together by pairs of sulphur atoms, so gathering up the chain somewhat. However, the chain is also extensively crumpled or folded into a compact form, rather as one might crumple up a piece of paper before throwing it away. However, in spite of the apparent irregularity of this form, it is always the same and adapted to perform a very special, a very specific, task.

Part of the bacterial cell wall is formed as a sort of string bag of glucose-like molecules linked together in chains, called polysaccharide chains. The lysozyme molecule cuts through these chains at a particular point, so breaking down the whole string bag and hence the bacterium itself. It does this by means of a stereospecific site, a deep infolding or cleft in the lysozyme molecule, the sides of which are linked with a particular grouping of amino acids. Some of the free ends of these amino acids establish chemical bonds with the six links of the polysaccharide chain which fit into the length of the cleft. In addition, the pull of these bonds strains the chain somewhat between its fourth and fifth links, where the break is made. This distortion weakens an internal bond between a carbon and an oxygen atom which join these links, and one free end of amino acid attaches to the oxygen, so breaking the bond. A further action involving another amino acid causes the two separate parts of the original polysaccharide chain to drop away, leaving the enzyme in its original state, ready to break another chain. Figure 4.27 gives a very free impression of the geometry of this mechanism.

All this depends on the configuration of perhaps 10 different free ends of amino acids within the cleft, and most of the complexity of the lysozyme molecule appears to be necessary to contrive this special arrangement in space. Even then, the stereospecificity is not as exclusive as might be desired. One small molecule, an amino sugar like one of those forming the polysaccharide chain, will prevent the lysozyme from working, and it is thought this is because it bonds within the cleft just as one of the links would do. However, without the rest of the chain, there is no bond breaking and subsequent separation of the amino sugar.

One of the many complications omitted from this simplified account is that all the parts of these molecules are vibrating furiously because of the randomly distributed thermal energy they possess. At times this may make it appropriate to think of these stereospecific mechanisms as restricting the degrees of freedom of the chemically-active components in such a way as to increase the probability of the desired event.*

The whole basis of reproduction and growth lies in stereospecific mechanisms of much greater sophistication and elegance, which we are just learning to understand and to use in limited ways for our own purposes. These mechanisms will be mentioned again in Chapter 9.

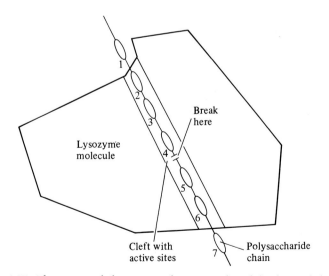

Fig. 4.27. *The action of the enzyme lysozyme (based freely, with kind permission, on Phillips, D. C. (1966). The three-dimensional structure of an enzyme molecule. In the* Scientific American, *vol. 215, No. 5, p. 68).*

* Blow, D. M. (1976). Structure and mechanism of chymotrypsin. *Accounts of Chemical Research*, 9, pp. 145–52.

5

Structures

The term structure is used very widely for any complex whole showing functional relationships between its parts, as, the management structure of a company, the structure of a language, etc. In this chapter its use will be confined to load-bearing structures, material (but not necessarily solid) bodies that sustain or resist forces. Most of the designed world consists of structures of one kind or another; aqueducts and arteries, boilers, bones and bridges, cathedrals and car bodies, teeth, termitaries, trees, tyres and tennis rackets are all structures whose design is chiefly dictated by the loads they must sustain.

Very often there is a conflict between the structural and other functions – the wing of a bird or an aircraft is best made thick for strength and thin for performance, a brick will be able to carry more load if it is dense but it will be a better thermal insulator if it is porous, and so on. Nearly always it is desirable to use as little material as possible both for reasons of economy, which apply no less in nature than in human affairs, and also usually for functional reasons. For example, more stone than necessary in the top of a cathedral would not only have been more laborious to raise, but might have led to collapse lower down, and bones, car-bodies, tyres and tennis rackets are all structures in which extra mass requires extra force to accelerate it, and is therefore undesirable.

5.1 Elements of structures: struts and ties

Many structures can be regarded as assemblies of elements having relatively simple functions. Thus cathedrals are built up from piers, columns, arches and slabs, all simple structural elements.

Two such elements, the simplest, we have already met in Chapter 3, the tie, which carries a purely tensile load, and the strut, or column, which carries a purely compressive load (see Fig. 5.1). Many large engineering structures of the nineteenth and twentieth centuries are built chiefly of struts or ties – bridges like the Forth railway bridge and the Sydney Harbour bridge, and the roofs of the great nineteenth-century railway terminals, for example.

As a representative type of strut, consider the column, formed of a stack of parallel-ended stones (see Fig. 5.2(a)). It can sustain a large compressive or squashing force, but it cannot support any stretching force. A tie, on the other hand, represented by a rope, can support a large stretching or tensile load but no compression (Fig. 5.2(b)). Note also that we can deflect a member in tension like the vertical rope in Fig. 5.2(c) by a strut or a tie acting on it at right-angles, and that the deflection produced is proportional to the force applied. The figure has been drawn for an applied force equal to half the tension P in the rope.

Figure 5.2(d) shows two compression members B, B similarly deflected and maintained in equilibrium by a lateral force of P/2. However like the tension case these may appear, there is an important difference. With the rope, we can deflect it by applying a force, but with the columns, if we deflect them we must apply a force in the opposite direction to prevent the deflection increasing of its own accord and the columns collapsing. Samson is conventionally represented as fetching down the temple of the Philistines

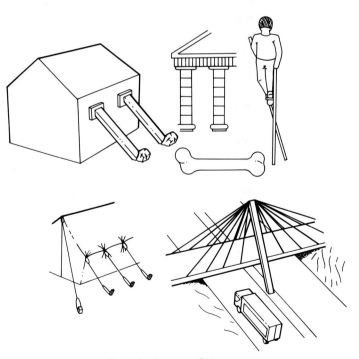

Fig. 5.1. Struts and ties.

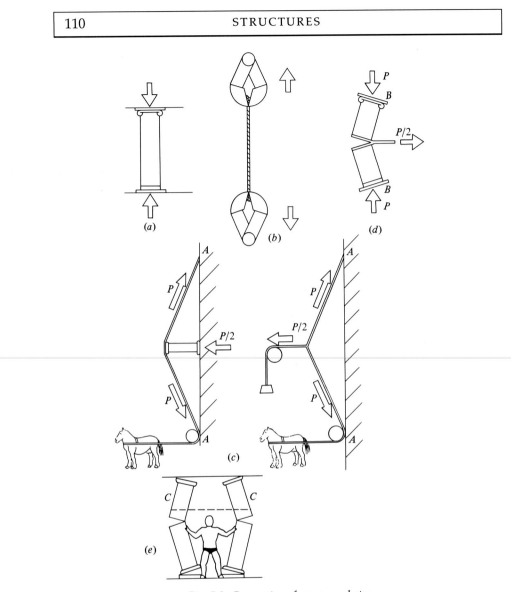

Fig. 5.2. Properties of struts and ties.

by forcing two pillars apart (see Fig. 5.2(*e*)). Once he had succeeded in deflecting them at all, they would have continued to bend outwards and collapse. A loop of stout enough rope round the pillars at *CC* would have stopped the motion, and if Samson had then stopped pushing, the structure would have remained bandy-legged in appearance but it would have held up, with each pillar in the state shown in Fig. 5.2(*d*). If Samson had forced apart two ties supporting a balcony, the ties would have moved easily at first but they would have pushed back harder and harder the further they were pushed, with no sudden collapse or instability. This difference between the inherently unstable strut and the inherently stable tie is very important in the design of structures.

5.1.1 Structures built from struts and ties

Figure 5.3 shows three simple structures cantilevered from walls and being used to hoist a load at A. In each case, each member is drawn as a tie (a rope) or a strut (a column), and the reader should be able to see that the right choices have been made, and that anywhere a change from rope to column or vice versa would result in collapse (although the character of DC in structure (c) may be less obvious than the others). For example, in structure (a), the rope BA would hang straight down the wall without the column AC to push it out. In structure (b), the rope DBA would straighten between D and A without the strut BC to push it out of line.

We can usually deduce whether the load in a member joining any two points, say X and Y, is compressive or tensile by imagining the member removed, and seeing if X and Y come closer together or separate. For example, if the member DB were removed from structure (c), it is clear that the triangle ABC would pivot clockwise about C, increasing the distance

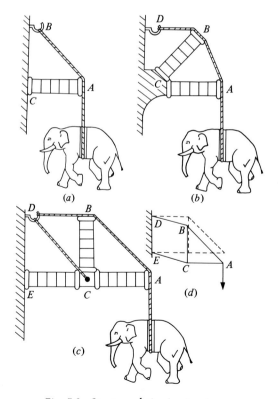

Fig. 5.3. Struts and ties in structures.

BD; it follows that the member joining B and D must be holding B back, and so is in tension. If member DC in this structure were removed, the members DB and EC would turn clockwise, the triangle ABC moving bodily downwards without rotating, as shown in Fig. 5.3(d); in this process DC would extend, so it is being stretched and must therefore be a tie.

This idea of introducing an imaginary motion to find the direction of the force in a member of a structure is closely related to the use of an imaginary motion to make deductions about the harnessing of ploughs in the last chapter.

5.1.2 Components of a force and equilibrium

In Fig. 5.2(c), the rope AA, under a tension P, is deflected by a force of magnitude $P/2$. To analyse this effect, which is central to the understanding of structures, two simple ideas are needed.

The first is the *vectorial representation of forces*, in which a force is represented by a line in the same direction and of length proportional to its magnitude (e.g., if we used a scale of 1 mm represents 10 N, we would represent a force of 300 N by a line 30 mm long).

The second idea (which was used in Chapter 2 in studying the sling) is that of *components of a force* and *resolution of a force into components*. Figure 5.4(a) shows the top of the strut and the middle of the rope from Fig. 5.2(c), regarded as 'cut away' from the rest and studied separately. Now this little piece of the structure is not accelerating, but remains stationary, so it must be in equilibrium under the forces acting on it, which are two sideways and slightly downward forces of magnitude P, from the rope, and an upward force of magnitude $P/2$, from the strut. Now it is clear the two P forces balance one another sideways, and combine together to produce a downward effect balancing the $P/2$ force. By means of the idea of resolving a force into its components, we can put these intuitive ideas on a formal footing.

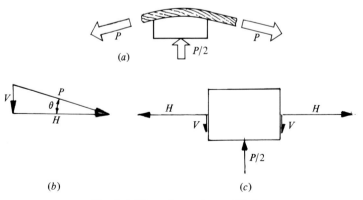

Fig. 5.4. *Three forces in equilibrium.*

In Fig. 5.4(*b*), the arrow *P* is the vectorial representation of the force in the right-hand part of the rope. By drawing the right-angled triangle shown, we obtain the vectorial representations of two other forces, *V* and *H*, whose combined effect is the same as that of *P*, so that we can replace the right-hand *P* in Fig. 5.4(*a*) by *V* and *H*, as in Fig. 5.4(*c*). *V* and *H* are called the vertical and horizontal *components* of *P*, and separating *P* into them is called *resolving P*.

In the same way, we can resolve the left-hand *P* into a vertical and a horizontal component, as has also been done in Fig. 5.4(*c*), and then the equilibrium of the parts considered as 'cut away' can be studied in the same way as that of the sledge and the aircraft in Chapter 2. Horizontally, the two components *H* balance one another. Vertically, the two downward components *V* must together balance the upward force $P/2$, so that the magnitude of *V* must be $P/4$. Thus in the right-angled triangle of Fig. 5.4(*b*), the hypotenuse is four times the shortest side, from which, by drawing or trigonometry, it can be found that the angle θ is about $14\frac{1}{2}°$.

Using these ideas, the functioning of a large number of structures can be explained.

5.1.3   The suspension bridge

Consider a cable hanging across a gap *MN* (see Fig. 5.5(*a*)) and carrying a number of weights. Each weight produces a kink in the cable, and a heavier weight will produce a sharper kink. Let us look at one kink, the second from the left, ringed in the figure, and consider the forces there using the idea of components.

There is a downward force, *W*, say, due to the weight, and there are forces at angles from the cable on each side. These forces are shown split into horizontal and vertical components in Fig. 5.5(*b*), where the components are shown as full lines and the cable forces themselves are shown as broken lines. Now the horizontal forces must balance, so the horizontal component of the force in the cable to the left of the kink, H_L, must be equal to the horizontal component of the force in the cable to the right of the kink, H_R, or $H_L = H_R$. Also the vertical forces must balance. The left-hand part of the cable pulls upwards with a force whose vertical component is V_L. The right-hand part pulls down with a force whose vertical component is V_R. The net effect of the kink is an upward force $V_L - V_R$, and this upward force supports the weight *W*, so $W = V_L - V_R$.

A suspension bridge can be regarded as a pair of parallel cables carrying a large number of equal and equally spaced weights, as in Fig. 5.5(*c*). Since all the weights are equal, the kinks must all be equal (more precisely, each involves an equal change of slope) and this requires that they all lie on a parabola. The horizontal component of the tension in each cable, *H*, is the

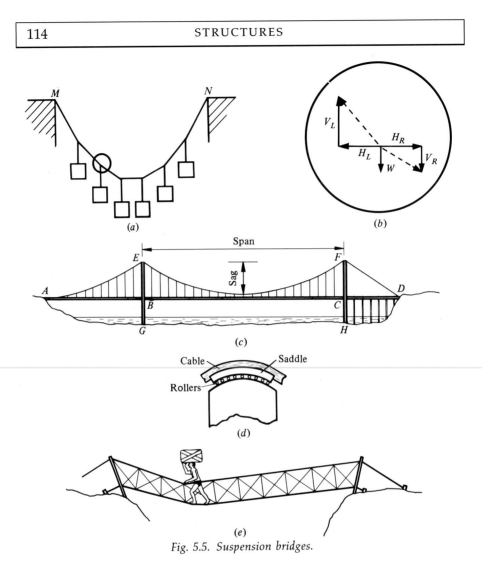

Fig. 5.5. *Suspension bridges.*

same throughout the length of the bridge, and from the geometry of the parabola it is easily found that

$$H = \frac{\frac{1}{2}\,\text{weight of span}\ \times \text{length of span}}{8 \times \text{sag of cable}}$$

As the sag is generally about one-twelfth of the span, H is about three-quarters of the weight of the span. Thus, if the span weighs 40 000 tons, each cable carries a tension of about 30 000 tons, or about 300 million newtons. At the towers, the force in the cable is about 5% higher, because of the vertical component, but we shall neglect this small effect.

The cables are made of high-tensile steel wire, simlar to the wire used for piano-strings, carrying a stress of about 500 N/mm^2, so that a combined area of about 600 000 mm^2 would be needed, corresponding to a diameter of nearly a metre. In practice, the cables are made from solid wires about

4 mm in diameter, of which some 50 000 or so would be needed in this case. They are all laid in a parallel bundle, without twisting.

The side-spans of suspension bridges may also be suspended, as at AB in Fig. 5.5(c), or they may be separately supported, as at CD. In either case, the anchorages for the cables at A and D must be able to carry the same enormous load as the cables themselves.

The cables at E and F must carry equal loads on either side of the towers. In fact, a balancing degree of freedom is required, such as is provided by the swingletree or the differential. In earlier designs, the cable was often fixed to a saddle which was free to slide on rollers on the top of the tower, as in Fig. 5.5(d). An alternative is to hinge the whole tower at the bottom so that it can swing to and fro and is only held up by the cables, as may be seen in the Chelsea bridge in London where the hinges are conspicuous. But nowadays a very simple method is usually used: the towers are made flexible so they can bend as if they were hinged at the bottom.

If the loads were unequal, the kinks would be unequal, with sharper kinks at the bigger loads, as may be seen in the primitive bridge in Fig. 5.5(e). This tendency is present even in the modern suspension bridge, but the deck ($ABCD$ in Fig. 5.5(c)) is made stiff so that it resists bending and spreads the load over a length of cable. Even then, the suspension bridge is better suited to evenly-spread loads than large single loads such as railway trains, with their heavy locomotives.

In the suspension bridge, a generally horizontally-tending tension member, the cables, supports a load distributed along its length. The small increments of load from the hangers supporting the deck gradually turn the direction of the tension member so it is concave upwards, as in Fig. 5.5(a),(c) and (e).

5.1.4 The arch

The arch bridge is the inverse of the suspension bridge. A generally horizontally-tending compression member, the arch, supports a load distributed along its length. The small increments of load W from the deck of the bridge and the arch's own weight gradually turn the direction of the compression member so it is concave downwards (see Fig. 5.6(a)). With equal, equally-spaced loads (which include the weight of the arch itself) there will be a parabolic *line of thrust* corresponding to the parabolic form of the cables in the suspension bridge, only now the forces are compressive and the parabola is concave downwards. There is a constant horizontal component of force H in the arch, given by

$$H = \frac{\text{weight of span} \times \text{length of span}}{8 \times \text{rise of the bridge}}$$

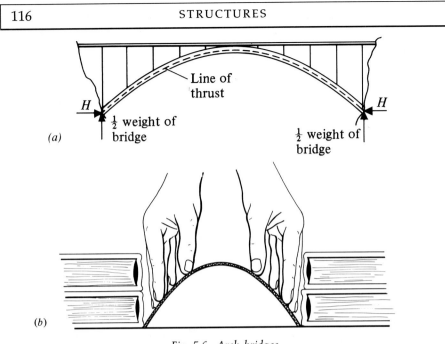

Fig. 5.6. Arch bridges.

However, while the H in the cables of the suspension bridge was tensile and pulled the towers together, this H is compressive and pushes the ends of the arch (or abutments) apart.

The action of the arch is easily studied by means of a model made from a piece of corrugated cardboard arched between heavy books lying on a table, as in Fig. 5.6(*b*). Supply a distributed load by pressing gently with the finger-tips, and it will be obvious that the books are being forced outwards by the abutment thrust H. Try to replace the finger-tip loading by weights placed on the cardboard, and you will discover that the arch is inherently unstable, as the earlier discussion of struts and ties would lead us to expect. Pressing with the finger-tips, you automatically preserve the arch form, whereas weights do not adjust themselves to the behaviour of the structure but cause it to go unstable and collapse, unless it is made stiff so it keeps its form.

A suspension bridge, however, is inherently stable and will simply change its form slightly to adjust to a different system of loads, as in Fig. 5.5(*e*).

The instability of an arch means it cannot be full of flexibilities like a suspension bridge, but must have only a few flexible points in it, or none at all. Many arches are made with hinges in them, although they are not usually conspicuous and are often difficult to recognise.

Figure 5.7(*a*) shows a famous arch bridge, the Salginatobel bridge in Switzerland, built by Robert Maillart in 1930. It has hinges at A, B and C, of the type shown in Fig. 4.23, which have the effect of making the line of

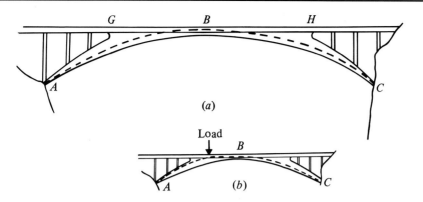

Fig. 5.7. *The Salginatobel bridge.*

thrust go through them. The dotted line shows the line of thrust with no load on the bridge. Fig. 5.7(*b*) shows the line of thrust with a heavy load on the left-hand side: the heavy load requires a large kink, concave downwards, in the line of thrust to balance it, just as a heavy load on the suspension bridge requires a large kink in the tensioned cable to balance it. The value of *H* increases, the line of thrust still goes through *A*, *B* and *C*, and between *B* and *C* the line of thrust moves down.

To see why the line of thrust must go through any hinges there are, consider the hinges in Fig. 5.8. If we squeeze the slightly bent one, it will fold up because the line of thrust *AA* does not go through the pivot point: only if the thrust *AA* goes through the pivot will it not collapse, and even then, it is unstable – disturb it slightly and it folds up! Similarly, if we pull the halves of a hinge outwards, the line of tension goes through the pivot.

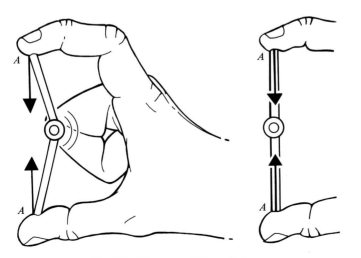

Fig. 5.8. *Hinges and lines of thrust.*

A cable is like a chain of hinges or pivots, so the line of tension must lie in it all the way.

In the Salginatobel bridge, all the possible lines of thrust lie in the shape *AGBHCBA* (Fig. 5.7(*a*)), and this is the underlying philosophy of Maillart's design, which is in a sense a wrapping of material round the whole family of lines of thrust.*

Note that three hinges is the most an arch can have: with four it would collapse, although a suspension bridge has, in effect, an infinite number of hinges.

5.1.5 Masonry arches

Masonry is virtually incapable of supporting tension. A well-built masonry arch with sound abutments, when new, will develop the right abutment thrust *H* to produce a line of thrust that will follow the centreline of the stonework fairly closely, and all will be well (see Fig. 5.9(*a*)). But after a few centuries the abutments may give slightly, *H* may decrease, a smaller *H* means the line of thrust becomes steeper near the abutments, acquiring a more heavily-arched form which moves towards the inside of the arch (intrados) at the springings, *A* and *C* in Fig. 5.9(*b*), and towards the outside (extrados) at the top, *B*. This movement of the line of thrust causes the joints of the masonry to open on the side from which it moves away, i.e., on the inside at *B* and on the outside at *A* and *C*; in effect, hinges form on the line of thrust, as in Fig. 5.9(*b*). If in that figure, *A* and *C* give and move apart, the hinges will bend slightly in the sense indicated by the opening of the joints. Cracks of this sort can often be seen in old arches – they may not be at just the same points, and if the abutments have moved together, which they sometimes do, the cracks will open from the other side of each 'hinge'.†

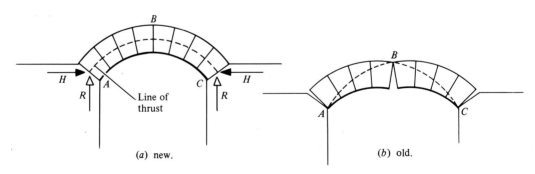

Fig. 5.9. Masonry arches.

* Billington, D. P. (1979). *Robert Maillart's Bridges.* Princeton University Press.
† Heyman, J. (1968). On the rubber vaults of the middle ages, and other matters. *Gazette des Beaux-Arts*, vol. 71, pp. 177–88.

5.1.6 Cathedrals

Because masonry cannot sustain a tension, stone buildings with wide covered areas present special difficulties which the mediaeval mason, lacking science, overcame by trial and error methods. Roofing in could be done only by arches or domes, which exert an outward abutment thrust which must be sustained either by very deep, heavy buttresses or by further arches. Since the buttress is a crude and clumsy solution, it is more elegant to use arches. For a cathedral, where only the central arch is really wanted, the side arches were made to slope sharply downwards, forming the well-known flying buttresses (see Fig. 5.10).

Except in the case of domes, which is discussed later, and a few details, the mason was limited to placing one stone upon another, to form a heap, in effect, but a very subtly ordered and daring heap, a heap with huge voids in it. The need was to keep the structure everywhere in compression, using the downward loads due to gravity and characteristic of heaps to ensure that lines of thrust remained well within the stone – not that the mason would have understood such a statement of his problem. Proceeding empirically, with the occasional collapse that told him he had gone too far, the mediaeval master mason built these extraordinary airy structures, which seem almost to defy gravity and challenge the density of stone, the very properties that give them their strength and stability. Moreover, these remarkable developments occurred in a relatively short time, so that the evolution was fast for a time when mechanics was not yet understood and the 'reproductive cycle' so long: a man does not build many cathedrals in a lifetime.

Except for timber trusses, sometimes with iron ties, architecture remained largely heap-based until the industrial revolution. The first real change came in the nineteenth century, with the extensive use of iron (see Fig. 5.11).

5.1.7 The brontosaurus and the Saltash bridge

Since an arch pushes its abutments outwards, and a suspension bridge pulls them inwards, a judicious combination of the two should produce no net horizontal force. This is the principle Brunel used in his Royal Albert railway bridge over the River Tamar at Saltash, built in 1859 (Fig. 5.12). The arches are wrought iron tubes of oval cross-section, and they push outward with a thrust calculated to balance the inward pull of the chains at the bottom, which are made of flat links of wrought iron pinned together. A similar combination of arch and suspension cable has been used by nature for hundreds of millions of years in the design of large four-legged land animals, perhaps most strikingly in the brontosaurus and other great prehistoric reptiles (see Fig. 5.13). The legs correspond to the towers, the spine to the arch and the tensile components of the belly to the chains.

Fig. 5.10. Flying buttress, Westminster Abbey.

5.1.8 The use of ties and struts

The simplest structural elements, struts and ties, have peculiar weaknesses and advantages which influence the designer strongly in their use. The disadvantage of struts is that if they are slender, they may buckle, as was seen in Chapter 3. The disadvantage of ties is that they have to be connected to other members at their ends by connections which are strong in tension. An interesting aspect of these weaknesses is that struts are most satisfactory

Fig. 5.11. Paddington Station roof.

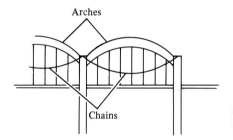

Fig. 5.12. Principle of the Saltash bridge.

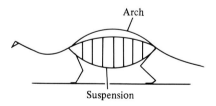

Fig. 5.13. Structure of brontosaurus.

when they are stocky, and ties when they are slender (for the amount of extra material needed for end-connections is a much smaller fraction for long, thin ties). These limitations are neatly avoided in tents and in suspended roofs for large areas like stadiums, such as the one shown in Fig.

1.1; these combine a few heavy struts with a very large number of long thin ties.

The problem of making end-connections to ties is a serious one for engineers (see Fig. 5.14). A screwed connection, such as (a) or a pinned, riveted or bolted one, such as (b), is weaker than the tie itself, as is a welded joint, unless the cross-section of the tie can be made larger at the end, which is often difficult. A convenient concept is the efficiency of a joint, defined as

$$\frac{\text{strength of connection}}{\text{strength of members connected}},$$

where the strength of the weaker member is taken when they are unequal. A joint of very high efficiency is the timber joint shown in Fig. 5.14(c), used in roof joists and only possible because wood is easily cut to this awkward shape and because modern adhesives are so good.

Living organisms, on the other hand, manage the end-connections of their ties very neatly – the insertion, as it is called, of ligaments and tendons on the bones is compact and efficient, and is typical of the advantages that growth processes often show over the methods open to engineers.

Most structural tasks can be achieved using only struts and ties, and because of their simplicity it is useful to study economy of design in structures in terms of these two elements.

5.2   Structural economy

Consider a structure consisting entirely of struts and ties, made in one material that is equally strong in tension and compression, and in which problems of buckling of compression members do not arise. Imagine that the structure has only to be designed for one particular system of loads, and that under it a particular member X, of length L, is subject to a force P. Then the required cross-sectional area of the member must be made proportional to P, the load it has to carry, and the volume of material in it will be its area times its length, which is proportionate to $P \times L$.

More precisely, if the design stress is f, both in tension and compression, the cross-sectional area of X must be P/f and the volume of material in it must be PL/f (ignoring end-connections). Note that while it is usually

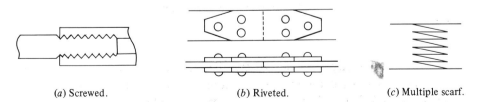

(a) Screwed.   (b) Riveted.   (c) Multiple scarf.

Fig. 5.14. End-connections for ties.

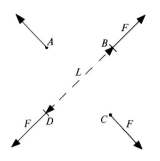

Fig. 5.15. A system of loads.

convenient for compressive stresses and forces to be regarded as negative, we can take both as positive in the present treatent. To find approximately the volume of material in a design of structure, we can simply add up the products of $P \times L$ for all the members and divide the sum by f.

Now consider the design of a structure, consisting of struts and ties only, to support a particular system of loads. As an example, the system shown in Fig. 5.15 consists of four equal forces F acting diagonally outwards at the corners A, B, C and D of a square of diagonal length L. Firstly, the forces do balance each other, without which the problem would not be real. Secondly, two simple diagonal ties AC and BD would serve, each of which would carry a tension F for a length L, so that for this particular structure, the sum of products of forces in members by their lengths would be simply $2FL$. Figure 5.16 shows a series of other possible structures: the question is, are any of these more economical?

The answer is, that the first three are just as economical, using exactly as much material. Structure (d), however, is less economical, because it contains compression as well as tension members. The structures $(a),(b)$ and (c)

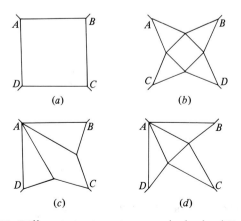

Fig. 5.16. Different structures to carry the loads of Fig. 5.15.

use the absolute minimum volume of material, $2FL/f$, and this will be true of any other structure designed for this loading if it contains only tension members.

This result can be deduced from a theorem of Clerk Maxwell (1869). In any structure to bear a given set of loads and composed of ties and struts working at the same design stress f, the difference between the volume of material in tension (T, say) and that in compression (C, say) is a constant for all designs. It follows that if we can reduce C by a certain amount, we shall automatically reduce T by the same amount, and that *if* we can reduce C to zero, no further reduction in material used is possible. Thus if, as in the case of the loading system in Fig. 5.15, we can find a design composed of ties only, that design will use an irreducible minimum of material, the same amount as any other 'ties only' design. If an arrangement is used, such as that in Fig. 5.16(d), which contains struts when it is possible to use ties only, then that design will use more than the minimum amount of material, to the extent of exactly twice the amount in the struts, C. The same argument applies when struts only can be used: then 'compression only' designs will all be optimum, using the irreducible minimum of material, and any designs including ties will use more material to the extent of twice the volume in the ties.

Unfortunately, Maxwell's theorem will not tell us the minimum quantity of material required except in those special cases in which a structure can consist only of struts or only of ties. But it is clear that there must always be a lower limit to the amount of material required and, indeed, that limit can always be found, sometimes exactly but always closely enough for practical purposes. One ideal minimum material design is shown in Fig. 5.17, and very strange it is. The spirals are mathematically akin to those of seashells, though there they are the outcome of growth processes, not of structural economy (see Chapter 9).

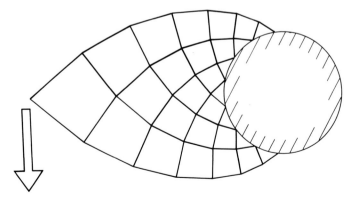

Fig. 5.17. An ideal minimum material structure.

While structures like that in Fig. 5.17 are not practical designs, if we can find practical forms which come close to them in structural economy – and we usually can – then we know that there is little to be gained by further study. In this way, work on optimum structures can give useful guidance to practical engineers, providing an absolute measure of their success, or otherwise, in saving material.*

5.3 Shear

Consider a rectangular frame (Fig. 5.18) such as a wooden work-bench, and imagine a force W applied to the top as shown. Now unless the joints are very strong and stiff, the frame will be relatively easily deformed in the way shown by the dotted lines. This kind of deformation is called 'shearing', and it is kinematically equivalent to the deformation in the shear failures considered in Chapter 3, where one piece of material slid or 'sheared' relative to the other.

Shearing deformation is readily seen in a cardboard box which has a bottom but no top (Fig. 5.19). Here the open top will shear very easily, but the bottom, being closed, remains rectangular. This example shows that one structural element which will resist shear is a flat plate. However, where the amount of shear to be resisted is small, a plate may use too much material, for if it is made too thin it will buckle and be useless. An alternative is to use diagonal ties or struts. For instance, in Fig. 5.18, the frame may be made stiff against shearing in the direction shown by connecting B to C by a tie. To resist shearing in the opposite direction, a tie from A to D is needed. Alternatively, a single diagonal member able to resist tension and compression will stiffen the frame in both senses.

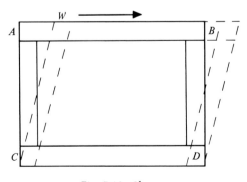

Fig. 5.18. Shear.

* For a difficult but excellent treatment of economy in structures, see Cox, H. L. (1965). *The Design of Structures of Least Weight*. Oxford: Pergamon.

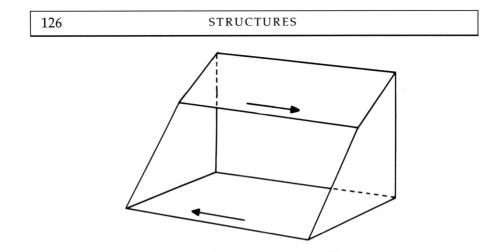

Fig. 5.19. Shear in an open-topped box.

5.4   Beams and bending

Imagine taking a twig in your hands and bending it till it snaps; it has not been subjected to tension or compression, but it does bend, corresponding to stretching of the material on the outside of the bend and shortening of that on the inside. In fact, half the material of the twig is in tension, and the other half is in compression: moreover, the material is not all equally stressed. The stress, whether compressive or tensile, is greatest on the extreme outside and inside of the bend, and decreases between.

The strength of a member in tension, or in compression if buckling is not involved, is proportional to its cross-sectional area and independent of the shape of that cross-section. However, if we take a strip of wood of rectangular section 20 mm by 10 mm and break it by bending, we know by experience it is easier to do so the flat way on than edgeways. It requires just half the force, in fact. The areas in tension and compression are just the same, but edgeways they are twice as far apart and so twice as able to resist bending. It follows that to make a member strong in bending we should concentrate as much of the cross-section as we can in two areas well-separated in the plane of bending.

Very often a member must be able to resist bending equally in any plane, as in a twig, an axle or a thigh-bone, and in such cases a circular section is appropriate by symmetry. It is still important, though, to separate the areas in tension and compression as much as possible, so that a tube will be better than a solid round rod, a conclusion which experience confirms.

A beam is a structural element designed to resist bending. A common form used for steel beams, called an I-section, is shown in Fig. 5.20. The top and bottom of the I, called the flanges, are designed to take the longitudinal tension and compression we saw in the twig. They are joined by the web, which takes a rather small share of the tension and compression. The

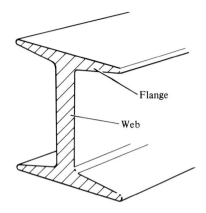

Fig. 5.20. I-beam of joist (rolled steel).

main function of the web is to join the flanges, and it will appear later that this is a structural function too, in a very precise sense.

Suppose that one end of such an I-beam is fixed firmly in a wall, and a weight W is attached to the other end, as in Fig. 5.21(a). Such a structure consisting of a beam fixed only at one end is called a 'cantilever'. The elephant has been drawn much too small, to save space, but the bend in the cantilever has been made much too large, for clarity. The beam bends downwards, the top flange is stretched and the bottom flange is compressed.

To find the tension in the top flange, imagine the cantilever hinged to the wall at the bottom, at P (see Fig. 5.21(b)) and prevented from collapsing by a *pull* on the top flange, at Q. To fix ideas, let us suppose the length of the cantiliever is just four times its depth d, as shown in the figure. Then the force W due to the elephant has just four times the leverage about P of the force at Q, which must therefore be $4W$ to prevent the cantilever collapsing.

Now imagine the cantilever hinged at the top corner, at Q (Fig. 5.21(c)), and we find the push required at P is $4W$. If we neglect any longitudinal forces in the web – and they are not very important – the pull of $4W$ at Q and the push of $4W$ at P are the tension in the top flange and the compression in the bottom flange, respectively. This may prompt the question 'are not the flanges then simply a tie and a strut?' but it will be seen they are not.

If the length of the cantilever had been only twice the depth, then the flange forces would have been only $2W$. But if we consider a section halfway along the cantilever, AB in Fig. 5.21(c), the part to the right is just such a cantilever, so that at A and B the flange forces must be $2W$. If we take a section $2.5d$ from the elephant end, then the flange forces there must be $2.5W$. At the point of attachment of the load, they disappear. Thus the top flange carries a tension which increases from zero at the extreme right

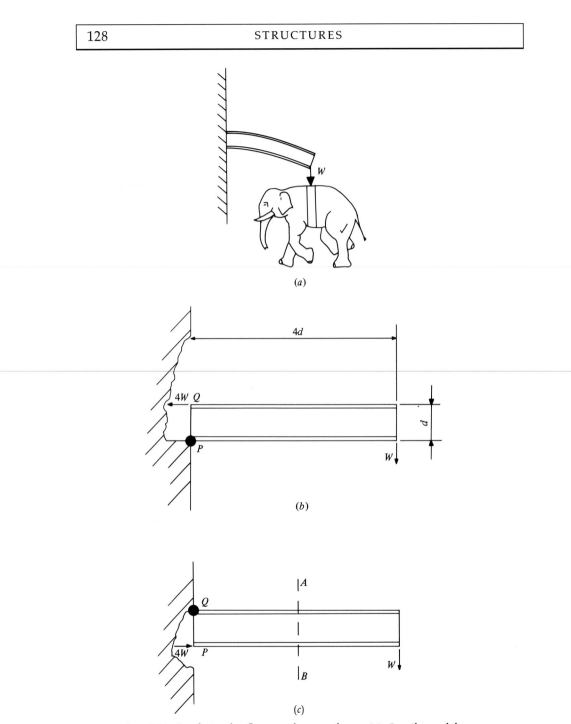

Fig. 5.21. Loads in the flanges of a cantilever. (a) Cantilevered beam loaded at end. (b) Load in top flange. (c) Load in bottom flange.

to $4W$ at Q, and it is this variable tension which distinguishes it from a simple tie. Now this increasing load must come from somewhere, and the only place it can come from is the web.

To see more clearly the function of the web, imagine what would happen if it were removed (Fig. 5.22(a)). The beam would collapse in *shear*. Even more insight is given by considering a thin horizontal slice of the web, at the top just under the top flange, to disappear. The collapse would be as in Fig. 5.22(b), the web sliding along the top flange. To restore the beam to the horizontal, pairs of points like Q and Q', A and A', R and R' which were opposite one another could be pulled back together by forces to the right on the flange, and equal forces to the left on the web. The sum of all the forces either way would be $4W$, and in the flange they would build up the tension from nil at R to $4W$ at Q.

The function of the flanges is to resist bending. Although it does help the flanges slightly in this task, the chief function of the web is to resist shear. It follows that a beam which has no shear ought not to need a web. Let us see if this is so.

Figure 5.23 shows a beam supported at two points $2d$ in from the ends. The portion between the supports is free from shear, and so no web is required in that stretch. It can be seen that there is no shear, because the arrangement is entirely symmetrical so that there can be no tendency to shear to left or right (provided the elephants are of equal weight). The tension in the top flange is $2W$ from end to end of the central, webless section, since there is no web to transfer load in or out of it. Similarly, the compression in the bottom flange is $2W$ over the central stretch.

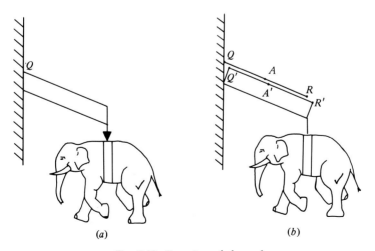

Fig. 5.22. Function of the web.

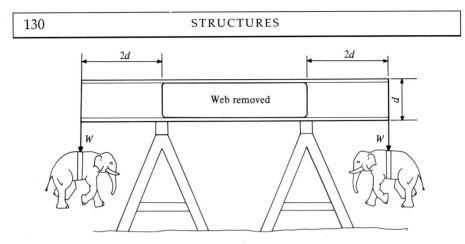

Fig. 5.23. Beam in pure bending.

5.4.1   Bending moments: hogging and sagging

The two equal and opposite flange forces, each of $4W$, at the fixing of the beam in Fig. 5.21 balance each other except for a turning effect, or moment, whose magnitude is the product of either force and the distance between them. Thus, in the cantilever of Fig. 5.21, we can say that the moment at the fixing is $4Wd$. Such a moment in a beam is called a *bending moment*: if it tends to bend the beam convex upwards, as in Fig. 5.21(*a*), it is called a 'hogging moment', if to bend it convex downwards, it is called a 'sagging moment'.

The terms 'hogging' and 'sagging' arose in connection with ships. If a long, light ship is heavily loaded in its midlength, it will tend to be bent convex downwards, to sag, in fact.

If it is loaded heavily at bow and stern, it will tend to droop at the ends, bending convex upwards or 'hogging'. Very often small wooden ships would be permanently deformed in one of these ways, and then they were said to be 'sagged' or 'hogged'.

The ancient Egyptian ships, perhaps because of a shortage of long pieces of timber, perhaps because of a tradition derived from reed-boats, but most probably because it worked well, were strengthened by a bundle of ropes running from stem to stern and passing over vertical posts standing on the keel (Fig. 5.24). The bundle of ropes was tightened by twisting it by means

Fig. 5.24. Egyptian ship, Vth Dynasty.

of a bar thrust through it and then lashing the bar, as may be seen in the figure (this device, of a bar used to twist and so tighten ropes, is called a Spanish windlass). The general effect would have been to bend the ship convex downwards (sagging) so balancing the usual tendency of such vessels to hog.

The Egyptian ship is a little reminiscent in its structure of the brontosaurus or the Saltash bridge, only inverted, with the compression member, the keel and planking, at the bottom, and the tension member, the rope bundle, at the top. It is interesting that in the ship of Fig. 5.24, of the Fifth Dynasty, the vertical posts are graduated in height, whereas in later ships they were all of the same height (for instance, in the ships sent by Queen Hatshepsut to Punt, in about 1500 BC). Structurally, the older pattern would appear more suitable. However, the great virtue of the rope bundle may not so much have been the provision of the tension flange of a beam, as the putting into compression of the wooden hull, the 'prestressing' of it, in fact, as in prestressed concrete.

5.4.2   The economical design of beams

The cantilever of Fig. 5.21 has flange loads which increase from zero at the right-hand end to $4W$ at the support. It would be possible, therefore, to taper the flange away to nothing at the right-hand end and then its cross-sectional area would everywhere be appropriate to the load it had to carry (Fig. 5.25(a)). This form appeals to our intuitive appreciation as being a fitting one, and so it is.

Another way we could reduce the material in the beam is to taper it in depth, as in Fig. 5.25(b). The flange load in the parallel beam falls off to the right because the depth remains the same while the leverage of the load decreases. However, in the beam of Fig. 5.25(b), the depth of the beam falls off in proportion to the decrease in leverage of the load, so that the flange

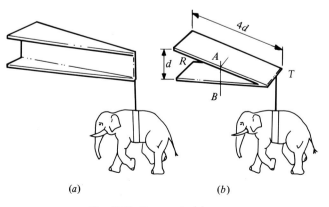

(a)   (b)

Fig. 5.25. Economical beams.

loads are constant at 4*W*. Since no load has to be transferred into the flange except at the extreme right, we can deduce that no web is needed.

The top flange carries a uniform tensile load of 4*W* from end to end, so it could be, ideally at least, a cable, and the bottom flange carries a uniform compressive load, and could be replaced by a pillar. In short, the beam has been reduced to a tie and a strut, like those in Fig. 5.3(*a*).

Which is the more economical in material of these two designs of beam? That in Fig. 5.25(*a*) has less material in the flanges, but that in Fig. 5.25(*b*) has no material at all in the web. The answer depends on the nature of the material and whether the beam is relatively heavily loaded or not. As a general rule, however, long thin beams will be more economical with tapered flanges and short deep beams will be better with taper in depth, the transition being at a length/depth ratio of about two to one.

In nature, beams are usually tapered both in depth and breadth. A branch of a tree is a good example, with a round section tapering more or less uniformly, although in some kinds, such as cedars, the depth may be substantially greater than the width near the trunk, a shape which improves the structural efficiency. The design of a branch has to meet many requirements besides bending strength, and one of these, the need to grow, is very well-suited by the round section. However, there is a structural reason why an I-form would be ill-suited to a branch.

5.4.3 Beams that must also resist twisting

If we took the I-beam of Fig. 5.21 and extended it to the side by another strong beam, so that looked at from below the two would appear as a letter L, then very little load could be supported at the extremity of this extension (see Fig. 5.26) because the I-form is extremely weak in twisting, or torsion, typically, many times as weak as it is in bending.

This weakness is well-illustrated by a long narrow cardboard box, such as those in which kitchen foil or wrapping films are sold. Try to bend one, and it is quite stiff up to the point where the compression side collapses by buckling – but do not spoil the box yet. Try twisting it, as in Fig. 5.27, and you will find it very weak indeed. The twisting process involves shear, as can be seen by the way the edge of the lid *CD* slides along the adjacent edge *AB* of the box part. Suppress this sliding or shearing action by taping the edges together along *AB* and *CD*, and the box will be found to become very stiff in twisting: in an example which I have just tried, the stiffness increased 200 times.

The sliding of the edges *AB* and *CD* along one another is of exactly the same kind as the sliding of *Q'R'* along *QR* in Fig. 5.22(*b*) – it is a shearing action. When the box is taped up, each of its four sides acts just like a shear web in a beam, but to do this it needs to be fixed to another along both sides. To resist twisting the most efficient structure is a tubular one.

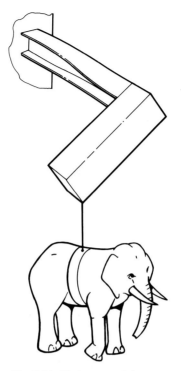

Fig. 5.26. Twist in an I-beam.

Fig. 5.27. Twist in a box.

If a bicycle were made with a frame made of I-section material in place of the usual tubes it would twist out-of-plane very easily and be very unserviceable. The bones of birds and other creatures have to resist both bending and twisting, and so they are generally tubular, a form which is excellent for both purposes.

While in nature tubes present no difficulties of production, this is not true of engineering; an I-section is easily made by rolling steel between suitably-shaped rollers, but tubes are not so easy. The chassis of many early cars and modern vehicles were and are made from I-beams, one along either side. If these were to be joined by further I-beams running across the chassis, the resulting structure would twist very easily and give poor road-holding. However, if tubes of large diameter are used as cross-members, the chassis cannot twist without twisting them, and as they are stiff in twisting, so is the whole chassis (Fig. 5.28).

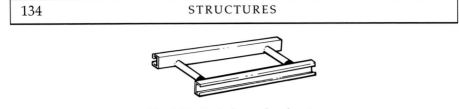

Fig. 5.28. Basic frame for chassis.

5.5 Feathers

The large flight feathers of birds' wings are remarkable examples of design, being light, strong and stiff. They also provide good thermal insulation, and, above all, their ability to resist damage is outstanding.

Figure 5.29(*a*) shows the construction of a feather, with a central spine or *rachis* from which numerous smaller *barbs* spring to form the aerofoil surface or *vane*. These barbs are closely spaced and joined by comb-like fringes of barbules (see Fig. 5.29(*b*)) which intermesh with each other, clinging together by minute hooks, much as burrs or teasels cling to hair or fabric. The ends of the barbules nearest to the attachment to the bird are curled under to form tunnels which trap air for thermal insulation.

The heaviest loading on the feather occurs during the downstroke of the wing, when it acts like a cantilever subject to a fairly uniformly distributed upward load over its whole surface. The barbs, each acting as a cantilever, transmit the load to the central rachis, and for this purpose they are made narrow and deep.

The feather is an example of a 'tiered' structure, where the loading falls in the first place on the barbules, which transmit it to the barbs, which in turn transmit it to the rachis and quill, which transmit it to the wing itself. Early aircraft wings worked like this, the load falling first on a fabric covering, which transmitted it to ribs which transmitted it to spars running along the wing. The fabric spanning between the ribs acted rather like suspension bridges, but all the 'tiers' of the feather are cantilevers. In a wooden floor in a house, there are two tiers – first floor-boards, which act as continuous beams, rather like cantilever bridges without hinges, and then a second tier of joists. Such tiered structures are well-suited to the concentration of small loads from a large area or the distribution of large loads over a large area, problems of what may be described as 'load diffusion'.

The rachis is a cantilever loaded from underneath, and so it is subjected to a sagging bending moment which throws its top flange into compression and its bottom flange into tension – for indeed, it has flanges. It is of box section (see Fig. 5.29(*d*)), rather wider than it is deep, for depth tends to spoil the flow of the air over the feather. Here is a case of conflict between the structural function, for which more depth would be desirable, and the aerodynamic one. The small depth of the rachis means that the shear load

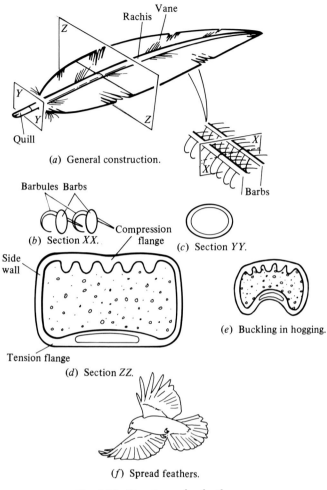

(a) General construction.

(b) Section *XX*.

(c) Section *YY*.

(d) Section *ZZ*.

(e) Buckling in hogging.

(f) Spread feathers.

Fig. 5.29. *Structure of a feather.*

in the webs will be small compared with the loads in the flanges. Consequently, the side walls of the box which constitute the web or webs of this beam can be very thin, so that they are in danger of buckling: this is prevented by filling the interior of the box with a foam-like structure of feather material, which is only about 5% solid. This filling supports the webs sideways and stabilises them against buckling.

The top or compression flange has a mean thickness of about 10% of the depth of the beam and is also in danger of buckling under flight loading. However, besides being supported by the foam, it is formed with longitudinal ribs integral with it (see Fig. 5.29(d)) which stiffen it. The same purpose, preventing buckling, would be served by transverse ribs from side wall to side wall, but as these would not contribute to the flange area as longitudinal ones do, they would represent extra weight. The bottom flange has no stiffening ribs but being normally in tension is in no danger of buckling.

The chief provision against damage lies in the structure of the aerofoil surface. The barbs may readily be separated, say, by the passage of a pencil between them, but will reunite by re-engagement of the hooks on the barbules, simply by passing the vane between the fingers, restoring the aerofoil in its original smooth entirety. The bird carries out this reforming process by preening with its beak.

The structural material of feathers is a protein called *keratin*, which is also the chief constituent of finger-nails and hair. Keratin is not only strong for its weight – it will sustain its own weight freely hanging to the extent of 30 km (Fig. 3.8) – but it is also stiff for a substance based on chain molecules.

It has been suggested that feathers may need to be stiff in order to avoid flutter, self-induced vibrations of the kind which destroyed the Tacoma Narrows bridge (see Chapter 3). It is difficult to see this happening within the serried ranks of ordinary plumage, but some birds, for aerodynamic reasons, spread the wing-tip feathers in some regimes of flight (see Fig. 5.29(f)) and in such cases flutter could occur. If so, the closed section of the rachis and the quill would be important in contributing torsional stiffness.

The hulls of some ships, the box girders of some bridges and the wings of some supersonic aircraft all resemble the rachis of the feather in their basic structural philosophy.

5.6 Hoops, arteries and domes

A number of elephants standing in a circle and all pushing outwards could be restrained by a single rope, as in Fig. 5.30(a). Similarly, an internal 'pressure' in an artery, a hosepipe or a sausage-balloon inflates it to a circular cross-section, as in the other examples of hoop action in Fig. 5.30. The tension in a hoop can easily be calculated from the internal pressure by considering a short length L of hose, of radius r, as shown in Fig. 5.30(e), inflated to internal pressure greater by p than that of the atmosphere. Engineers called this excess of pressure above atmospheric pressure the *gauge* pressure, because it is what will be indicated on a pressure gauge: clearly, it is only the excess which has to be carried by the structure, in this case, the hose. Imagine the hose divided in two by a rigid board, as in the figure: then one half of the hose can be attached to the board and the resulting tube of semicircular cross-section will work perfectly well providing the board can stand the bending stresses it will have in it. In particular, the pressure forces on the hose material will be the same as before and will

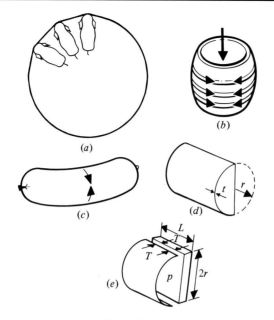

Fig. 5.30. Hoops.

be in equilibrium with those on the board – otherwise such a semicircular tube would be a convenient perpetual motion machine, exerting an unbalanced force which could be put to good use.

The pressure forces on the board, which measures L long by $2r$ wide, all act in the same direction and so add up to

$$p \times L \times 2r,$$

and they must balance the two tension forces T in the two edges of the hose-pipe, so that

$$T = pLr$$

or, more conveniently, the hoop force per unit length of hose is

$$\frac{T}{L} = pr.$$

If the hose is t thick, then the area of material carrying load T is Lt and so the hoop-wise stress in it is

$$\sigma_{\text{hoop}} = \frac{\text{force}}{\text{area}} = \frac{T}{Lt} = \frac{pr}{t}$$

By imagining a circular hose closed off at its end by a circular board of radius r, we can likewise show that the force on this board is $\pi r^2 p$, and as the circumference of the hose is $2\pi r$ and the material is t thick, the *longitudinal* stress in the wall is

$$\sigma_{\text{long}} = \frac{\text{force}}{\text{area}}$$

$$= \frac{\pi r^2 p}{2\pi r t} = \frac{1}{2}\frac{pr}{t}$$

i.e., just half the hoop stress. This result will be useful later.

An interesting use of the hoop principle to resist bursting forces round a circle is in the construction of domes. A dome may be regarded as a large number of half-arches radiating from a centre, each exerting an outward abutment thrust (see Fig. 5.31), just like the circle of elephants in Fig. 5.30(a). Large domes such as that of St. Peter's, Rome, and St. Paul's, London, were usually built with a chain round their circumference at the bottom to hold them in. During its construction in 1743, the dome of St. Peter's began to crack, and as a result of a celebrated correspondence between the Pope and some of the leading experts of Christendom, heavier chains were provided which cured the trouble.

The ring or tube in circumferential tension – hoop stress, as it is called – is one of the most important structural elements in nature and engineering.

5.6.1   Ring carrying inward loads

Clearly, if the elephants in Fig. 5.30(a) were pushing inwards, then a rope would not serve to keep them apart, but a beam in the form of a ring would. This beam would be in compression, and the action of any portion of it would be rather like that of an arch.

The tie, a structure in tension, is inherently stable but the strut, in compression, is inherently unstable, being prone to buckle out sideways and collapse. The suspension bridge, another tension structure, is inherently stable and the arch, a compression structure, is inherently unstable. A hoop in tension, say, the section of a sausage-balloon, is stable, tending to become circular as it is inflated, whatever its initial shape. On the other hand, a hoop in compression is unstable; any irregularity in its shape tends to increase and may eventually cause buckling and collapse.

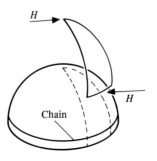

Fig. 5.31. Domes.

When an aircraft flies at altitude the pressure inside the cylindrical fuselage is kept higher than that outside so that the air inside is thick enough to breathe: the cabin is pressurised. The differential pressure between inside and outside produces tensile hoop stresses in the skin of the fuselage, increasing its stability.

In a submarine, the air inside the cylindrical pressure hull, in which the crew spaces are, is maintained at atmospheric pressure, while the outside pressure increases due to the depth of seawater above the vessel. This differential pressure, acting inwards, throws the skin of the hull into compressive hoop stress, tending to make it buckle. To prevent collapse, stiffening frames have to be provided.

5.6.2 The bicycle wheel

The bicycle wheel is an elegant structure that illustrates several principles of design. First, consider the naive (and useless) version shown in Fig. 5.32(a), in which the spokes are radial and lie in a single plane. The spokes are tightened up, that is, thrown into an initial tension or *prestressed*, by means of the threaded sleeves which secure them to the rim. When the rider mounts the bicycle, so that a heavy load W falls on the hub, the tension in the spokes at the top increases, and that in the spokes at the bottom decreases, so the load is carried partly to the top and partly to the bottom of the rim. If the spokes were not prestressed, the bottom ones would not share the load since they are too thin to bear even a small compressive load, which would simply buckle them: the load would all be carried

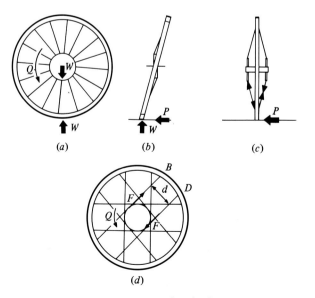

Fig. 5.32. Bicycle wheel.

to the the top of the rim, which would be subject to about twice as much bending moment, and would have to be reinforced. It is true that the pre-stress in the spokes throws the rim into a heavy compression, but it is a compression whose line of thrust will follow precisely round the section just as in a perfect arch bridge, producing no bending and therefore only relatively low stresses.

The wheel design of Fig. 5.32(a), with prestressed spokes, would be well-adapted to take loads such as W lying in its own plane. But when cornering bicycles have to take a side load, such as P in Fig. 5.32(b), and the design with all the spokes in one plane would collapse in such circumstances in the mode of failure shown in the figure; this is prevented by using a long hub, AB (Fig. 5.32(c)), and connecting half the spokes to each end, so 'triangulat-ing' the rim from the hub just as T in Fig. 5.25(b) is triangulated from the support at R.

Before the design is satisfactory, however, one more provision must be made. Suppose in Fig. 5.32(a) the rim is held securely and the hub is twisted with a moment or torque Q. Because turning the hub produces very little extension in the radial spokes, they are very ill-fitted to resist this torque, by the principle of work we saw in Chapter 4. Just as the work-bench of Fig. 5.19(a) was strengthened in shear by a brace running in such a direction that shearing would lengthen it considerably, so in the bicycle wheel it is strengthened against torque by aligning the spokes so that twisting the hub relative to the rim would change their length considerably. This is done by making them run *tangentially* to the hub, as in Fig. 5.32(d), instead of radially. Now if we apply the torque Q, a spoke such as B will be stretched, so that the tension in it is increased, while the spoke D will be shortened, reducing the tension in it. Thus the pair of spokes B, D will exert a resisting couple equal to the magnitude F of these changes in force times the distance d between them, as shown by the arrows in the figure. Every other such pair of spokes will develop a similar resisting moment, so that the total of them all is equal and opposite to Q.

Some appreciation of the economy of the design of the bicycle wheel is given by imagining the material of the spokes converted into a thin disc of the same mass. The thickness of this disc at the outside would then be about one-fifteenth of a millimetre, or rather less than most sheets of paper.

5.7 Prestressing

The bicycle wheel is a good example of *prestressing*, the deliberate introduc-tion of self-balancing forces into a structure before it is loaded. The tension in the spokes is balanced by the compression in the hoop of the rim; it enables the very slender spokes to act as struts and carry a compressive load due to external forces applied to the wheel, up to the point where that

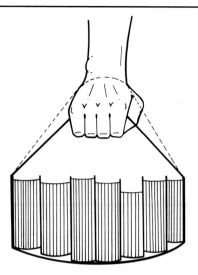

Fig. 5.33. Carrying books.

compressive load is equal to, and cancels out, the initial tension or prestressing load put in by tightening the end-fittings.

The tyre of the bicycle is another example of prestressing in tension, the 'antagonistic' or balancing member here being a fluid, the air under pressure in the inner tube. But one of the most important cases of prestressing is in concrete, where the object is to produce an initial compression in a material which cannot support tension.

Consider a bundle of books being carried by means of a strap (Fig. 5.33). The tension in the strap clamps the books together endwise so that they behave, within limits, as one solid block. On the other hand, if there is too much slack in the strap, as in the dotted line in Fig. 5.33, there will not be enough endwise compression on the books, which will fall out.

Figure 5.34 shows a kind of concrete bridge which works on this basis. The section AB hangs from the cables (it does not rest on the tower at all) and its own weight produces a large endwise compressive force in it which

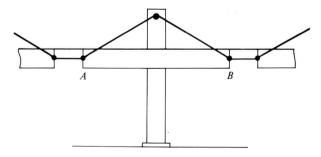

Fig. 5.34. Bridle-chord bridge.

throws the concrete into compression. There is a sagging bending moment in *AB* which will tend to produce a tensile stress in the concrete at the bottom of the bridge section, a stress which concrete cannot sustain. But as long as this tensile stress does not exceed the compressive stress produced by hanging *AB* from the cables, there will be no net tensile stress, so the concrete will be satisfactory.

In *reinforced* concrete, steel rods or wires are laid in the tension side of the concrete to take the tensile stress. But the concrete is still stretched and may crack, and it will not contribute to the stiffness of the beam nor protect the steel from corrosion. By using *prestressed* reinforcement, the concrete can be thrown into compression and will not crack, and stiffer, lighter structures can be built.

Consider the concrete beam shown in Fig. 5.35(*a*), which has a steel rod through the middle tightened by means of nuts screwed on its ends (nuts are not used in practice). The beam carries a uniformly distributed load so that it has a sagging moment in the middle and no moment in the ends. Now the uniform compressive prestress produced by the steel is needed by the concrete at the bottom of the beam, but not that at the top. We could obtain the same result as far as the bottom concrete is concerned by a thinner rod nearer the bottom of the section and exerting a smaller compressive force, but if it lay outside the middle third of the depth of the beam, it would actually throw the top concrete into tension. This would be prevented in the middle of the length of the beam by the compression in the top due to the sagging bending moment, but it would crack the top of the beam near the ends.

We can achieve this promised economy without cracking the top concrete at the ends by making the wires follow a curved path, near the bottom in the middle and central at the ends, as in Fig. 5.35(*b*). It may occur to you that the beam is now really a kind of suspension bridge, with the rod acting as the cable and the concrete serving to carry the abutment pull *H* from end to end, and you would be quite right.

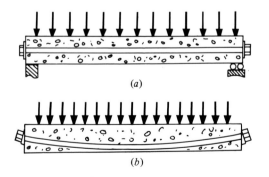

Fig. 5.35. Prestressed beam.

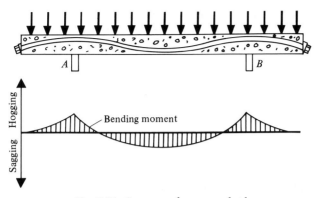

Fig. 5.36. Prestressed concrete bridge.

Note that this design is only possible if the beam never has to cope with the case of no load – and it never does, because its own weight is an important proportion of that load. But if suddenly all the load were removed from such a beam, e.g., by standing it upright, it might well crack under the unopposed bending stresses caused by the deviation of the prestressing rod from the centreline.

Prestressing steel can either be stretched while the concrete is poured round it and sets, gripping it and holding it stretched, or it can be put through a tube cast into the concrete and tightened when convenient. The former method is called *pretensioning*, the latter, *post-tensioning*. The prestressing wires of post-tensioned structures are often called *tendons*.

5.7.1 Prestressed concrete bridges

Figure 5.36 shows a section through a prestressed concrete bridge together with the bending moment under its own weight and that of the traffic over it. The principal prestressing tendons undulate along their length, lying in the bottom of the section where there is a sagging moment, as in the beam just studied, and in the top of the section over the supports at A and B, where there is a hogging moment, as there is in a rolled-up rug cast over the back of a chair so that it hangs down on either side. Indeed, this analogy leads to a vivid way of visualising the working of such a bridge as a whole. Figure 5.37 shows a long, saggy object, such as a roll of carpet, laid across two supports at A and B. If a rope C through the middle, represented by

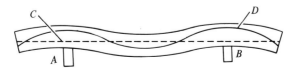

Fig. 5.37. Visualisation of tendon action.

the dotted line, is tightened, it will tend to straighten and stiffen the whole 'beam'. If, however, instead of going straight down the middle, the rope follows the wavy path D, it will be very much more effective when it is tightened.

An interesting aspect of the prestressed concrete bridge in its more daring and elegant embodiments is the approach to an organic form. Because of the subtlety of the considerations governing the path given to the tendons, they have a liveliness of outline not found in the straight lines and conics of simple designs. Their exterior simplicity disguises the internal textural intricacies, the maze of reinforcement, often curiously shaped, the local areas of stronger concrete with special reinforcement that bear comparison with the elaborate internal structure of bones.

A final point of interest is that such bridges often have hinges in them, of the type shown in Fig. 4.23, which are introduced to ensure that the designed bending moment distribution is achieved.

5.8 Fluids as structural elements

In discussing Maxwell's Theorem, it was remarked that nearly always neither T nor C was zero – in other words, the vast majority of structures need both tension and compression members. But while tension members must be solids, a compression member can be fluid, and a great number of living organisms use a liquid as a major structural component.

For example, most flowers droop if plucked and not put in water, because their stems are of very flexible fibres prestressed by the internal pressure of sap, and evaporative losses destroy the pressure and cause them to wilt.

Earthworms, slugs, sea-anemones and many other creatures rely on liquids for the compressive part of their structure. Motion is obtained, not by the operation of muscles on kinematic link mechanisms, as in the jaw and the knee, but by the reaction of muscles against fluid pressures. For example, an earthworm is a tube full of viscera and fluids, provided with circumferential belts of muscle and longitudinal muscles (Fig. 5.38). Since the contents of the tube are virtually incompressible, contracting the circumferential muscles causes the worm to extend, and, because of the internal pressure, the extension is forcible and can push the creature's sharp end into the earth. If the longitudinal muscles contract and the cirumferential ones relax, then the worm becomes shorter and fatter. By contracting the longitudinal muscles along one side, the worm bends powerfully to that side, but only because it has the fluid pressure inside to react against – otherwise it would simply develop wrinkles. The worm develops an internal bending moment, with the tensile force in the longitudinal muscles and the compressive force in its liquid contents (Fig. 5.38).

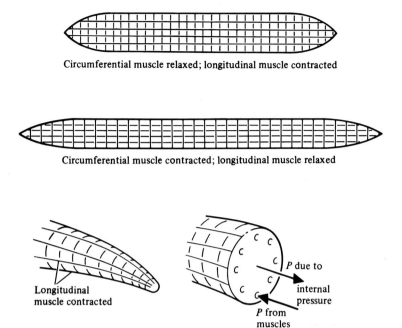

Fig. 5.38. *Structure and motion of an earthworm.*

A simple example of a structure with solid tension and gaseous compression elements is a sausage-balloon. The skin of the balloon is in tension both longitudinally and circumferentially, and both these tensions are balanced by the pressure of the air inside. If the balloon is bent, the longitudinal tension is reduced on the inside of the bend and increased on the outside of the bend, until eventually the tension disappears on the inside of the bend. The skin, being unable to support a compressive load, buckles (see Fig. 5.39). A very similar form of buckling may be seen in a bent flower stem.

Some spiders have leg joints rather like the bent balloons of Fig. 5.39 (see Fig. 5.40) in which there is a soft fold in the hard tubular exoskeleton.

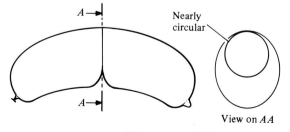

Fig. 5.39. *Buckled sausage-balloon.*

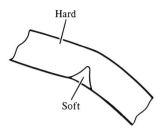

Fig. 5.40. Spider's leg joint.

Muscles are used to flex the joint, but it is straightened by pressure applied to the interior of the tube, which springs out just as the balloon would if released.*

Man has been rather slow to use fluids for structural purposes, although tyres are an important case and the thin structural skins of large rockets are stabilised by internal fluid pressure.

Structure is a vital aspect of all organisms, trees, horses, humans, birds, of all buildings and machines, and all vehicles, especially aircraft. All structural design is based on combinations and variations of a few basic elements, struts, ties, beams, arches, hoops, shells, and so on, most of which have been touched on, but a vast range of interesting and elegant forms can be devised from these few elements. The subject is also enriched by the influence of materials and production methods. Thus a stalk of grass, a feather, an aircraft wing and a supertanker are all structures based on a similar plan and similar needs – the ability to resist bending, extreme economy in the use of materials, very small loads relative to the size leading to problems of stability, and so on – and complicated by difficult non-structural requirements, aerodynamic form, growth, manufacture, transmission or storage of liquids, requirements of access for inspection, repair or maintenance. Structures like the human pelvis or the components of aircraft undercarriages are less simple in their load-bearing role, because they form parts of mechanisms and move relative to the co-operating parts, and so they lack the direct appeal to the understanding of a feather, but they are full of subtleties.

Mechanism and structure are two characteristics common to most design. A third, present in all living things and all but the simplest designs of man, is that of systems.

* Parry, D. A. and Brown, R. H. J. (1959). The hydraulic mechanism of the spider leg. *Journal of Experimental Biology*, 36, pp. 423–33.

6

Systems

A third manifestation of design is in systems – combinations of components or organs for performing particular functions. Thus, the digestive system of an animal consists of jaws and teeth, salivary glands, gullet, stomach, and so on, and its function is to process food and extract from it the substances which support life. A railway system consists of railwaymen, rails, rolling stock, signals, and so on, and its function is to move people and goods.

These two examples of systems are of characteristic types. The digestive system is a *process plant*, in effect, where material flows through and is processed to yield some useful product or products. It has a series or line arrangement of components, one after another, and a flow from one end to the other. The railway system is a *service* system, which provides a function, transport in this case, over an area. It has a *network* of components, with flows in both directions down any branch.

Not all systems will fit into these two categories, but they do provide the key to a number of important aspects of design in which the ideas of flows, lines, loops and networks are central.

6.1 Process plants

A radio receiver displays many of the features of the process plant type of system. It consists of a large number of components, resistors, capacitors, coils, transistors, etc., connected into circuits by circuit boards and wires. It processes an incoming radio signal and puts it out as sound from a loudspeaker. It is organised into a sequence of stages, or *sub-systems*, in series, each of which processes the signal in sequence and most of which are stages of amplification. It has within it a service system of its own which supplies the necessary power to the sub-systems, and it has numerous *feed-back loops*, the nature and purposes of which are a characteristic of the design of many kinds of systems.

147

The amplification stages are designed both to perform well in themselves and to *match* the stages on either side. Furthermore, they are designed so that each sub-system is less variable in performance than the components of which it is built. All these are characteristics shared by many systems.

Another system of this sort is a paper-making machine. It has inputs of water, wood-pulp, the white clay kaolin, and various other substances, electrical energy, heat in the form of steam, and so on. Each stage or section is *matched* to the one before – for example, once the paper has been formed it must proceed at the same speed down the rest of the line: if a later stage were to run faster, the paper would be torn in pieces, and if it ran slower, the paper would accumulate in a great mass, be ruined and jam up the machine.

The paper-making plant has feed-back loops also, and of a kind easier to understand than those of the more familar, but more recondite, radio receiver. The water, wood-pulp and clay are mixed into a creamy slurry at an early stage, and at points down the line measurements are made of the consistency, density and moisture content of the paper being produced. These measurements, in the form of electrical signals, are processed and *fed back* to the mixing stage where they adjust the composition very slightly. At other points the tension in the paper may be measured and very small adjustments made to the speed of the rollers.

Both in processing radio signals and making paper, the object of feed-back is often to stabilise the system in spite of disturbing influences, just as our digestive system will try to cope with whatever food we send down, and our body temperature remains much the same through frost and heat-wave.

6.2 Service systems

Service systems distribute or collect some thing or things or provide some function throughout a region. Thus the sap of a plant collects water and important dissolved salts (of nitrogen, potassium, phosphorus and other elements) along with waste products of the living processes in the roots, it takes them out to the leaves and it bears back the products of photosynthesis which sustain the life and growth of the entire organism. It also carries signals in the form of substances that control growth and other functions.

The control system of an aircraft provides power at a number of points, to operate the rudder, elevators, undercarriage, and so on, and also controls the application of that power. It too has feed-back loops, some of a very special kind designed to match the aircraft to the pilot's own feed-back loops. It is one of a number of systems spreading throughout the aircraft.

A railway system provides transport for people and goods throughout its network. Its power system, as it were, the rails and trains which carry out the prime function of transport, are controlled by a superimposed signalling system, which not only controls the trains but also detects where they are and modifies its instructions accordingly – it has a feed-back system, in fact.

The human body is full of systems of all kinds – the blood system, the nervous system, the lymphatic system – mostly very finely ramified and, in the three cases mentioned, extending almost throughout the entire body. The blood carries all the three resources which design deals in; energy, in the form of fuels for the muscles, materials, to build up and repair the organism, and signals, in the form of chemical substances like adrenalin, which readies the body for violent action. For a few special functions it even carries energy in the form of its own pressure, like the hydraulic fluid in an aircraft. But its function in carrying signals is mainly confined to messages where speed is not required – it is the postal system of the body; messages which must go fast – and that is the vast majority, and covers nearly all the instructions and an even greater preponderance of the information – are handled by the nervous system and that as yet largely incomprehensible processor, the brain, the pinnacle of design to date, which can itself design.

6.3 Interaction of components in systems: matching

Some systems can be regarded as consisting of major components or sub-systems which interact in relatively few, simple and sharply-defined ways with each other. For example, in the radio receiver, the interaction between the different stages of amplification can be represented in simple mathematical terms which can be studied and which form a basis for the design of the whole system. The stages must be matched to one another, so that the performance of the whole is optimised. In the case of the radio receiver the technical basis of this optimisation is beyond the scope of this book, and so other examples will be used.

In the paper-making machine, matching the various sections is simply a question of making them all run at the same speed and have the same width. Matching in other systems is often not much more difficult a matter, though it may well seem so because the quantities to be matched are less easily recognised than the speed and width of paper.

Consider a windlass which is used for raising a bucket of water from a well (Fig. 6.1) by winding a rope up on a drum. If we make the radius r of the drum too large, a man will not be able to exert enough force on the handle to raise the bucket of water; if we make it too small, the bucket will be easy to raise but at each turn of the windlass it will rise very little, so

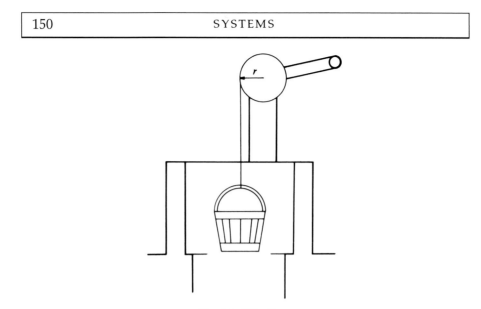

Fig. 6.1. Windlass.

that it will take too long to reach the top. An ideal radius is one such that the force he must exert on the handle is large, but not uncomfortably so; here the quantity to be matched is a force.

Finally, consider a gas turbine (Fig. 6.2). Air, after being compressed in a compressor and heated by burning fuel then drives a turbine. The turbine must be designed to suit the quantity and pressure of hot air entering it, just as a section of a paper-making machine must suit the width and speed of the paper. However, the gas turbine may have to operate at a variety of speeds and powers, so that its components have to match each other suffi-ciently well over a wide range of conditions, which makes the case much more difficult.

6.4   Characteristics of systems

6.4.1   Characteristics: the man/bicycle system

Many systems include a power-developing device – a motor, an engine, a set of muscles – driving a power-absorbing device, a pump, a propeller, a

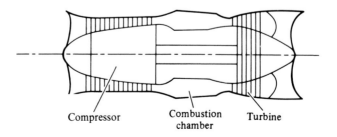

Compressor   Combustion   Turbine
                     chamber

Fig. 6.2. Gas turbine.

windlass, swings. When such a system is operating steadily, the power developed by the one must exactly equal that absorbed by the other. If the motor power exceeds that required by a pump, then the excess power causes the system to speed up. If the pump requires more power than the motor is supplying, perhaps because the load on the pump has suddenly increased through a tap being opened somewhere, then the system will slow down until a balance is reached again, or will perhaps stop altogether.

We have already seen a case like this, that of the singer exciting the wineglass into resonant vibration, in Chapter 3, only in that case amplitude of vibration corresponds to speed (see Fig. 3.11). If the amplitude is low, say, when the singer has just hit the note at which the glass resonates, the power input from the sound waves exceeds the power absorbed by damping in the glass, and the amplitude increases, until either the point A is reached where the two powers balance, or the glass breaks.

Consider the case of a cyclist and his machine. On the level, the force at the back wheel required to drive the bicycle along is relatively small, and so he uses a high gear, one in which one turn of the crank turns the wheel about three times. In this gear, there is little force at the back wheel but a given crank speed gives a high forward speed of the bicycle. On a slight hill he will change into a lower gear, producing more force but less speed. The high gear corresponds to a large radius of drum on the windlass, the low gear to a small radius.

Figure 6.3(a) is the *characteristic* of a gear/rider combination. It shows the force he can comfortably produce at various speeds at the back wheel in two gears, a low one, A, and a high one, B. With the low gear, he produces more force at lower speeds, but as he goes faster and the cranks turn more quickly a speed is reached, at point P in the figure, beyond which he could produce more force in the higher gear.

Figure 6.3(b) shows the resistance characteristic of the bicycle and rider, the force resisting forward motion (which would be the drag in an aircraft)

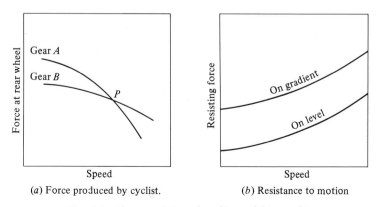

(a) Force produced by cyclist. (b) Resistance to motion

Fig. 6.3. *Characteristics of cyclist and his machine.*

plotted against speed. Two such characteristics are shown, one for travel on a level road, the other for climbing a gentle gradient. The resistance increases with speed in both cases, but is naturally higher for the uphill case.

Now when the bicycle is travelling at a steady speed, neither accelerating nor decelerating, then the force at the back wheel produced by the cyclist exactly balances the resisting force, as we saw with the sledge and the air-craft in Chapter 2. If we plot both characteristics on the same graph, as has been done in Fig. 6.4, then any intersection, such as Q, between a curve of the force at the back wheel and a curve of resistance represents such a point of equality between the two forces. Thus at Q, the force at the back wheel in gear B is equal to the resistance climbing the gradient, so the speed in that gear or that gradient may be read off as v_1 on the speed axis at the bottom of the graph. Similarly, v_2 is the speed climbing the gradient in gear A, and it will be noted that v_2 is higher than v_1. On the level, however, a higher speed is obtained for the same apparent effort in the higher gear, B.

In practice, a cyclist knows by experience that the higher gear suits him better on the level and the lower gear is better for climbing. He may have a choice of 10 gears, when it becomes quite difficult to decide which is best. The gears in cars and other vehicles serve the same basic purpose of match-ing the prime mover, the engine, to the speed and the gradient.

Curves like those in Fig. 6.3, of force against speed, are called character-istics, and they are an important aid in the design of systems. In this case, we plot the force/speed characteristic of the cyclist, the power producer, and that of the bicycle, the power absorber, on the same graph and to the same scales. The intersection point (Q or R, according to the gear, in Fig. 6.4)

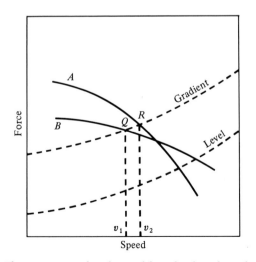

Fig. 6.4. Characteristics of cyclist and bicycle plotted on the same axes.

gives the speed at which the system will operate. It also gives the power, because the power is speed times force, as was seen in Chapter 2.

6.4.2 Characteristics

Some characteristics are not simple curves like those of Fig. 6.3. A transistor or a pump requires a family of curves, because both require two quantities to be specified in order to define a working point. At one speed, the bicycle has one resistance. But at a particular speed, a pump will develop a pressure which varies according to how much liquid we draw from it. Figure 6.5 is a characteristic for a centrifugal pump, where the curve marked 1000 rev/min, say, shows how the pressure varies with the flow rate delivered when the pump is driven at a fixed speed of 1000 rev/min.

Consider the arrangement of Fig. 6.6, where a motor drives a centrifugal pump drawing water from an ornamental pond and ejecting it through a fountain. The fountain has a single-curve characteristic, like the bicycle: for any particular pressure in the supply, it delivers a particular flow of water, giving a roughly parabolic characteristic like the dotted line in Fig. 6.5. If the pump is supplying the fountain, then the flow into the fountain is the same as the flow from the pump, and the pressure into the fountain must be the same as the delivery pressure of the pump – i.e., the two are represented by the same point on the superimposed characteristics of Fig. 6.5. Thus if the pump is supplying the fountain and running at 1000 rev/min

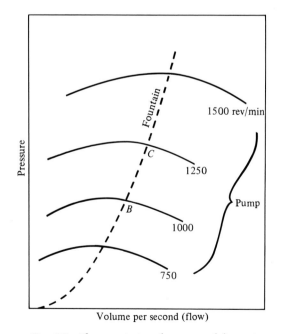

Fig. 6.5. Characteristics of pump and fountain.

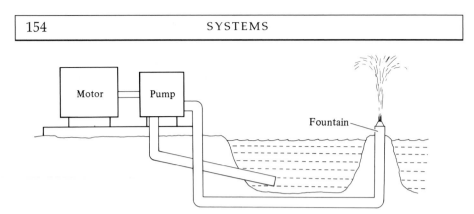

Fig. 6.6. Motor/pump/fountain system.

its flow and pressure must be represented by a point on the 1000 rev/min curve in Fig. 6.5, and this point must be *B*, the intersection with the fountain characteristic. Similarly, if the pump is running at 1250 rev/min, its working point when supplying the fountain is *C*. We can now construct a speed/power characteristic for the pump/fountain combination from Fig. 6.5, as in Fig. 6.7, for if we pick any speed, say, 1000 rev/min, we can pick off the flow from point *B* in Fig. 6.5, and the power is given by the product of pressure and flow.

6.4.3   Degrees of freedom

A pump may be regarded as having two degrees of freedom – a slight broadening of the very important idea we applied to mechanisms in Chapter 4 and to structures in Chapter 5. To specify its working point, we must have two pieces of information, say, the speed and the pressure against which it is working; if we know these two, we can locate the point on the characteristic curves (Fig. 6.5) and hence find the flow. The characteristic of the pump is a set of curves, because it shows the relation between the three quantities pressure, flow and speed.

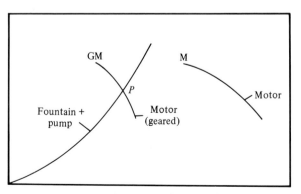

Fig. 6.7. Motor/pump/fountain working point.

The fountain has one degree of freedom – if the pressure is fixed, that determines the flow, and the characteristic is a single curve relating pressure and flow.

Thus the pump and the fountain between them have three degrees of freedom. But when we couple them with a pipe, we introduce constraints which reduce the degrees of freedom. First, the flow from the pump must be equal to that into the fountain, which removes one degree of freedom. Second, the pressure leaving the pump must be equal to the pressure at entry to the fountain, which removes another degree of freedom, leaving only one, i.e., the characteristic of the pump/fountain sub-system has one degree of freedom and is a single curve, as we saw (Fig. 6.7).

A transistor has two degrees of freedom, so that its characteristic is a set of curves. Coupled to a voltage source and a load resistor, however, the characteristic reduces to a single line (the load line) and so this sub-system (which may be the basis of an amplification stage in a transistor radio) has only one degree of freedom.

From a further characteristic of the pump, we can find the power required to drive it at any speed and any flow. Now when the pump and the fountain are connected together, the speed fixes the flow and hence the power input, and so we can draw a single curve of power against speed for the combination, as in Fig. 6.7.

6.4.4 Gearing

The next step is to look for an electric motor to drive the pump/fountain combination. Now a motor has a power against speed characteristic which can be plotted on the same axes as the pump/fountain characteristic. The intersection of the two will then give the speed at which the pump/fountain/motor system will run.

In Fig. 6.7 such a motor characteristic has been plotted as a line from M on the fountain/pump characteristic, and the two do not intersect. At M, the motor is at the lowest allowable speed, where it takes the largest permissible current, and yet its power is much less than is necessary to drive the pump at that speed. However, if gearing is used to reduce the speed of the motor to one-half, the characteristic of the geared motor becomes GM, intersecting the pump/fountain characteristic at P, a practicable working point. In this particular system, gearing can probably be avoided by using a lower-speed motor, or a higher-speed pump, or both, and as gearing is rather undesirable in such an installation, this is what would normally be done. In a ship driven by gas turbines, however, the efficient speed of the propellers may be one-hundredth of that of the turbines, so that gearing is unavoidable, practically speaking. The problem arises because the gases in the turbine, moving at very high speed and so having very great kinetic energy, have to transfer that energy to a much greater mass of seawater, moving at a very much lower speed: there is a very large mismatch of speed between the original driving medium, the gas, and the ultimate driven one, the sea.

The same fundamental mismatch, but in a more extreme form still, occurs in ships driven by steam turbines. However, attempts were made in the early days of turbine ships to avoid the use of high-speed, high-power gearing, which at that time was an unknown field, and to mount the turbine and the propeller on the same shaft. By providing a very large number of stages in series in the turbine, the steam converted its thermal energy into kinetic energy in smaller fractions at a time, so reducing its speed. The propeller blades were designed with very small pitch, that is to say, so that if the propeller were screwed into wood like a wood-screw, the distance it would move in per turn would be relatively small: this meant that to make the ship move at a given speed, the propeller would be turning relatively fast. Also, the turbine blades were put as far out from the centre of the shaft as possible and the propeller blades were made as short as possible, so that the former would move as fast as possible compared with the latter.* In the 'Turbinia', Parsons' vessel with which he demonstrated the potential speed of turbine-driven ships so spectacularly at the Royal Review of the Fleet in 1904, this need to keep the diameter of the propellers small led to him putting three small ones on each of the three turbine shafts (see Fig. 6.8).

But the fundamental mismatch between steam and sea was too great to be overcome efficiently with turbine and propellers on one shaft. Electrical transmission was tried – that is to say, the turbine was used to drive an electrical generator which provided power to drive a motor on the propeller shaft, but that arrangement, although it solved the matching problem and was very flexible and quiet, was also very expensive and rather inefficient.

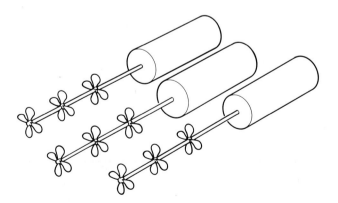

Fig. 6.8. Turbine and propeller arrangement, 'Turbinia'.

* Stodola, A. (1945). *Steam and Gas Turbines*, pp. 809 *et seq.* New York: Smith.

Gearing was meanwhile developed which was capable of absorbing the very high powers and producing the very large torques required, and gearing has for a long time provided an efficient and relatively cheap solution to the mismatch problem of turbine and propeller.

Gearing is rather noisy because of the continual engagement and disengagemet of the teeth. It is made quieter by using helical teeth, so that engagement takes place progressively along their length rather than all at once, as with straight teeth, but they remain noisy. In recent years work has been done on making them quieter by the simple expedient of leaving the teeth off the gears, and pressing the smooth cylinders that remain hard together so that the drive is transmitted by friction between them, just as a bicycle is propelled by friction between the road and the rear tyre. These 'traction' drives, as they are called, are efficient and quiet, but have not yet reached the stage of full practicality.

6.4.5 Piston-engined ships

Most ships nowadays are driven by internal combustion engines (oil or diesel engines). The best speed for these in the large powers used is probably about 300 rev/min, while the best speed for the propeller may be about 80 rev/min. This is not too big a mismatch, so that at present the usual solution is to drive the propeller directly from the engine, both running at 100–110 rev/min. This does mean that the engine is nearly three times the size and weight it need be, but gearing presents difficulties because of the fluctuating torque of the engine and the torsional vibrations which are set up.*

6.4.6 Other matching problems

Other matching problems arise which cannot be solved practically without the introduction of 'gearing' of some kind. For example, computers and other microelectronic devices require a much lower voltage than the mains supply, and so a transformer is used to 'gear' the voltage down to the right level (in this case, another kind of change, from alternating to direct current, is also needed).

There is a very close analogy indeed between the transformer and mechanical gearing. Gearing down an electric motor reduces the speed and increases the torque in the inverse ratio, while speed times torque, which is power, remains the same except for the small losses in the gearing. Stepping up an electrical supply by means of a transformer reduces the current and increases the voltage in the inverse ratio, while current times voltage, which is power, remains the same except for the small losses in the transformer.

---

* Thompson, R. V. (1985). Marine technology – present and future. *Proceedings of the Institution of Mechanical Engineers*, 199, no. 45.

For some systems, no practical form of 'gearing' or 'transformer' exists. For example, we might have a pump and a fountain which did not match because the pump pressure was not high enough although the flow was too great. In this case no practical 'gearing' device exists.

The list below gives examples of some simple systems consisting of a driver and a driven component, together with the quantities which might be used for the axes in plotting characteristics. In each case the product of the two quantities is power and the one on the vertical axis has the character of a force or voltage, a measure of intensity, while the other is a speed or flow rate, a measure of quantity per unit time (e.g., speed is quantity of road passing per unit time).

Man	Bicycle	speed	force
Engine	Car	speed	force
Engine	Propeller	rotational speed	torque
Pump	Fountain	flow rate	pressure
Heart	Body	flow rate	pressure
Muscle	Wing	speed	force
Mains	Computer	current	voltage

In the first three, gearing is used to achieve matching. In the next three, the components must be designed to suit. In the last, a transformer must be used.

In the muscle/wing case, matching is achieved as far as possible by the selection of the point of attachment of the muscle to the bones. However, this method has limitations, as will be seen in the next section.

6.5   Matching in flapping-wing flight

It was remarked in Chapter 2 that small size is an advantage in flying, because a lower power/weight ratio will suffice. However, for flapping flight using muscles, there is a matching problem at large sizes which is an even more serious handicap, and without which, it may be speculated, much larger flying creatures might have evolved.

If we double all the linear dimensions of a bird, its length, its wingspan, and so forth, it will have four times the wing area and eight times the mass, as with the fish in Chapter 2. The load to be supported by unit area of wing is thus doubled, and to sustain this extra load the speed must be increased by $\sqrt{2}$ times, or about 40%. To keep the wing working properly, its speed relative to the bird must also increase by $\sqrt{2}$ times. However, since the wing is twice as long, its stroke also will be twice as long. The wing-tips, then, are moving $\sqrt{2}$ times as fast but have twice as far to go, so that they take $2/\sqrt{2}$ or $\sqrt{2}$ times as long per stroke, or say 40% longer.

Let us look at the power requirements of the bird. It now weighs eight times as much and so requires eight times the lift. Unless its lift/drag ratio (see Chapter 2) has been improved, the drag will be increased eight times, and so at $\sqrt{2}$ times the speed it will require $8\sqrt{2}$ times the power. However, the wing muscles are only working $1/\sqrt{2}$ times as often, which below a certain frequency of operation reduces their power output. Indeed, the *force* the muscles are capable of exerting becomes the limit, and under that condition the mass of muscle required will have gone up $8\sqrt{2} \times 2 = 16$ times! Thus a bird of eight times the mass would require twice as large a fraction of its body-weight to consist of flying muscles.

This is not what is found, because small birds have roughly the same proportion of their weight devoted to flying muscles. However, this is because it is valuable for them to be able to fly faster and also to have a reserve of power for manoeuvring and carrying loads so that they have more than the minimum amount of flying muscle. Also the aerodynamic performance of the large bird should be slightly better. Nevertheless, large birds have much less power margin – look at a swan taking off, for example, and the length of runway it requires, compared with the take-off of a sparrow – and the highest mass of any flying bird is about 12 kg (swans, pelicans and condors all reach about this weight). Because of their small power margin also, large birds will have a limited range of flight, being unable to take off with a very high fat content (see Chapter 2).

6.5.1 The propeller-driven swan

Men have flown across the English Channel and further powered by their own muscles. The specific power (power per unit mass) of a man is only about one-sixth of that of a bird, and after adding the mass of even a very light aircraft, it falls to about one-tenth. It is all the more astonishing therefore that man can do so well, especially in view of his further disadvantages of high absolute mass, when we have seen that birds cannot fly at masses above one-tenth that of the man and machine combination. The enormous design gap these ratios of one-to-ten represent is bridged by various devices, chief of which is the low flying speed of the man-powered aircraft, achieved by a very large wing area. Drag is roughly proportional to weight, so that the specific power required for flight is roughly proportional to speed. But this low speed (about one-quarter that of a bird) is not enough to explain this remarkable achievement. There is also the superior propulsive efficiency of the propeller and the perfect matching that can be achieved between it and the pilot's muscles by the use of gearing in the transmission.

For much larger birds to be possible, nature would need to evolve either a new kind of muscle, or else a means by which the wing muscles could contract several times for each wing beat, doing useful work each time. It is possible to imagine such a mechanism using ratchets, but it is difficult to imagine nature ever evolving it. More efficient, as well as better matched, would be a bony crankshaft turned by direct-acting muscles and driving a propeller. The idea of a giant propeller-driven swan, weighing as much as a lion, is a remote one, but the power side of it looks workable. Apart from the problem of an evolutionary route to such a creature, the unlimited rotation required of the crankshaft and propeller is a design problem that does not seem soluble in large living organisms.

The 'muscles' of aircraft used to be pistons in the cylinders of petrol engines, which drove a propeller through step-down gearing, so that each 'muscle' operated several times for each revolution of the propeller.

6.6 Other forms of matching

Gears and transformers are the simplest forms of matching, and much more subtle forms exist. Figure 6.9 shows a familiar design of reading lamp, the Terry 'Anglepoise', in which there are three springs at C. One of them balances the moment of the weight W of the lamp about the pivot at A, and the other two keep the whole in equilibrium about the pivot at B. As the lamp is moved about, the forces in the springs alter by just the amount necessary to keep it in balance. (It is interesting that this balance is mathematically exact, and requires that the tension in the springs should be proportional to their length. You might like to think about how such a spring is made – it has to be such that when released it shortens until the coils press themselves tightly together and a finite force is needed to separate them at all.)

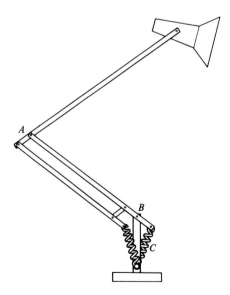

Fig. 6.9. Terry 'Anglepoise' lamp.

When a car is standing at the kerb, its weight rests on the springs. When it is jacked up, the springs extend and take less of the load, the remainder falling on the jack, which becomes harder and harder to raise if it has a constant mechanical advantage. The car jack shown in Fig. 6.10 consists of a rhombus of four equal links pivoted together at their ends. A screw pulls *B* and *C* together so causing the diagonal *AD* to extend and raising the jacking point of the car at *A*. To begin with, when there is little load at *A* and the rhombus is rather flat, a shortening of one centimetre, say, in the distance *BC* causes a much larger increase in the distance *AD*. However, when the jack reaches the position shown in dotted lines, this effect is reversed and a given shortening of *BC* causes a smaller increase in *AD*. By the principle of work, this means that the mechanical advantage has increased to suit the greater weight now acting on *A*. Thus the effort needed to turn the screw is made more uniform by matching the mechanical advantage of the jack to the increasing load.

6.6.1   The matching of powder grains

A rather unusual design problem occurs in artillery. The principal components involved are a gun barrel, a shell and the powder whose explosion drives the shell from the barrel. The 'explosion' is just the very rapid burning of grains of explosive (rather low explosive, or propellant, as it is called) which quickly evolves large quantities of hot gas in the chamber of the gun. The resulting high pressure accelerates the shell, which travels down the barrel, so enlarging the space occupied by the gas and reducing the pressure. The burning which evolves the gas takes place at the surface of the powder, and the rate of gas production is therefore proportional to the total surface

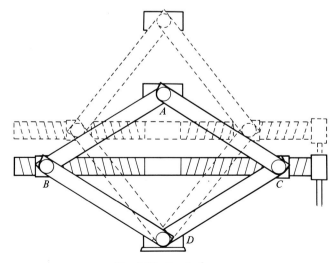

Fig. 6.10. Car jack.

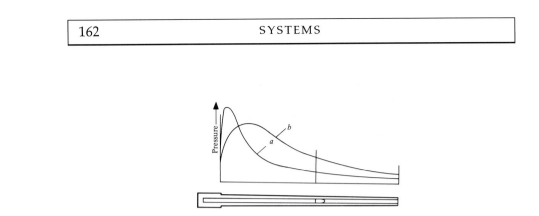

Fig. 6.11. Pressures in a gun barrel.

area of the grains. If the grains are shaped as small cylinders or cubes then the area falls as they burn away, and so does the rate of evolution of gas.

There are two problems with this system. Firstly, the shell is (relatively) slow to gather velocity, so that very high pressures are built up in the early stages. Then, when the shell has travelled a little way, the lower rate of burning is inadequate to keep up the pressure behind it. Curve *a* in Fig. 6.11 shows the kind of variation of pressure behind the shell with the distance travelled by the shell typical of solid grains, with a very high peak near the beginning and tailing away rapidly as the shell nears the muzzle.

The force on the base of the shell is proportional to the pressure, and so the area under the force/shell travel curve is proportional to the work done on the shell, which becomes its kinetic energy. If a curve like *b* could be obtained with a bigger area under it, the maximum pressure the barrel had to withstand would be reduced, and yet the kinetic energy, and so the muzzle velocity, would be increased. To do this, the initial rate of burning, and so of gas generation, must be reduced, to reduce the initial peak pressure, and yet the later burning must be at a higher rate than before. This requires that the surface area of the grains should increase as they burn, and this can be achieved by using a shape with holes in it, such as the one shown in Fig. 6.12. As burning proceeds, the outside surfaces decrease in

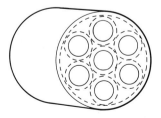

Fig. 6.12. Powder grain.

area but the holes increase in diameter, giving an overall increase in area. With grains of this sort, a curve is obtained like *b* in Fig. 6.11. This is not entirely without disadvantage – for example, the bursting pressure to which the muzzle end is subjected is increased, and there is a tendency to produce a bright flash from tiny fragments of grains still burning as they leave the gun.*

6.6.2 The matching of fish muscles

Most fish swim by bending their bodies from side to side in an undulating pattern. In steady swimming, they do this by means of a thin outer layer of red muscle (see Chapter 2) but in flight or pursuit they bring into use the underlying white muscle, which extends to the backbone. Figure 6.13 is a plan view of a fish bending to the left, which it does by contracting the muscle on the left side of its body. Suppose first that all the muscles lay longitudinally in the fish. Now in such a motion, muscle fibres far from the backbone would contract fully, perhaps by 20% of their length, while those half-way out would contract by only 10% and those close to the spine, hardly at all. Now, since the work done by a muscle is the product of the force it exerts and the amount it contracts, this means that muscle near the spine would do little work, and only that near the surface would be fully used. The average utilisation would be only about a half.

However, the fibres of fish muscle do not lie longitudinally, but along roughly helical paths such as are shown in the upper half of Fig. 6.13. Thus any path is near the surface for part of its length and near the spine for another part, and the average contraction along all the paths is much the same. By this arrangement all the muscle fibres can contract fully and make their full contribution to the speed of the fish, which can also bend more sharply.†

The same principle operates in the helical stranding of a rope – if all the strands lay parallel, then when passed over a small pulley, all the tension would fall on the outermost strands. Interestingly enough, there are cases where strands are laid up straight, with good reason, in suspension bridge cables, for example, and in some bowstrings.

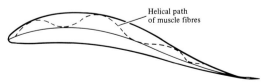

Helical path
of muscle fibres

Fig. 6.13. Swimming of fish.

* Hayes, T. J. (1938). *Elements of Ordnance*. New York: Wiley.
† Alexander, R. McN. (1967). *Functional Design in Fishes*. London: Hutchinson.

6.6.3 Matching in human muscles

An example of matching is found in the human elbow. The action of muscles on a limb bone is sometimes used as an illustration of a lever, but it is easy to see that the straightening of the arm is is not a good example. Figure 6.14(*a*) shows a person doing 'press-ups', with the bones of the arm represented by lines AB and CD, with the pivot between them, in the elbow, at B. The muscles straightening the arm are indicated by the broken line AD. The attachment of the muscle at D is to a smooth rounded end of bone (the olecranon process) round which the tendon wraps as the arm is bent, rather than directly to the side of the bone as in the scrap view (Fig. 6.14(*b*)). It can be seen that with a direct attachment, the 'arm' of the force in the muscles about B would be very small, and it would not be possible to produce the force necessary to perform a press-up.

6.6.4 Matching in bows

An example in which the muscles which bend the human arm are concerned is the matching of a bow to an archer. The 'weight' of the bow, the force needed to draw it, is limited by the strength of the archer and the distance it can be drawn is limited by his or her stature: within those limits it is usually desirable to make the stored energy as great as possible.

If the force F at the archer's hand is plotted against the distance x which the bow has been drawn, as in Fig. 6.15(*a*), then the area under the graph is the stored energy. Clearly, with the same weight of bow and the same draw, there will be more stored energy with a fuller graph like *b* than with one like *a*, and the composite bow described in Section 3.4.2 would have a fuller graph than a long-bow. But suppose that we were to sit down and think about what would be an ideal graph – clearly, it would be very full, but it would also be an advantage if F peaked and then fell off, as in graph *c*, for then, when the bow was fully drawn, the archer could easily hold it so while he or she took aim.

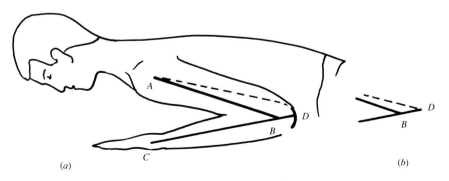

Fig. 6.14. Matching in the elbow.

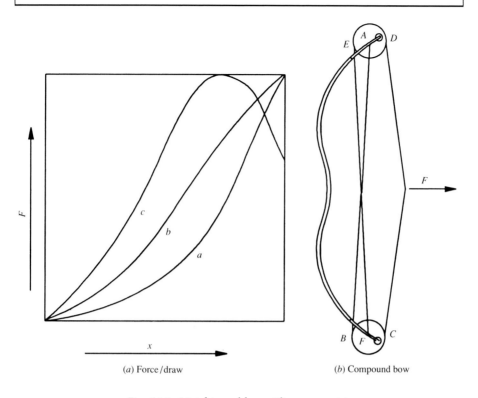

(a) Force/draw   (b) Compound bow

Fig. 6.15. Matching of bows (diagrammatic).

Such bows are made: they are called compound bows, or 'wheelies', and
they have the form shown in Fig. 6.15(*b*). The string is attached to one tip
of the bow at *A*, goes round the wheel on the other tip in a groove in the
rim from *B* to *C*, across to the rim of the other wheel at *D*, round it to *E*
and back to the second tip at *F*, where it is fastened off. Thus the bow
appears to have three strings, *AB*, *CD* and *EF*, although it really has only
one. The arrow touches only the part *CD*, the parts *AB* and *EF* remaining
straight. The wheels are pivoted to the tips, but they are eccentric on their
pivots, and they may not even be circular.

When the bow is drawn, the increasing force in *CD* tends to turn the
wheels, increasing the tension in *AB* and *EF* and pulling the tips powerfully
together. The tension in *CD* rises very rapidly, giving a full graph. However,
the rotation of the wheels on their eccentric pivots alters the ratio of tensions in such a way that the force *F* eventually falls, giving the desired
characteristic, *c*.

All these examples of matching – bicycle gears, ship propulsion, powered
flight, electrical transformers, a reading lamp counterbalance system, the
form of propellant grains, fish muscles, elbows and bows – illustrate components being given the characteristics they need for their task. This is a

very central idea in design, and as for evolution, 'survival of the fittest' might be rendered as 'survival of the best matched'.

6.7 Control systems

Mention has been made already of feed-back and controls. A simple example is the thermostat of a central heating system, which may switch on a pump circulating hot water when the temperature in the building falls below a certain level, and switch it off again when the temperature reaches another, higher, level. This is called a 'closed-loop' system, because the heat from the radiators warms the house which affects a temperature sensor which controls the pump which pumps the hot water to the radiators which . . . and so on (see Fig. 6.16).

The temperature of the hot water in turn may be monitored by a boiler thermostat, which forms part of another closed loop and controls the burner of the boiler.

More typical, but less familiar, is the positional servo, which is simply a device which positions an object, say, the cutting tool of a lathe, in accordance with the setting on a dial or the information mechanically read from a tape or received from a computer. This system is illustrated in Fig. 6.17(a), where a lathe is shown fitted with a positional control of the distance Y of the saddle carrying the cutting tool from a datum D. The saddle can move to left or right along the lathe bed, driven by the motor M which turns a long screw, the lead-screw, which carries a nut fixed to the saddle. There is a feed-back system which measures the actual position, Y, of the saddle and compares it with the desired position X. If X is greater than Y, as in the figure, meaning that the saddle should be further to the right, then a voltage is applied to the motor which makes it turn in such a direction as to move the saddle to the right. If Y is greater than X, then the voltage is applied the opposite way to the terminals of the motor, which then moves the saddle to the left.

This form of positional servo may be represented by the diagram in Fig. 6.17(b), which is a simplified version of those used by the designers of control systems. The arrow at the left labelled X represents the input from

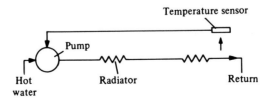

Fig. 6.16. Feed-back in central heating.

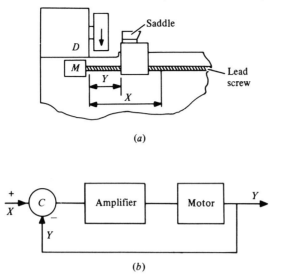

(a)

(b)

Fig. 6.17. Positional servo.

the operator, the computer, the fluidic system, etc., which is controlling the lathe. The arrow at the right labelled Y represents the output, which is the actual position of the saddle. The comparator C compares X and Y, which in this case simply means 'subtracts Y from X to obtain a signal proportional to $X-Y$'. This signal is then amplified, if necessary, and fed to the motor. The quantity $(X-Y)$ is called the *error*, for the obvious reason that it represents how far the saddle is from where it should be.

While these arrangements may sound simple enough, they present certain difficulties in practice, most of which may be regarded as questions of matching. For example, in the positional servo, too powerful a response to error can result in a system which goes into self-excited oscillation (see Chapter 3) drawing its power supply, not from the wind, as with the Tacoma bridge, or from a machine-tool spindle, as with chatter, or from the violinist's right arm, but from its own driving motor. These problems can be overcome by judicious selection or design of the components forming the control system. It is also desirable that the system controlled should itself be designed with a proper regard to these aspects.

6.7.1 Design of control systems

The control engineer uses a powerful symbolism to help him in designing systems: Fig. 6.18 is a very simple example by way of illustration. The servo-amplifier and motor of Fig. 6.17(b) have been represented by a box in which stands a mathematical expression called the 'open-loop transfer function' of these components. The meaning is, that the output from the box is

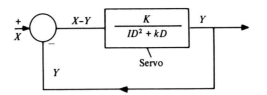

Fig. 6.18. Diagram of closed-loop system.

equal to the input to the box times the transfer function, i.e., if we call the output Y, as in the figure, but leave the input unlabelled for the moment, we have,

$$Y \text{ (output)} = \text{transfer function} \times \text{input}$$
$$= \frac{K}{ID^2 + kD} \times (\text{input})$$

Here K is a constant derived from the characteristic of the amplifier and the characteristic of the motor and k is a measure of friction in the system. The D stands for a difficult idea to do with the differential calculus, but briefly, the whole term ID^2 represents the effect of the *inertia I* of the motor, the lead-screw and the saddle, the fact that when a voltage is applied it takes time for the motor to speed up.

By writing $(X-Y)$ for the input in the equation, the control engineer has a mathematical model, albeit an oversimplified one, of the system, from which he can predict its behaviour and design out the problems of instability (self-excited oscillation), offset (stopping the saddle just before $Y = X$), sluggishness of response, etc. To do this, he may need to modify the transfer function.

For example, he may introduce a network to change the input to the servo from the error, $(X-Y)$, as in Fig. 6.18, to something more complicated, such as the error plus a multiple of the speed at which the error is changing, in order perhaps to prevent unstable oscillations or to reduce the 'settling' time of the system, the time it takes to settle down in a new position to which it has been directed.

6.7.2 Reliability of systems

An important aspect of design generally is reliability, and it is one which is illustrated well by systems. Very complicated systems are more prone to unreliability than simple ones, simply by virtue of the numbers of components they contain. For example, if a system consists of 10 components each of which has a likelihood of being defective of one in 100 000, the likelihood that the system is defective is about one in 10 000; if it contains 1000 components of the same reliability, however, its likelihood of being defective is almost 1% which may well be unacceptable.

There are two possible ways round this difficulty. The first is to strive for very high component reliabilities, and this is always an aim. On the other hand, the second way, which is to build in redundancies, is much more promising in most cases. If we duplicate, triplicate or even quadruplicate components in systems, very high reliabilities can be obtained, provided we make sure that those components cannot all fail from a common cause. Some aircraft control systems, for example, have some of their channels quadrupled, and sometimes different modes of operation are used, e.g., hydraulic and electrical actuators for one control surface.

Nature also provides redundant channels and components, some of which may be incomplete or immature, developing fully only when the need for them arises. Thus, the human brain, though injured and seriously reduced in its functions, is sometimes able to recover some of its faculties, presumably by bringing into service parts of itself which were not performing any essential function before.

Another aspect of reliability is fail-safe design, as when the wing of an aircraft is designed to show cracks before it is in danger of collapse, or when cars are provided with two braking circuits, each operatng on one front-wheel and the diagonally-opposite rear-wheel.

6.7.3 More advanced control systems

One simple control system has been sketchily described, but many systems are much more complicated. Moreover, the control systems of living organisms are more sophisticated still, and we understand as yet very little of how they function.

An important distinction in man-made control systems, practically rather than fundamentally, is that between 'analogue' and 'digital' processing of information. In an analogue system, the actual position, Y, of the saddle may be presented by a physical quantity, a voltage or an air-pressure, say, and the position, itself a quantity, is represented by another voltage or air-pressure. In a digital system, the position is represented by a series of digits, generated by a device called an encoder and coded in binary fashion.

The advantage of the digital system lies in the speed and accuracy with which operations such as subtraction, multiplication, and so forth can be performed by modern microprocessors, which enable the designer to build complicated and subtle rules into the control system. For example, in the sea-wave energy converters discussed in the next chapter it may be desirable to use control rules that take the apparent height of the oncoming wave, the position and velocity of the device itself and signals from adjacent converters in the row, make calculations and finally decide whether to open or shut certain valves. This calculation, which might take several minutes by hand, will be repeated several times a second, and action taken only when the result of the calculation has the right value. This complex sort of control is needed if the most energy is to be collected from the sea, and the relatively small cost of such a system would be repaid many times over by the increased output.

The positional servo dealt with just one variable, saddle position, while modern process plants may handle dozens of variables. It becomes worth while to digitise all the signals and to use a computer to process the information they provide and dispatch the necessary instructions to servos. This idea of information processing links control systems with communications systems, and also with the nervous systems of living creatures.

6.8 Communications

Space will allow only a very brief mention of this important but uncharacteristic and highly-specialised area of design. Some of its subtleties arise in the problem of conveying as many messages, or, looked at another way, as much information, as possible through a single channel, and to do this in the face of 'noise', that is, random 'signals' arising adventitiously in the circuits and blurring the useful ones. The extent of the noise problem is often expressed by the term 'signal-to-noise ratio', which has crept into colloquial use among engineers. It turns out that information theory has a great deal in common with thermodynamics. Indeed, one kind of noise is actually thermodynamic in origin, so that some instruments dealing with very weak signals need to be operated at very low temperatures, not much above absolute zero, where thermal energy is almost absent.

As an example of the design problems characteristic of telecommunications, consider the transmission of television, in the first case, that of a monochrome picture. To avoid a flickering appearance, the pictures should follow each other at about 50 per second. The picture is produced a line at a time by a bright spot which 'scans' or travels across the screen of the television set from left to right, varying in brightness as it goes, being nearly extinct in the dark areas of the image and very bright in the highlights, with all levels between (see Fig. 6.19). After completing one line the spot jumps back to the left-hand side, and starts another line, a little lower down, and so on, perhaps 600 times, until the whole picture is built up. Now ideally, the spot should be able to change intensity as it travels across the screen in about the width of a line, so that the image is as sharp from side-to-side as it is up-and-down. Thus, with a square picture made of 600 horizontal lines, each line ought to consist of 600 adjacent 'dots' of differing brightness. As the picture is not square, but wider than it is high, the 600-line picture ought to have about 800 dots for the definition to be equally good vertically and horizontally.

Now let us look at the total number of intensity signals, or dots, we have to transmit per second. Each picture takes 600 lines of 800 dots, or, say,

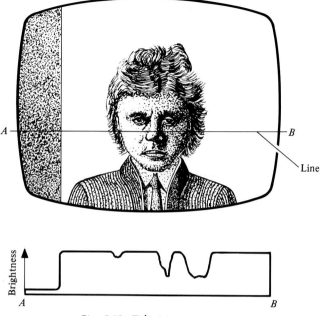

Fig. 6.19. Television picture.

500 000 signals. At 50 pictures per second, that would be 25 million signals per second, almost five for every letter in the Bible.

If we now turn to colour transmission, each coloured dot should be built up from three dots, one in each of the primary colours, so that, ideally, just three times as many signals are required. In addition, other signals must be squeezed in – the sound transmission, the synchronizing signals that switch the spot back from right to left to start a new line and those that make it start a new picture. Altogether a prodigious flow of information is needed, on the face of things, of the order of 20 bibles per second. Let us see how the designers of television systems have dealt with the problem.

6.8.1 Television bandwidths

In Chapter 3 it was seen how a pendulum had a resonant frequency at which it offered very little resistance to motion, so that a relatively small alternating force at that frequency would build up a large vibration in it. The television receiver contains circuits with resonant frequencies which can be tuned to the incoming radio waves from the station required, so that they respond to that particular stream of signals and not to any other – this is the property called 'selectivity'. But if the waves were at a constant frequency and a constant volume, no picture, nor any other form of message or information, could be transmitted, so the strength, or amplitude, of the waves, is varied – this is called 'amplitude-modulation', or AM, for short.

Thus, the signal stength, or amplitude, can be made high when the spot is to be bright and low when it is required to be faint, or vice versa. Figure 6.20 shows such an AM signal, where it will be seen that one spot width corresponds to a large number of cycles or waves – say, about 50. This may seem uneconomical, because it looks as if one cycle, by the height and depth of its peak and its trough, could carry the information for the density of two spots, so that our 100 million signals per second could be carried by 50 million waves per second, or a frequency of 50 megahertz (50 MHz), whereas at 50 waves per spot we should need a frequency 100 times as great (5000 MHz) (actual TV broadcast frequences lie in between these extremes).

The reason it requires 50 or so cycles to transmit one 'spot' is concerned with selectivity. Varying the amplitude of the waves means that they will excite resonant circuits tuned to other frequencies, slightly above and slightly below the basic or carrier frequency, and so each broadcast requires a *bandwidth* or range of frequencies, all to itself. If the variations of amplitude are more rapid then the spread of frequencies above and below the carrier frequency will be greater, and so a bigger bandwidth or slice of frequencies will be occupied. Indeed, the 50 MHz we estimated above is roughly the bandwidth required, not the frequency.

It is a measure of the skill of the designers that excellent high-definition television can be transmitted and received using a bandwidth of only about one-eighth as much. Roughly speaking, this reduction of the bandwidth can be regarded as dividing it by two three times over. One of these three halvings is achieved simply enough by accepting a lower standard of definition horizontally to vertically: the other two are where the ingenuity has been displayed.

One reduction by a half is achieved by transmitting, not 50 pictures, but only 50 *half*-pictures per second. Instead of starting at the top and tracing all the 600 or so lines in succession, the spot traces only every other line, and then comes back and traces the lines it missed out before. Because these alternating half-pictures every fiftieth of a second cover the scene, even if only partially, no flicker is apparent even though a *complete* picture takes one twenty-fifth of a second. This device is called interlacing.

The methods by which further reductions in bandwidth are achieved are too difficult to explain here, but they hinge essentially on sacrificing

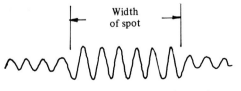

Fig. 6.20. Amplitude-modulated signal.

information which does not contribute much. Interlacing does this, sacrificing in effect some detail of movement which the eye is insufficiently fast and sharp to follow. Similarly, because the eye is unable to appreciate colour very exactly in small patches, the degree of definition in colour can be sacrifced without perceptible loss, and the television designer takes advantage of this to reduce the bandwidth still further. There are also many smaller savings which have been introduced to pack the essential signals more tightly into the bandwidth, which have the general effect of enabling more channels to be broadcast simultaneously without interference.

There is still some information-transmitting capacity left. To see that this is so, view a television screen through a vertical slit and move your head from side to side fairly fast: a slit about 7 mm wide about 400 mm in front of the eyes is about right, and the crack at the hinges of a partly open door is suitable. You will see a number of dark bars, almost vertical but slightly leaning at the top to the side to which you are moving your head. The faster you move your head, the fewer and thicker the bars you see. These dark bars correspond to those strips you happen to see through the slit at those instants when the screen is dark. The slope occurs because the picture is built up on the screen from the top downwards so there is a shearing effect, the dark appearing first at the top so that by the time it reaches the bottom your eye, and hence the strip of screen in view, have moved.

6.8.2 Data-processing

Remarkable as is the process whereby an image of the scene in the studio is built up on the screen of the television set, it is as nothing to the process that is applied to the image of that image formed on the screen at the back of the eye, the retina. A great number of tiny receptor cells, the rods and cones, receive the image and transmit signals through a bundle of nerves to the rear of the brain. However, for every such nerve there are about 200 individual light-sensitive cells: the retina has at the back a structure of nerve cells which reduce the inputs of all the receptor cells to about one two-hundredth as many outputs. In the words of the engineer, the back of the retina *processes* the signals, reducing the amount of information but retaining the essential part. This processing does much more than just reduce the quantity of signals; it takes some steps towards its codification or interpretation. Such steps may sharpen contrast, e.g., make a spot of a different colour more strongly differentiated from its background, so it springs out and draws the attention of the brain, or begin to render shape, e.g., react selectively to a long, thin patch and transmit a corresponding signal. A great deal more of this 'data-processing' goes on in the part of the brain to which the optic nerves lead, enabling us to recognise objects regardless of which way up or round they are. Very little is understood about these functions yet, but some interpretation of the elaborate systems of interconnection between the large and complex cells involved is possible by ingenious experiments.

Much data-processing is done by those human devices, computers. For example, a computer may be connected to a large number of transducers, that is, components which produce a signal which is a measure of some quantity in their environment such as temperature, pressure, aceleration, and so on. It will then read them all in rapid succession at prescribed intervals, perform calculations upon them and print out results in a convenient form. In experiments on obtaining power from sea-waves, for instance, the computer may read the height of the water at a dozen places 100 times in succession, and hence calculate and print out a few concise results, such as the power in the waves.

Very much more sophisticated tasks than this can be achieved with instrumentation connected directly to a computer. In one modern X-ray technique, a very narrow beam is projected through the patient's body and the amount absorbed is measured. By doing this at many angles and along many lines in a plane a great number of absorption measurements are obtained, from which a picture can be constructed. To see that this is possible, consider the four-region picture in Fig. 6.21(a) which is to be constructed using the absorptions measured along the lines indicated by arrows. Let a, b, c and d be the absorption in the four regions of the picture. Then the ray A passes through the top two regions, and $a + b$ units are absorbed. Let us suppose the measurement shows 5 units are absorbed; then we can write an equation,

$$a + b = 5.$$

At the end of each arrow, representing a ray, the absorption along that line is written in a circle. You will find that it is possible to determine that in this case $a = 1$, $b = 4$, $c = 2$ and $d = 3$, so that in our picture b is most

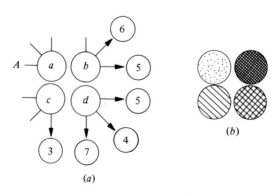

(a)

(b)

Fig. 6.21. Scanning with X-rays.

absorptive, or darkest, and so on, as indicated pictorially in Fig. 6.21(*b*). With many more spots, many more parallel rays in each direction and many more directions are needed but the definition of the picture wil be improved. The computations become very much heavier, of course, but a computer is ideally suited to this low-grade, high-volume, high-speed work and is able to present a picture on a television-type screen of the thin section, of a human head, say, which is being viewed; by moving this section relative to the head, the entire brain may be scanned. Not only is this method a great advance on X-ray photography in the information it provides, but it also reduces the exposure of any piece of tissue to radiation which can damage it.

6.9 The human brain

The most remarkable product of nature to date, on the earth at least, is the human brain. We do not understand, as yet, the data-processing which goes on in it, but there can be little doubt that it is much more advanced than anything man can do with computers. Indeed, it can be speculated that no task we have as yet attempted is so difficult as that of understanding how the brain does it. Even then, it is important to distinguish between this data-processing, which can at least be envisaged as purely mechanical, and thought. It is not appropriate here to discuss whether thought is itself purely a physical process, or whether the brain is the interface between an elaborate mechanism, the body, and some supernatural mind (the writer regards the first as disproved by consciousness, and finds the second too difficult to reconcile with what we think we know of the physical world). Suffice it to say, we have a long way to go before we understand the brain, even in its simpler aspects, this system that can itself create other systems, advanced enough, for example, to use to look at itself in an extraordinary way with a thin beam of X-rays ceaselessly moving in a plane.

6.9.1 Neural networks

The studies that have been made of the brain reveal structures of special cells, called 'neurones', connected together in a complex fashion. From the deductions that have been made about how these structures function, computers have been programmed to operate in a similar fashion. Such artificial *neural networks* (ANNs) have proved to be useful for certain tasks, particularly those which require the recognition of some generalised feature. For instance, they are suited to tasks like reading handwriting, because they can recognise a T shape, for instance, whether large, small, thin or fat, erect or sloping, and in spite of waviness or kinks in the strokes. So far their uses are limited and specialised, but parallels with the human brain suggest that when we understand them better they will prove capable of great things.

6.10   Systems, mechanisms and structures

The quality of a system depends on the quality of the components which form it, as well as the excellence of its own organisation. On the other hand, systems can improve the performance of components such as mechanisms and structures.

Suppose we wish to build a very large, very accurate structure, a giant radio telescope of the bowl type which is 'steerable', i.e., can be pointed at any part of the sky. Such a structure deforms slightly under its own weight and the action of the wind, and the deformation varies as the bowl is tilted to different angles and the wind rises and falls. Two approaches are possible to maintaining the accuracy at an acceptable level: firstly, we can use a material of high specific stiffness (see Chapter 3) such as carbon-fibre, in a structural design of the highest efficiency, and accept the small distortion that will still occur. We can 'build out' a certain amount of the distortion (so that if we suddenly moved the bowl out into space, with no gravity forces, it would be slightly out of shape) so that in one chosen position the form will be perfect, but as the angle changes to above or below that one value, the *change* in distortion will mean slight errors. As an alternative policy, however, we could build adjusting screws into the bowl, operated by servo-mechanisms controlled by a central controller which continually measured the distortion and removed it by changing the adjusters. Thus, in anthropomorphic terms, the bowl would observe and correct its own imperfections, enabling better results to be obtained with less material. In effect, a system of transducers and servo-mechanisms replaces a great deal of structure.

In a similar way, some manufacturing processes which would have used specialised mechanisms in the past can now be performed with a less specialised but more highly-automated processor. In gear manufacture, machines are used in which very accurate and stiff mechanisms cause the workpiece and tool to move in the correct relation to one another to generate the required gear tooth forms. To achieve flexibility and enable a great variety of gears to be made, many additional motions have to be built in, which tend to reduce the accuracy and the stiffness unless construction is made even more massive and precise, so that the machines become complex, clumsy and expensive.

An alternative, not much explored as yet, is to use a rather simple machine of refined construction in which the relative motions of workpiece and tool are governed by high-precision positional servos under numerical control. In this way the necessary accuracy resides in relatively cheap precision components, such as glass plates with very accurate markings on them, rather than heavy elaborate mechanisms. In addition, the new machines are likely to be easier and quicker to set up and more versatile.

In the future, we are likely to see more and more use of the alliances which are possible between control systems and mechanisms or structures. With the advent of the very large-scale integrated circuit (VLSI), the microprocessor or 'silicon chip', new ways of solving old problems will become attractive, leading to better products made at lower cost.

In these cases, particularly that of the radio telescope with a 'steerable' bowl, we choose the design problem to be one we can manage more easily. Instead of building a very rigid accurate bowl, we build an accurate measuring system to check the amount an inaccurate bowl is out of shape, and use a control system to adjust it; an accurate measuring system is cheaper to make than a very rigid bowl.

Another interesting alliance between structures and control systems has appeared recently in aircraft. The severest loadings on civil aircraft wings arise because of gusts of wind. Some aircraft now have control systems which quickly adjust the control surfaces on the wings when a gust comes, so as to reduce the additional loading and hence the maximum bending moment in the wings. Provided there is sufficient confidence in the control system, it is then possible to reduce the strength, and hence the weight, of the aircraft wing, with economies in operation.

However, when an advance is made, like this one of reducing gust loadings, we need to consider carefully how to take the best advantage of it. Instead of reducing the strength of cross-section of the wing, we could keep it much the same but make the wing longer and thinner. This would increase the lift/drag (L/D) ratio, and hence reduce the fuel consumption. The fuel carried would then be less, and once again there would be more lift to spare for extra passengers or freight.

In design it is always desirable to examine carefully the way in which any favourable change or advance is applied, so as to obtain the maximum benefit. Nearly always, the ideal is to cut the cake several ways, e.g., in the aircraft with controls that react to gusts, to make the wing a little longer, a litle narrower, a little shallower and a little lighter. On the other hand, if it is not a new aircraft to be designed but an old one to be modified, a simpler approach may save retooling costs.

6.11 Systems design in nature

Nature is the greatest system designer and we are stumbling in her footsteps most of the time. Sometimes, as in the case of flight, we have consciously imitated nature, perhaps too closely at first in that case, for some inventors thought that flapping wings must be the right way to go (see Section 6.5.1). Sometimes we have discovered later that we were following behind nature, as in the case of sonar, where it was a long time before we realised that the tank of oil in the head of some whales was a lens for sound waves. The simpler control systems of nature are still very complicated and subtle by human standards; we do not yet understand all the variables influencing human body temperature, for example. In heating our buildings, we have progressed from the simple thermostat reacting only to the temperature of its single sensor, to ones which operate on several pieces of data, for instance, the temperature outside as well as that at several places.

Where we are most evidently behind, however, is in the comparison of our most advanced robots with, say, a pianist or a tennis player. We can make a mechanically-played piano, a pianola, but it can only play the punched cards it is fed, or we can make, with great difficulty, a robot that can play a piano. We might be able to build a tennis-playing robot, but if we were restricted to the sort of power–mass ratios that nature disposes of, I imagine it would be easy enough to beat them. Humans can still beat computers at chess, even though the computer can carry out 'mental' operations thousands of times faster than the human, because the man's thinking is of altogether a higher type. But the same human might be able to play the piano, and tennis, and chess, and, moreover, speak several languages (while no computer can speak one), drive a car and ride a bicycle, do real mathematics and a thousand other things which are beyond the computer: we are still at an early stage of learning.

With learning, however, we touch the sort of area where the gap between our efforts and natural design is widest. We can make computers 'learn', but only in very limited ways which have been built into the software. We can build in the conditioned reflex, the response taught by association, like Pavlov's dogs salivating at the sound of the bell they had learned to associate with food. We can make adaptive control systems that 'observe' their own performance and improve it, rather like a human learning a skill of co-ordination, such as catching a ball. What we cannot yet do is to make a computer that can perceive a problem and teach itself how to cope with it. For all the talk of 'artificial intelligence', that objective still seems very remote.

It is perhaps some consolation that nature has only achieved much in the way of intelligence in our own species, so far as we know. Wallace, with Darwin, one of the originators of the theory of evolution through natural selection (Section 1.11), maintained that the human brain could not have evolved in that way because primitive man had no need for great intelligence. But Wallace underestimated, it may be suggested, the high level of intelligence needed to develop early technology. The bow seems only to have appeared about 20 000 years ago, and rapidly swept away the spear-throwers used before. Since it would have been a great advantage for millennia before and would surely have been hit upon much earlier otherwise, the human brain must have been only just good enough to invent it. To this argument it might be replied that it applies only to engineering ability and not to artistic ability, which was of no survival value, and that therefore Wallace's point is sound. In this book, however, the view is taken that intellectually the engineer and the artist are not far apart, as the works of Leonardo demonstrate so beautifully, and as much human experience witnesses. This view is developed in Chapter 8.

7

The practice, principles and philosophy of design

7.1 Introduction

Earlier chapters have shown some of the problems facing the functional designer and some of the solutions that have been found or copied from nature. Some areas of knowledge important in design have been touched upon – mechanics, materials, kinematics, structures – but nothing has emerged in the way of a systematic method of design. Indeed, a truly systematic approach is as unthinkable in such a field as it would be, say, in play writing. All there can be is guidance, generalisations and principles as broad as the unities of time, place and action in drama. However, there are three great sources of help to the designer, which will be studied in this chapter.

The first is *practice*, what has been done already, in living organisms, or, most often, by other designers. The second is *principles* of design, derived by reflection, or by abstraction from a wide knowledge of practice, or both. The last is the *philosophy* of design, which manifests itself as a single strong thread of sometimes rather abstract reasoning running through certain designs, usually having an apparent inevitability. It is rare, but perhaps should be more common; certainly, it is difficult to explain, even with examples, but it can be recognised.

7.2 Practice

The human designer in his work is able to draw upon a vast number of examples, mostly from the works of men but occasionally from the living world. The evolutionary process, on the other hand, cannot profit by experience, so that, for instance, flight has been developed from scratch on the earth at least four times – in the insects, in the reptiles, in the birds and in the bats. Only very recently, by man, has it been done using the examples of these earlier cases.

One of the many specialised forms of the English language is that peculiar to patents. Most of what is described in the average patent is standard design practice, and it would be burdensome to explain at length all the different ways in which some of these essential but non-original parts of the invention might be carried out. For example, two components may need fastening together, and this might be done by bolts, or by welding, or by an adhesive, or by a spring clip engaging in a groove, and so on, not indefinitely indeed, but at great length. In the interests of brevity, use is often made of the phrase 'other known means', as a more explicit alternative to 'et cetera'.

Design practice might be regarded as mostly the intelligent application of the formidable armoury of 'known means'. Not every designer knows every part of that armoury, but he is able to draw upon specialists who do. Frequently, he is able to incorporate general-purpose components which have not only already been designed but are available from stock, and about which there is much information, obtained by actual test. Often design reduces very largely to a matter of combining such known components in a suitable fashion, as we saw in the last chapter, which also showed that in most cases this is no simple problem, and may draw on intellectual skills of a high order.

At the other end of the scale, even the designer of an internal combustion engine still makes use of a large number of stock or standard items – bolts, nuts, washers and screws, some bearings and seals, the electrical components, much of the cooling system probably. He may be able to use an existing pattern of water pump, for which he will be able to obtain the *characteristics* (see Chapter 6) to see whether it will suit that system, and so on.

7.2.1 Standardisation

Associated with design practice is the important question of standardisation.* A designer requiring a bearing, say, has a bewildering choice at his disposal (although even then, for the most critical cases it is quite possible he will find none to suit his need). When he chooses one, he may well have several sources of supply of interchangeable bearings from other makers.

Clearly, standardisation offers important economic advantages and reduces the complications in a world already too complicated for most of us. But its limitations should be observed: standardisation may be impractical

(a) where components are very large and used in small numbers
(b) where components are used in very large numbers
(c) where the technical problems are critical
(d) where the component is very heavily integrated into the design
(e) in very cheap and simple devices.

As an example of (a), consider the case of ships, which regarded as components of transport systems are very large but also are used in fairly large

* For a fascinating account of the standardised design of ancient artillery, see Marsden, E. W. (1971). *Greek and Roman Artillery: Technical Treatises*. Oxford: Clarendon Press.

numbers. Here both standardised products and single 'one-off' ships are available, and the standardised products can be modified considerably to suit the customer's particular needs or wishes. In the case of the less common types of ships, the economies of even a limited degree of standardisation are not sufficient to justify the restriction of customer choice.

Where very large numbers of components of one kind are needed for one particular purpose, all the possible economies of large-scale production may be gained without using a standard item. Furthermore, a special component may well be inherently less expensive than the nearest suitable standard component, which will usually have a higher performance than is needed, in some respects at least.

Where technical problems are acute, standardisation is rarely possible. For example, it might be possible to find a stock design with a suitable characteristic for a rocket fuel pump, but it would be unsuitable in other ways – it would not be made of suitable materials or have suitable seals for the liquids it had to pump, it would be much too heavy and probably its efficiency would not be quite as high as would be obtainable with a special design.

Above all, with highly interactive designs, standardisation is not feasible, because each component interacts in so many ways with the others, and therefore has to meet so many requirements that no conceivable range of standard items could cover it. This is the case with most of the larger components in engines, cars, aircraft, machine-tools and domestic appliances, to take a few categories. However, advantages similar to those of standardisation are often to be had by basing new designs on components of the old, or by designing a range of products using common parts. Thus, large compression-ignition engines are available with, say, 4–10 cylinders, depending on the power required, and these are able to use the same connecting-rods, the same pistons, the same bearings, etc., throughout the range, even though components like the crankshaft must be different for each size. Aircraft may be available in 'stretched' versions, in which many parts are the same as in the basic design, and so on.

Finally, in very cheap and simple devices very little standardisation of parts is possible. For example, an electric torch, including a switch, sells for about the same price as a standard switch alone. This good value is possible because the switch is of a very simple kind, only suitable for a very low voltage, and because it has no separate casing to support its working parts, this function being performed by the body of the torch itself. This combination of functions saves not only the cost of a separate switch body, but also the cost of fixing that body to the torch and making the necessary connections.

7.2.2   Bearings and pivots

In Chapter 4, hinge joints or pivots were discussed. A bearing is just a pivot that permits indefinite rotation in one direction, unlike a knee or a door hinge which turns only through rather less than two right-angles. Examples are the bearings on which car or bicycle wheels turn, or those in which the crankshaft of an engine rotates. In the average household there are likely to be some hundreds, in vehicles, clocks and watches, toys, sewing machines, fans, etc., and even more hinges and pivots (especially if we count those in humans and other animals).

For different purposes, bearings and pivots have to meet different requirements. A car wheel bearing must be strong, it must turn freely and it must run for long periods of time at heavy loadings without wearing out. On the other hand, it may, indeed, it should, have a little slack in it, which would be unacceptable in some machine-tool bearings where it would result in inaccurate parts being made. The bearing for the machine-tool would have very little slack but too much friction in it, although freely enough turning for most purposes, for certain uses in instruments requiring bearings free of both slack and friction. The roller and ball bearings which are very satisfactory in some large turbines may be useless in very small ones turning at 100 000 rev/min or more, and so on.

Two enemies of bearings and pivots are the friction and wear which occur when solid bodies rub on one another. In many critical cases, nature and the engineering designer do their best to ensure that no solid parts touch, by separating them with a fluid layer, or *film*. The most mobile joints of warm-blooded animals, for example, are of the synovial type, encapsulated by a membrane which retains the synovial fluid which lubricates the working surfaces. This alone would not prevent the bones coming into contact, but they are separated by a spongy cartilage out of which fluid squeezes as the bones press on it, creating a separating film between cartilage and bone and preventing solid-to-solid contact. In Fig. 7.1, *AB* shows the zone

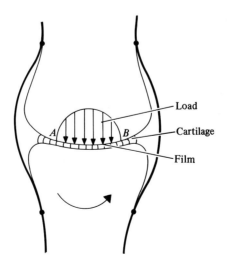

Fig. 7.1. Synovial joint.

carrying the load in such a joint. If the lower bone is turning anti-clockwise, then this zone may be moving to the right, so that fresh lubricating film is being squeezed out of more cartilage at B as it comes under load. Meanwhile, cartilage unloaded at A will be free to soak up more liquid.

It may be seen that there is a defect in this film-generating scheme – under continuing load without movement of the joint the film will squeeze out and will not be renewed. When there is no movement, this does not matter, but when the joint starts moving again there will be solid contact to begin with. This difficulty appears to be overcome by nature by the remarkable properties of the lubricant, synovial fluid.* Engineering designers have the same difficulty, very nearly, in their equivalent of the synovial joint, which is called a hydrodynamic bearing. This is the kind of bearing used in most places in a car engine, for example. Figure 7.2 shows how it works. A shaft runs in a bearing of slightly larger diameter; the difference between diameters, called the clearance, has been greatly exaggerated in the figure, to show that unless the shaft runs in the middle, there will be a point C where the gap between it and the bearing is a minimum. With the shaft turning anti-clockwise as shown, if oil is pumped into the gap at I, it will be dragged by the friction of the shaft through the gap ACB, which begins by narrowing to C and then widens again. The effect of this is to develop a pressure in the oil, rising to a maximum at C, which enables it to keep the shaft and the bearing apart in spite of a considerable load forcing them together. (This is quite different from the venturi effect, where the narrowing of a wide passage causes a *drop* in pressure in a fluid of relatively low viscosity.) Thin *hydrodynamic* lubricating films of this kind can develop very high pressures, but they do rely on relative motion of the parts concerned, and they disappear when the shaft stops turning. When the shaft starts again, there will be no film for an instant and more wear may take place in such instants than in the rest of the life of the bearing.

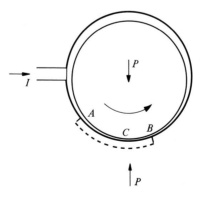

Fig. 7.2. Hydrodynamic and hydrostatic bearings.

* McCutchen, C. W. (1962). Animal joints and weeping lubrication. *New Scientist*, 15, pp. 412–15.

With some very heavy machinery, or in cases where relative motion is too slow to build up a hydrodynamic film, a *hydrostatic* bearing may be used. Imagine that in Fig. 7.2, instead of oil being pumped in at *I* under low pressure, it is pumped in at *C* at very high pressure, so that it pushes its way towards *A* and *B* and lifts the shaft in the process. To help the pressure to spread, a shallow depression is made in the wall of the bearing around *C* (shown by a broken line in the figure). The principle is essentially the same as that of the hovercraft, with the perimeter of the shallow depression being equivalent to the hovercraft skirt.

A very simple demonstration of a hydrostatic bearing may be had by rinsing a tumbler with a flat rim in hot water to make it both hot and wet, and then quickly upending it on a smooth level surface like a kitchen work-top. The water seals the rim and the hot glass raises the temperature and therefore the pressure of the air trapped inside, so that often the tumbler is lifted and will float for a few seconds, travelling down any slope there may be on the surface, even though that slope is too small to be appreciable to the eye.

The hydrodynamic bearing relies on the viscosity of the oil to drag it into the narrowing zone of increasing pressure. If a very low-viscosity lubricant is used, like a gas, the film can still be generated but it will be very thin, and when the thickness of the film becomes comparable with the bumpiness or roughnesses of the shaft and bearing surfaces, solid contact will occur. Both hydrodynamic and hydrostatic bearings can be, and are, made to work with gas as a lubricant, but the parts must be very smooth and accurately shaped, the gas must be dust-free and the loads that can be carried are not high.

In roller and ball bearings, instead of separating the shaft and its bearing by a fluid film, a row of rolling-elements is used, which, ideally at any rate, slide on neither. In fact, there is a little sliding, particularly in ball bearings, but it is at very low velocities and so involves little wear or friction.

In many bearings it not practical to provide rolling-elements or a hydro-dynamic or hydrostatic film. Even then lubrication helps to reduce friction and wear. Where no fluid lubricant is possible, solid lubricants like graphite, molybdenum disulphide or polytetrafluorethylene (PTFE), may be applied to or incorporated in the material of the bearing.

In unlubricated or slightly lubricated bearings, the physical and chemical nature of the materials used has a profound influence on the rate of wear. The balance staff of the watch the author is wearing at the time of writing has oscillated about 2000 million times so far, and has worn very little because it is of hardened steel running in bearings of synthetic ruby (crystals of aluminium oxide). Both materials are very hard, but because one is an oxide the very tiny roughnesses of the smooth steel surfaces do not continually weld to it and break away as they would with another steel surface.

7.2.3 The repertoire of bearings

The designer has available to him today a vast range of bearings based on these principles, mostly with known performances. If none of these suit, he can either design special bearings himself or have them designed to his requirements by experts. A wealth of experience has been accumulated on the behaviour in use of these bearings and bearing types, and much of it has been set down in convenient written form.

There are rolling-element bearings in which the elements are balls, rollers, long thin rollers (called needle rollers), cones, barrel-shaped bodies and cones with barrelled sides, with and without cages to space them apart, with tracks or races of different forms, designed to withstand loads at right-angles to the shaft, or along the shaft in one or both directions, or various combinations of such loads. A few kinds are shown in Fig. 7.3. One company in the UK makes 35 major types, each in many variations and sizes, a total of over 10 000 different bearings, but even that vast range does not include three of the varieties shown in the figure.

The range of hydrodynamic bearings is even larger. Many of them are available as shells, sometimes quite thin, to fit into holes the customer makes himself, and these generally have a backing of some strong material, say, steel, covered with a surface of special bearing alloy, of which a variety are available to suit different uses. On top of this there may be one or two further, very thin, layers of other materials, often including rather rare or unusual metals. Different patterns of oil holes and oil ways (channels cut

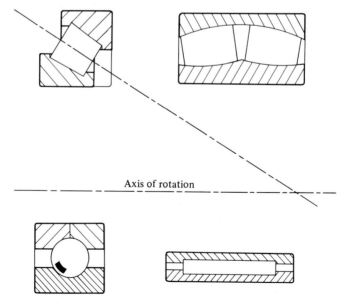

Fig. 7.3. A few types of rolling-element bearings.

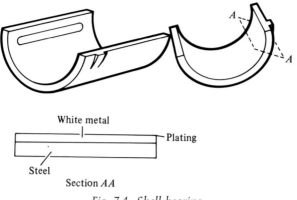

White metal

Plating

Steel

Section *AA*

Fig. 7.4. Shell bearing.

in the surface of the bearing – see Fig. 7.4) are provided to admit oil at the right place, as at *I* in Fig. 7.2, and to distribute it through the bearing. Figure 7.5 shows another type of hydrodynamic bearing, in which the working surface consists of pads which can tilt slightly under the oil pressure to conform to the shaft, a feature which makes it possible to use thin films combining high load capacity and great stiffness, which is important in machine-tools (see Chapter 3).

Simpler bearings than these may be of porous bronze (made by compressing and heating powdered metal – sintering as it is called) with a lubricant or a slippery plastic filing the pores, or of rubber for use in water which also acts as a lubricant. Where only a pivot is required, i.e., where the rotation is limited, it may be possible to bend the material instead, as was discussed in Chapter 4 (see Fig. 4.23). The pivots of the pads of bearings of the type shown in Fig. 7.5 are often of this flexural kind.

With all these 'known means' at his disposal, the designer rarely meets problems of bearing design other than selection and routine calculations.

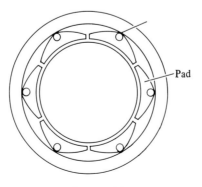

Pad

Fig. 7.5. Tilting pad bearing.

Even these may often be done by means of a routine program, on a computer.

7.2.4   Cylinders subject to high internal pressures

A common structural requirement is for a cylinder to contain an internal pressure, as in a boiler, a cylinder of an engine or an aerosol can. Where the excess of the internal over the external pressure (the 'gauge' pressure) is very high, say, of the same order as the strength of the material itself, a difficulty arises, which can be called 'the problem of the lazy layers' and may be explained by reference to Fig. 7.6. Consider a thin ring *B* of the material round the inside and a similar ring *A* around the outside. When the interior is pressurised, the entire cylinder, including the rings *A* and *B*, moves outwards due to the hoop stresses set up in it (see Chapter 5). However, the material between *A* and *B* is compressed radially by the pressure, so that the distance between the two rings is reduced. It follows that the amount by which the ring *B* grows radially must be greater than the growth in ring *A*, which means that the increase in circumference of *A* must be less than that in *B*. When this is taken into account together with the greater length of circumference of *A*, the circumferential strain (see Chapter 3) and hence the hoop stress, in *A* must be very much less than in *B*. In fact, if ring *A* is twice the diameter of ring *B*, the stress in it will only be about 40% as much. Thus the outer layers contribute much less to holding in the pressure than they might. This effect becomes important when the gauge pressure p is equal to about one-quarter or more of the allowable hoop stress f, and by the time $p = 1/2f$, on the basis of one commonly-used design criterion, the outside diameter becomes infinite! However, if the hoop stress could be made the same all the way through the cylinder wall, then for $p = 1/2f$ the outside diameter would only need to be twice the inside diameter, on the same criterion. But how can the hoop tension be more

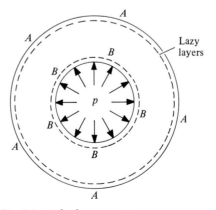

Fig. 7.6. Cylinder containing pressure.

evenly spread between the layers? One way is to make the tube in layers, with thin layers of fluid between which are maintained at intermediate pressures. Thus with $p = 1/3f$, we have an intermediate thin layer of fluid at a gauge pressure of $1/6f$ (see Fig. 7.7)a)), an arrangement which saves nearly 40% of the material needed for a single thick wall. A large number of layers could ideally save just over half the material, so the further gain is not large, even if the fluid layers are of negligible thickness. This type of situation is common in design; where improvement is possible by dividing a function (here, the containing of pressure) into a number N of stages, the advantage gained by increasing N rapidly decreases, often as the inverse square of N.

The arrangement of Fig. 7.7(a) has rarely been used; one problem is that of providing a very reliable source of the intermediate pressure. It has been used at the high-pressure end of steam turbine casings, where this intermediate pressure is automatically available. But a very similar effect can be obtained by using two concentric cylinders as in Fig. 7.7(b), and making the outside diameter of the inner tube slightly larger than the inside diameter of the outer one. They are fitted together by heating the outer tube to expand it till it will slip over the inner one. When the whole has become cool, the outer tube is left in a stretched state, that is, in hoop tension, and the inner tube in hoop compression. Thus the effect is much the same as when a thin layer of fluid at intermediate pressure is interposed in Fig. 7.7(a), stretching the outer and squashing the inner cylinder.

The shrinking of one cylinder over another is just one way of prestressing a thick-walled cylinder, so that the outside starts off in tension and the inside in compression. When the cylinder is pressurised, a large tension is added to the compressive prestress in the inner layers, and a small tension is added to the tensile prestress in the outer layers, leaving the combined stress a fairly uniform tensile one through the whole thickness of the wall, which is an efficient use of the material.

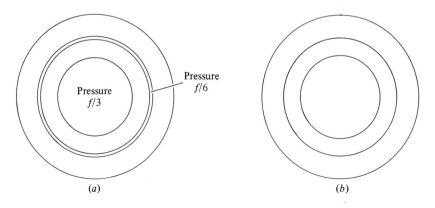

Fig. 7.7. *Economical thick-walled cylinders.*

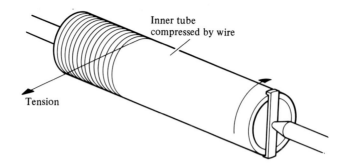

Fig. 7.8. Wire-wound cylinder.

A very simple way of producing the required pattern of prestress is just to subject a solid thick-walled cylinder to a high internal pressure, so that the tensile stress in the inside layers is enough to cause them to stretch plastically (i.e., permanently – see Chapter 3), a process called autofrettage. When the pressure is removed, the inside layers are too big for the outer layers, as it were, so that they are compressed and outer ones are stretched.*

Another way of producing a suitable prestress is to wind wire very tighty round a thin-walled cylinder until it is built up to the required total thickness (see Fig. 7.8). Here it must be remembered that the wire cannot take any load along the axis of the cylinder.

7.2.5 The evolution of gun barrels

The importance of cylinders to withstand high pressures first arose, alas, in the barrels of guns.

Almost the earliest guns were of the form shown in Fig. 7.9. Wrought iron bars were welded lengthwise (by heating and hammering together while very hot, like the steel and the iron in the scythe blade in Chapter 1)

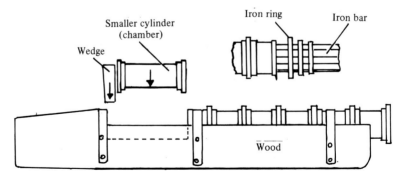

Fig. 7.9. Early artillery piece.

* Maning, W. R. D. and Labrow, S. (1971). *High Pressure Engineering*, pp. 31 *et seq*. London: Leonard Hill.

to form a tube, a tube which would have been very weak if it had been left at that stage. Wrought iron rings were then shrunk over this tube. A trough was formed at one end, the breech, into which was placed another, smaller, cylinder containing the powder. This smaller cylinder was wedged up against the breech and the gun was fired. The famous gun, Mons Meg, in front of Edinburgh Castle has this kind of wrought iron, built-up barrel, and so did the one which burst and killed James II of Scotland in 1460. Although there would be some prestress in these guns, the forge-welding would be very unreliable, as indeed would the whole structure.

With improvements in foundry practice, it became possible to cast guns in brass or bronze, and later in cast iron. From about 1860, guns began to be made of steel, and then shrunk and wire-wound constructions were adopted. Figure 7.10 shows the kind of elaborate arrangement of short tubes shrunk on to a central tube that was used in large-calibre guns around 1900.

It is interesting to note that at the end of the nineteenth century guns were made of wrought material (that is, material shaped by hammering and forging, as against casting) of built-up construction, prestressed and loaded at the breech end, as was the fourteenth-century gun in Fig. 7.9. In between, for four centuries, guns were almost universally cast, in one piece, without prestressing, and were muzzle-loaders. In this century, there has been a return to one-piece (or monolithic) construction, prestressing being achieved by subjecting the cylinder to an excessive hydraulic pressure, or autofrettage.

7.2.6 Containing pressure in string bags

One advantage of the wire-wound gun is that steel wire is readily made very strong, and the multiplicity of turns and layers makes for reliability. Other materials, such as glass, are very strong in fibre form, so that the principle of holding in pressure with windings is an attractive one. However, the gun is a particularly suitable case, because there is no longitudinal stress in the walls. In Chapter 5, it was shown that in a closed cylinder the pressure produces a longitudinal stress equal to half the hoop stress. If a wire winding is to be used on a closed cylinder, the turns must be laid helically so that they can resist longitudinal as well as hoop tensions. Moreover, the layers must be of both hands of helix, i.e., the turns of one layer must spiral the

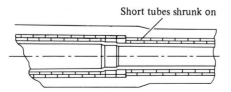

Fig. 7.10. Section of built-up gun barrel, about 1900.

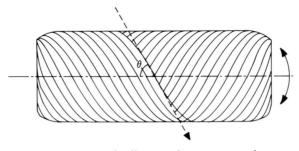

Fig. 7.11. Helically-wound pressure vessel.

opposite way to those of the one before, as in Fig. 7.11. Since the longitudinal loads are only half the hoop loads, it is to be expected that the turns
ought to lie rather closer to the circumferential than to the longitudinal
direction, so that the angle θ in Fig. 7.11 should be greater than 45°; it turns
out that θ should be a little under 55° – for the mathematically-minded θ
is arc tan $\sqrt{2}$. An alternative arrangement is to use circumferential fibres
togther with longitudinal ones (see Fig. 7.12(b)).

To contain a fluid or a gas under high pressure with what is, in effect, a
system of strings, requires that some solid fills the spaces in between, the
holes in the string bag, as it were. In the case of vessels like that in Fig.
7.11, used for containing gas under high pressure, where the strings are of
glass fibre, the 'proofing' or matrix is a synthetic resin. In the case of the
large pressure vessels used in gas-cooled nuclear reactors, the spaces
between the steel wires which form the strings are filled with concrete.

7.2.7 Nemertine worms

The angle θ mentioned in the last section can be deduced from considerations of equilibrium, but it can also be deduced in the following way. Imagine a cylinder, loosely wound in the fashion of Fig. 7.11, being stretched
axially, so that θ becomes smaller. The diameter of the cylinder will
decrease, but its length will increase. The decrease in diameter will reduce
the volume of the cylinder, but the increase in length will increase it. When

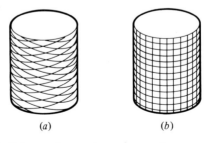

(a)   (b)

Fig. 7.12. Schemes for prestressed concrete pressure vessels.

the angle of the helix is 55°, then these two effects balance, and the volume of the cylinder is a maximum; if the angle is made greater or less than 55°, the volume decreases. Now since any fluid under pressure inside it will try to increase its volume, it will naturally tend to make the angle 55°; if the angle is more than 55°, the lengthwise forces will overcome the radial pressure, and the angle will decrease as the cylinder grows longer. Vice versa, if the angle is less than 55°, radial pressure will dominate and the cylinder will grow fatter and shorter.

A kind of worm, called a nemertine, has a tough skin reinforced with helical fibres in two sets, left- and right-handed. It will not by now surprise the reader that in the relaxed state of these animals, the fibres make an angle of about 55° with the axis. However, the object is not to resist internal pressure, since in this state the volume enclosed by the skin is a maximum and the worm is only half-full, as it were. If the worm contracts lengthwise, so becoming shorter and fatter, the volume enclosed becomes less until eventually it is equal to that of its innards and further contraction is impossible. If the worm extends, the volume will also contract until the creature becomes 'skin-tight', and further extension is impossible. By starting with an angle of 55° in the relaxed middle state, the worm combines a considerable range of extension in length with the protection of a strong jacket of nearly inextensible fibres.*

7.2.8 Fusion power

The most interesting of all containment problems is associated with fusion power, that is, atomic energy derived from the fusion of light atoms rather than the fission of heavy atoms. The advantages of fusion power, if it can be achieved, would include a relatively unlimited source of fuel, lower risks to the public at large and, probably, very much smaller amounts of radioactive material to be disposed of at the end of the life of the power station.

Unfortunately, the fusion reaction is only self-sustaining at very high temperatures, around 100 million degrees Centigrade, so that no solid material can possibly be used to contain the plasma, which is the name given to the very hot, strongly electrically charged 'gas' which matter becomes at these temperatures. In the depths of the sun, which is a giant fusion reactor, the plasma is confined by gravity acting on the immense mass of the outer layers. In devices of a more domestic scale, gravity is useless, and we are left with magnetic forces and inertia as possible containment means.

Magnetic containment is difficult to understand, but it works because the plasma particles are charged and the effect of a magnetic field on a charged particle is to make it travel in a helical path, an effect which has been used in the focusing of the 'cathode rays' in some television receivers. In

---

* Gray, J. (1968). *Animal Locomotion.* London: Weidenfeld & Nicholson.

experimental fusion reactors (which still consume much more energy than they produce) this effect is used to make a so-called 'magnetic bottle' to confine the plasma.

In inertial confinement the fuel is in the form of a small target, usually a tiny glass balloon, which is suddenly subjected to a very concentrated bombardment by laser or other very energetic beams. The resulting outward explosion produces a very powerful inward reaction on the core, which is violently compressed.

To make fusion power a practical possibility, the very difficult design problems associated with one of these two methods of containment must be solved. Magnetic bottles tend to be leaky and too much energy escapes: beams of sufficient energy and sufficient concentration are very difficult to produce. But the prize for solving the problems is possibly the greatest there has ever been for any activity of man, a lasting solution of the coming energy crisis. We need other sources to tide us over, but fusion power is the most attractive long-term prospect we have.

7.2.9 Pumps and turbines

As examples of the repertoire of known means available to the designer, we have studied bearings, which are components of mechanism, and stong cylinders for containing fluid, which are structures. The third and last example concerns a group of energy-converting devices, pumps and turbines. Pumps are used to do work on a fluid, increasing its pressure, while turbines extract energy from the fall in pressure of a fluid.

A designer very often needs a pump (or pumps) in a design. Sometimes he can integrate the pump into other parts very neatly, but often the pump is an addition or something to be bought off-the-shelf, and then he has a vast range to choose from.

Pumps can be classified according to the way in which work is done on the fluid. Unlike the sledge of Chapter 2, we cannot just push a fluid: we must first trap it in a closed space, as we do with the air in a bicycle pump, and then make the space smaller. Pumps of this kind are called 'displacement pumps', and have the characteristics that if we pump more slowly, we can still achieve the same pressure, and that if we stop pumping, the fluid cannot escape backwards through the pump.

A fan is a kind of pump, although the work it does on the air passing through it becomes apparent as kinetic energy not pressure: it does not trap the air in a closed space, but acts on it with a blade that moves fast enough to exert a force on it before it can escape out of the way.

Such pumps are called kinetic (or hydrokinetic). A windmill or a turbine is a kinetic motor (using the word 'motor' in its general sense of a source of mechanical work).

The pressure rise (or, in the case of a fan, the velocity rise) produced by a kinetic pump increases rapidly with the speed of rotation (the pressure rises roughly as the speed squared, just as we saw the fluid drag force on the fish in Chapter 2 varied as the square of the speed). If the pump stops, the fluid can flow backwards through it. Finally, if we forcibly stop the flow through a kinetic pump, say, by shutting a tap on the outlet, then the pump can continue turning, but if the flow from a displacement pump is stopped, either the pump must stop, or something will break, or a safety or relief valve must open to let the flow continue.

These two categories of fluid pumps and motors, displacement and kinetic, cover the vast majority of cases, but not all. For example, that type of waterwheel (Fig. 7.13) in which buckets or tanks in the rim are successively filled with water which weighs down one side, belongs to neither group, nor does the pumping action that raises the sap in trees. But gas, steam and water turbines, fans, windmills, helicopter rotors and ship's propellers are all kinetic machines, while bicycle and village pumps, hearts and reciprocating steam engines are all displacement machines.

Rotating machines appear to occur in nature only in the form of continuously rotating flagellae in some bacteria, and they are not truly kinetic but belong to yet another class of fluid machines relying on the viscosity of the fluid (shear pumps). However, wings, fins and the sinuously-moving bodies of fish and other swimming creatures are kinetic machines, though not rotary ones.

Because of the importance of these machines, a vast amount of human ingenuity has been expended on them and a large number of different types have been invented and evolved. This has been particularly true of the displacement category which offers more scope for variations, so that the kinetic one is easier to study. Figure 7.14 shows some pump and turbine rotors.

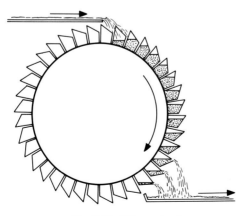

Fig. 7.13. Waterwheel.

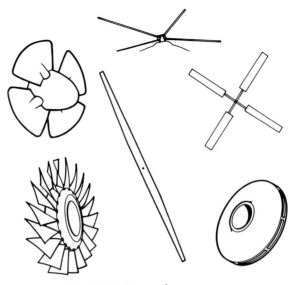

Fig. 7.14. Rotors of pumps, etc.

7.2.10 A classification of kinetic pumps and motors

If the work done on the fluid passing through a pump were all converted into kinetic energy, the fluid would reach a velocity called the 'spouting' or 'characteristic' velocity: if you were rash enough to drill a small hole in the delivery pipe of a running pump, the liquid that hit you in the eye would have this characteristic velocity, c. Now to do this amount of work on each unit mass of fluid ($\frac{1}{2}c^2$, if you remember Chapter 2) the blades of the rotor must travel with a speed u comparable with c – not less than $1/2c$, and generally not more than $10c$. Thus we may classify kinetic pumps by the ratio u/c, the ratio of the blade speed to the characteristic velocity; a low value means that the maximum work is being done on the fluid relative to the blade speed, while a high u/c means a relatively fast blade doing relatively little work. The blade with a low u/c will be strongly curved, so as to deflect the fluid strongly while the relatively fast blade will be nearly flat and only deflect the air slightly. Figure 7.15 shows some blade forms and their associated u/c values.

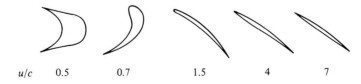

| u/c | 0.5 | 0.7 | 1.5 | 4 | 7 |

Fig. 7.15. Curvature of blades of kinetic machines.

Kinetic motors, i.e., turbines, can be classified using the same ratio u/c, only here the work is done by the fluid, and $\frac{1}{2}c^2$ is the energy extracted from unit mass.

Incidentally, do not imagine that a fast blade doing (or extracting) relatively little work is necessarily extravagant. For instance, economical windmills must have blades of this kind which move several times faster than the wind, in order that they should sweep through, and extract energy from, as much air per second as possible. Thus economical windmills have narrow blades moving fast, rather than broad blades moving slowly: both extract about the same energy from the air passing through, but the first uses less material and can be supported by a lighter structure. Moreover, a fast windmill will require less and lighter gearing to drive an electric generator than a slow windmill.*

To characterise pumps and motors more fully than u/c values alone can do, a second number can be used, which is a measure of the size of the flow through the machine relative to the size of the machine. Imagine all the fluid passing through the machine to pass also along a pipe having a diameter D equal to that of the machine itself. Let the velocity the fluid would have in this imaginary pipe be p. Then we can use the ratio p/c as our second measure (for the mathematically-inclined, if Q is the volume flow rate, since the area of the cross-section of the pipe is $D^2/4$, then $p = 4Q/(\pi D^2)$ and $p/c = 4Q/(\pi D^2 c)$). A high value of p/c indicates a machine handling a large volume of fluid for its size and speed, and vice versa.

Using these two numbers, u/c and p/c, we can represent all the many kinds of kinetic machine on a single plot, as in Fig. 7.16. Here machines nearer the top have a high blade speed compared with the characteristic velocity they induce in or remove from the fluid, and machines nearer the right are smaller relative to the volume of fluid they handle. By the use of such a chart, it is possible to see quickly what kind of kinetic machine will meet any particular need, although experienced designers in the field will know this without any chart and in some cases are likely to use a single criterion, which is called 'specific speed', which works nearly as well as this two-dimensional plotting.

Given a problem, say, to produce a given flow of fluid at a given pressure having available a driveshaft turning at a given speed, a point may be plotted on the diagram representing the ideal solution – this is simply a matter of a little algebra. If this point lies in the region of centrifugal pumps, then that is the solution.

Figure 7.16 gives only a slight indication of the vast range of kinetic pumps, turbines and fluid motors which have been designed and developed, and the problem of selection has many more dimensions than two. For

* Taylor, R. H. (1983). *Alternative Energy Sources*, pp. 22–4. London: Hilger.

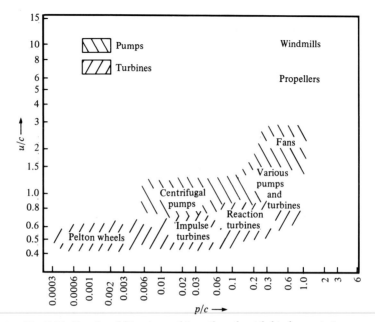

Fig. 7.16. Family of kinetic machines (based, with kind permission, on a figure in French, M. J. (1985). Conceptual Design for Engineers. London: Design Council).

instance, there are important differences in design between pumps for liquids and pumps for gases. Very viscous liquids and slurries like freshly-mixed concrete or coal mixed with water can be pumped with specially-designed machines. Very hot gases, liquid metals and liquids near their boiling points all present severe problems requiring special designs. The absolute values of the velocities u and c and the density of the fluid make important differences, which cannot be shown in a simple two-degree-of-freedom plot like Fig. 7.16. But over all this range, there is a wealth of design experience, proven designs and off-the-shelf machines to draw upon. Moreover, Fig. 7.16 shows only the kinetic family: the displacement family is even larger, and then there are several smaller families, like the shear pumps, important in very small creatures and in vacuum pumping (the hydrodynamic film bearing can be regarded as a hydrostatic bearing supplied by a shear pump which is part of the bearing), magneto-hydrodynamic pumps, thermosyphons and bubble-pumps, jet pumps, and so on.

7.3　The repertoire of known means

Three categories of known means have been looked at – bearings and pivots, cylinders to withstand high pressures, and pumps and turbines. From this small sample, itself only sampled, can be grasped some idea of the great wealth of design experience already accumulated by man and available for his use in further creations. Some other categories are very much larger – for example, that of fasteners: most of them represent the outcome of the ingenious and searching thought of hundreds of talented inventors and designers, vast tracts of intellectual enterprise into which scarcely a scholar has ever ventured.

7.4 Some design principles

7.4.1 Biasing

Besides the repertoire of known means that form the body of design practice, there are a number of wider principles, methods or approaches which a good designer will use automatically, perhaps without recognising their generality, because he has so often come across the kind of situation in which they are useful. Several of these principles have already appeared in previous chapters.

One such is biasing. A transistor is able to amplify an electric current, but it only works when the current flows in one direction; it cannot amplify a current which alternates in direction, which is what it is often required to do. But if a relatively large steady current is superimposed on the alternating current, so that the combined current always flows in the same direction, then the difficulty is overcome.

Another illustration of the biasing principle has already been seen in the bicycle wheel spoke (Chapter 5). Without prestressing, the thin spokes would only be able to take a tensile load, but by tightening them up to an initial locked-in tension, they can be made to take compression, which is essential for this elegant structure to work properly. The function of prestressing in concrete is similar, to bias the range of working stresses in the material, but into the compressive region this time.

These two cases, biasing of a transistor and prestressing, are both examples of one general idea, that of shifting the mean position of a working range to suit the limitations of a device or a material. Biasing is often called for in design, and a good designer will usually spot this because of his experience and his practice in solving problems of this general kind. He is unlikely to call it biasing, however, unless he happens to be an electronic designer.

7.4.2 The principle of least constraint

This principle may be stated as follows: 'when guiding one body relative to another, or securing one body to another, use the least number of constraints that will do'.

A simple example is given by stools and chairs. A two-legged stool will fall over. A three-legged stool will stand firm. A four-legged stool will generally rock if stood on a hard floor, because all of its legs will not be exactly level, or the floor may not be quite flat. The minimum number of legs (or constraints) is three.

Another example is the Salginatobel bridge (Fig. 5.7). It has three hinges in it. Now each hinge removes a constraint, because it lets the bridge bend freely at that point, and so the principle of least constraint says that we should put in as many hinges as possible, so as to leave the least constraints. The maximum possible number is three, for with four the bridge would become a four-bar chain, a mechanism with one degree of freedom, and it would collapse. The car suspensions discussed in Chapter 4 also follow the principle of least constraint: there are just sufficient links to leave the necessary up-and-down freedom of the suspension.

When the minimum number of constraints is used, the forces in the parts are all determined by equilibrium alone. When there are more than the minimum number of constraints, the forces in the parts depend critically on errors in manufacture, temperature difference between one part and another, and so on. For example, when someone sits on a three-legged stool, the load on each can ideally be calculated from equilibrium considerations, whereas this is not the case with a four-legged chair. If we slip a thin piece of wood under one leg of the chair, the load in that leg will be increased. If we do the same with a three-legged stool, the load will not change.

Design which follows the principle of least constraint is often called 'kinematic design'. However, kinematic design is not always to be preferred. For example, a kinematic design of ball bearing would have only three balls, and would be large and clumsy, and a four-legged chair is more stable than a three-legged one. But in most cases where the principle of least constraint can be applied, it is helpful.

A whimsical example is as follows. A man has a pair of trousers which will fall down round his ankles if he does not keep them up with either a belt or braces (US suspenders). However, he uses both together, thus contravening the principle of least constraint since either alone would suffice. Bending down to pick up a dropped coin, the braces need to move up at the rear and down in the front, they cannot do so because of the conflicting constraint of the belt, and they snap. But the strain has told on the belt too, and when he straightens up, it gives way and his trousers fall round his ankles. Thus the principle of least constraint or kinematic design involves avoiding fights between one component and another.

Sometimes, as in the ball bearing and the chair, it is better to forget the principle of least constraint. Then another principle may be invoked, which I call 'elastic design'. In the case of the braces (or suspenders), if they were sufficiently elastic, they would simply stretch, letting the less yielding belt hold the trousers where it would. The principle of elastic design might be stated: 'if there is going to be a fight, let it be a very uneven one and ensure that the loser is not hurt'. Thus the braces give way to the belt, but are so stretchy they are not hurt.

In the case of suspension bridges, we saw that sometimes the loads in the cables either side of a tower were allowed to equalize by using a roller saddle on top of the tower or a hinge at its base, both examples of kinematic design. In more modern designs, however, the towers are simply made to bend until the cable loads are nearly equal, an example of elastic design. The cables easily overpower the tower, but the tower is flexible enough not to be harmed.

7.4.3 The regenerative principle

The avoidance of irreversibilities, mentioned in Chapter 2, is a principle of great breadth, often affording only obvious, and so valueless, insights, but of great assistance in some very recondite problems, and also giving rise to some more specific and narrow principles, one of which is the very important and beautiful idea here called the regenerative principle.

As a first example of this principle, consider the feet of a waterfowl, say, a swan, living in a cold climate. If its feet were at its body temperature, then the loss of body heat to the cold water would be large and would require the swan to find extra food, a serious handicap. To overcome this loss, its feet are allowed to become cold. If, however, the blood supply to the feet gave up its heat to the surrounding water, then the loss would still be serious, since during its passage through the finely branching capillaries, the temperature of the blood would fall practically to that of the surrounding tissues.

To avoid this, the blood gives up its heat before it reaches the feet, and it gives it up to the returning cold blood, which is thereby raised nearly to body temperature. Thus the blood enters the feet cold, and there is little loss of body heat.

A little fable may be a help to explain the regenerative principle. A certain greedy king made all his subjects bring all their money every year to the court, where the chancellor had to assure himself that they had indeed brought a sum that he thought about right, and then send them on to the king. The king would than take half the money from each subject. This extortionate system caused great hardship to the subjects, until one of them devised a scheme whereby they might reduce their tax. Between the chancellor's office and the king's audience room there were three antechambers through which the subjects passed on their way in and out: the suggestion was that each man as he went through each of these rooms should share his money with the man going the other way that he met there. The subjects tried the scheme, and were very pleased with the results. How this

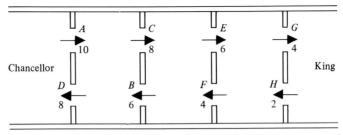

Fig. 7.17. Regenerative tax evasion.

'regenerative' tax evasion system worked is shown in Fig. 7.17. Suppose that at *A* a subject enters the first antechamber from the chancellor's office with ten gold pieces and meets a subject who enters from *B* on his way out, with six gold pieces. They share their money and leave with eight pieces each by doors *C* and *D*. The first subject, entering the second antechamber by door *C*, meets a subject with four gold pieces entering by door *F*, and so on. When the first subject reaches the king, he has only four gold pieces left, and so he leaves the king with just half of this, or two pieces. In passing through the three antechambers on his way out, he acquires two more gold pieces in each, leaving with eight. Thus the scheme effectively reduced the rate of taxation from 50% to 20%.

The parallel with the swan's blood system is clear, and is shown in Fig. 7.18. The flow of subjects represents the flow of blood, and money represents heat. The antechambers represent blood vessels (*AGHD* in the figure) in which the 'in' and 'out' (arterial and venous) streams of blood come into close thermal contact and exchange heat. The sharing of money represents the warmer 'in' stream giving heat to the cooler 'out' stream, and it is limited to sharing rather than interchanging sums, because to do more than bring the two streams to the same temperature would require heat to flow from a cooler to a hotter body, which is impossible (see Chapter 2). The grasping king is, of course, the foot, which can only remove a proportion of such heat as the blood brings in.

Some birds have a similar system for a different purpose. In flight, they tend to overheat because there is insufficient cooling to keep the body temperature down in the face of intense muscular activity and the heat it generates. While most of the body can stand some overheating, the brain is permanently damaged by quite a small temperature rise, and so it is protected

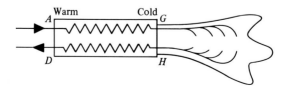

Fig. 7.18. Regenerative heat exchange in swan's leg.

by a regenerative system which keeps the head cooler than the rest of the body. There are many other examples of thermal regenerative systems in nature, particularly in the noses of arctic animals.

7.4.4 Man and thermal regeneration

An early use of thermal regeneration by man was in steel-making where the air entering the furnace was preheated by passing through a first stack of hot bricks, so making the furnace gases much hotter than they would otherwise have been. On leaving the furnace they were passed through a second stack of bricks to which they gave up most of their remaining heat. Periodically the air and gas flow through the system was reversed, so that the second stack, now very hot, replaced the first, which had fallen in temperature, and vice versa.

A difference here is that the heat is not transferred directly between two parts of a stream folded back on itself but is stored temporarily in a solid matrix between intermittent flows or 'blows'. Such a matrix is called a regenerator, and can replace the heat exchanger *AGHD* of Fig. 7.18, which is called a *counter-flow* or *counter-current* heat exchanger.

The regenerative principle is a very subtle one and arises in many guises, some difficult to understand. For example, in both nuclear and fossil fuel power stations the prime movers are steam turbines. The water entering the boilers, the *feed-water* or *feed*, as it is called, is heated first, in *feed-heaters*, by steam extracted from the turbine at a number of different pressures, on a pattern fundamentally similar to that of Fig. 7.17.

The turbine consists of a number of separate turbines, or 'stages', all on one shaft, through each of which the pressure and the temperature of the steam fall in passing. Since the steam extracted to heat the feed-water does not pass through the later stages, less work is obtained from the turbine, but also, since the feed-water enters the boiler hot, less fuel is needed. It is not obvious that on balance this regenerative feed-heating, as it is called, produces a net gain and it is a very complicated arrangement. In fact, it improves the overall efficiency substantially: the full explanation is too lengthy to give here, but the general grounds upon which it rests are explained later.

7.4.5 Side-ponds

An interesting historical invention that demonstrates the regenerative principle in a simple form, not in connection with heat, is that of side-ponds. These are shallow basins adjacent to a canal lock which reduce the quantity of water which flows from the reach above the lock to the reach below when a barge moves from one reach to the other.

First, consider the process of moving a barge from the reach above the lock to the one below, in which the water level is perhaps two metres lower. Imagine first that the lock is full, that is, on a level with the upper reach.

The gates to the upper reach are opened and the barge is moved into the lock (Fig. 7.19(*a*)). The upper gates are shut and the sluices in the lower gates are opened, letting water out of the lock into the lower reach (Fig. 7.19(*b*)). When the level has fallen to that in the lower reach, the lower gates are opened, and the barge moves out (Fig. 7.19(*c*)). If the next barge approaching is from the upper reach, then the lower gates and sluices are shut, and the lock is refilled from the upper reach, restoring the state shown in Fig. 7.19(*a*). One barge has moved down through the lock, and one lock-full of water has flowed down from the upper to the lower reach. In the case of a barge being locked up, a lock-full of water also flows down the canal. In a dry summer with many barge movements, the problem of water supply to the upper reaches may become serious.

Now consider the locking down of a barge when a single side-pond is used. The barge moves into the lock, as in Fig. 7.19(*a*). The upper gates are shut, and then, instead of opening the sluices in the lower gate, a sluice in the wall of the lock (*S* in the figure) is opened, letting water into the large shallow side-pond, which is at a level half-way between that of the upper and lower reaches. When the level of the water in the lock has fallen to *S*, the flow will stop. The sluice *S* is then shut and the slices in the lower gates are opened to lower the water the rest of the way. When it is required to fill the lock again, the first step is to open *S* and let the water from the side-pond in, so providing the first half lock-full. The remaining half lock-full or so is then let in from the upper reach. As this is the only water leaving the upper reach in the cycle, it is now possible to lock a barge up or down for only half the water.

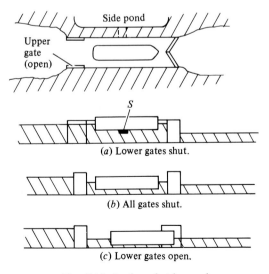

(*a*) Lower gates shut.

(*b*) All gates shut.

(*c*) Lower gates open.

Fig. 7.19. *Lock and side-pond.*

By using two side-ponds, the water consumption can be reduced to about one-third of a lock-full, and so on. An infinite number of infinitely shallow side-ponds would make the amount of water used infinitely small (if we neglect the effects of the displacement of the barges, but then we are studying an ideal case: the operation would also take an infinite time).

The side-pond arrangement illustrates several points which are typical of regenerative systems. Firstly, the remaining loss (or remaining irreversibility, as it is often is) is inversely proportional to the number of stages, N. In the case of side-ponds, N must be taken as one more than the number of ponds, because a lock without them is the one-stage device. Indeed, adding side-ponds is nearly the same as adding more locks – two locks of half the change in level will work on the same lock-full of water, and their lock-full is only half as much; but multiple locks are slower and dearer, requiring more barge movements, more lock basins and gates, etc. The same 'one-over-N' rule applies to the tax avoidance scheme, but there N is the number of antechambers plus two.

The one-over-N rule is a precise example of the wider, less precise principle of 'diminishing returns'. The first side-pond saves (nearly) 1/2 lock-full of water per cycle, the next (1/2–1/3) or 1/6 lock-full and the third (1/3–1/4) or 1/12 lock-full. Thus in terms of value for money, the second is three times, and the third, six times as expensive as the first: as far as the writer knows, three side-ponds was the practical limit.

So far, the side-pond has been regarded as of very large area. This was so that during filling and emptying the changes of level in it were negligible. With one side-pond of practical area, however, as it is filled, the level in it will rise, so that it is not possible to save quite half a lock-full of water. For example, if the single side-pond has twice the plan area of the lock, only two-fifths of a lock-full of water can be saved, and the bottom of the side-pond must be at two-fifths of the way up from the level in the lower reach to that in the upper.

In a comparison with the bricks of the steel-making process, the limited area of the side-pond compares with the limited heat capacity of the bricks: both reduce the savings that can be made.

7.4.6 Avoidance of irreversibilities

The regenerative principle is just a special case of the more general one of avoiding irreversibilities (see Chapter 2) which in turn is largely subsumed in two common-sense propositions:

(a) use the cheapest that will do
(b) throw away nothing of value.

Using these two articles of the wisdom of the home and the market place, much good design may be systematically arrived at, e.g., side-ponds and regenerative feed-heating.

Let us take the side-pond first. Without a side-pond, the first water to be let out when the lock is emptied is very nearly as valuable as water in the upper reach: it should not be thrown away, but stored in a side-pond, where it will have a value (with one pond) about half that of water in the upper reach. Conversely, when starting to fill the lock, the half-value water in the side-pond is the cheapest that will do, and it will only do until the lock is about half-full, so we use it first and only use the more valuable upper-reach water when we must.

In terms of irreversibilities, there is an irreversibility whenever water flows from a higher to a lower level without the potential energy being converted into some form (other than useless low-grade heat, which is what happens in the case of the lock; see Chapter 2). The greater the drop in level, the greater the irreversibility, so the use of the single side-pond roughly halves the irreversibilities.

Now let us consider the steam power plant in the power station and the regenerative feed-heating system. If cold water is put into the boiler, it will be heated from cold by the hot furnace gases. In this context, valuable is hot and cheap is less hot, so that the dearest thing we have is the hottest, the furnace gases, and they are certainly not the cheapest that will do to heat cold water. The cheapest is steam from somewhere in the low-pressure end of the turbine, at perhaps 80 °C, and so some is drawn off to heat the water from the condenser. Then some more steam from further up the turbine, at perhaps 120 °C, is used to heat it further, and so on.

One interesting aspect of the application of these principles, both to locks and to steam power systems, is that we can always be sure that an economy will result, even though the form it will take may not be apparent at first.

7.4.7   The cascade principle

Let us turn now to what looks like a very different problem, the fixing of a row of turbine blades to the rim of a disc by dovetailing, as in Fig. 7.20. The rim has been drawn straight, for simplicity, but it should be curved, being part of a circle. Each dovetail joint has to be fitted into a width P and should be as strong as possible to resist the centrifugal load F from the blade. Again, for simplicity, we assume both materials are of equal strength. The strength of the joint depends on the width of the dovetail necks x and y, and on having a good overhang z. All these three widths have to be fitted into the overall width P (the designer would call P the pitch), so that

$$x + y + 2z = P$$

A reasonable sharing out of the width is given by

$$x = y = 0.3P, \quad z = 0.2P$$

As a standard of comparison, we can consider an ideal welded joint, where the weld is equal in strength to the parent metal, extending from A to B

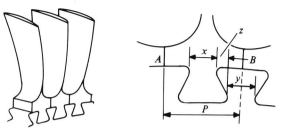

(a)

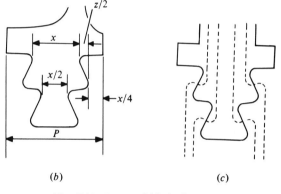

(b)   (c)

Fig. 7.20. Dovetail blade fixings.

and so of width P. The strength of the dovetail and this ideal joint would be in the ratio $x:P$, or roughly 0.3:1. This ratio may be regarded as a kind of efficiency, joint efficiency, defined generally as

$$\text{joint efficiency} , \; \eta = \frac{\text{strength of real joint}}{\text{strength of ideal joint}}$$

In the case of the single dovetail, then, $\eta = 0.3$. Now by using a two-stage dovetail, like that in Fig. 7.20(b), we can raise η to about 0.46. Since each pair of overhangs should take an equal share of the load, we can visualise the 'flow of force' in the blade as in Fig. 7.20(c), where each dotted line represents a quarter of the load F. If the wider neck is still of width x, in both blade and disc, then as the narrow neck carries half the load, its width should be $x/2$. Also, since the overhangs carry only half as much load each as the single dovetail, they should have width $z/2$.

The sharing out of the overall width P is now governed by the equation

$$x + \frac{x}{2} + z = P$$

and if z is $2/3x$ as before, this gives $x = 0.46P$. Going to a three-stage root gives $x = 0.56P$, to four stages, $0.63P$ – again, diminishing returns, approximating to the $1/N$ rule for a large number N of stages. The term 'stage' is appropriate because in the force flow diagram in Fig. 7.20(*c*), the load in the blade is shed via the overhangs in equal stages. Equally, the load in the disc neck builds up in equal stages, until all the load is in the disc. Ideally the joint efficiency approaches unity as *N* approaches infinity, as in the imaginary joint of Fig. 7.21, where the serrations of the dovetail have been made very fine and so cannot be seen. Where a strong enough adhesive is available, as in the case of wood, a glued joint realises this ideal form (as in the joint of Fig. 5.14(*c*)).

By transferring the load in stages from blade to disc, we can achieve a higher performance than when the transfer is in one step. Such an arrangement where a desired result is achieved by the cumulative contributions of a number of steps or stages is often called a 'cascade'. The regenerative feed-heating system can be regarded as a cascade.

The cascade idea can also be applied to a process like the liquefaction of natural gas, which requires heat to be removed over a wide range of temperature. A refrigerator could be designed which was capable of pumping all the heat to be extracted from $-160\,°C$, the temperature at which the gas will liquefy at atmospheric pressure, up to the ambient temperature of, say, $15\,°C$. But the ideal work required to extract one joule of heat at $-160°$ is 1.55 joule, whereas at $-120\,°C$ it is 0.88 J and at $-60\,°C$ it is only 0.35 J. Much less energy will be required if we design a refrigerator which extracts heat over several ranges of temperature, $-160\,°C$ to $-120\,°C$, $-120\,°C$ to $-60\,°C$ and $-60\,°C$ to $15\,°C$. This is easily done, since to reach such low temperatures anyway we need to use several refrigerants in a cascade, each pumping heat up through a stage and each working basically just like the refrigerant in a domestic refrigerator or freezer.

It is possible to design such a cascade refrigerator in which all the refrigerants are kept separate, but it is also possible to mix them all together. The effect is as if the cascade had a infinite number of stages, with heat being

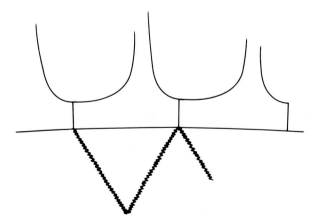

Fig. 7.21. Limiting case of dovetail fixings.

extracted at all temperatures, and a corresponding saving. However, a new kind of loss appears, called a mixing loss, whenever two streams of refrigerant of different composition mix together. This loss is another example of the second law of thermodynamics (Section 2.5). So great is the mixing loss where the freshwater of the Amazon enters the sea that it could provide all the world with electrical power, could it only be harnessed.

One of the alternatives to the chlorohydrocarbons (CFCs) that were used in refrigerators and have been banned because of the damage they do to the ozone layer, is a mixed refrigerant, a mixture of butane and pentane.

Rockets with separate steps or stages are another example of the cascade principle, with much in common with cascade refrigerators. Here the *first* stage lifts itself and all the subsequent stages, just as the *last* stage of the cascade refrigerator lifts the heat rejected by all the earlier stages and that which must be extracted in its own step.

7.4.8 Number of stages

The cascade and regenerative principles can be applied to a great many design problems besides the few which have been discussed, for example, distillation and separation processes, gear-boxes and transmissions, and electronic and electrical systems. There is not a single underlying mathematical scheme, but some relationships recur very frequently – for example, the 1/N rule is obeyed sometimes from $N = 1$ onwards, in other cases only very roughly at low values of N but more closely with large Ns, and in rare cases, not at all. They are closely linked with other ideas such as matching and avoiding irreversibilities, and can often be developed on the homespun principles, use the cheapest that will serve and never throw away what may prove useful later.

The cascade and regenerative principles are just particular applications of the wider principle of matching looked at in Chapter 6. In side-ponds, in all transfers of water we match the height of the supply and the height of the point of delivery as closely as we can: in the fir tree root, we match the widths of the necks to the loads they carry as closely as we can: in the refrigerator, we match the temperature of the coolant to the temperature of the thing cooled as closely as we can.

7.4.9 Matching

One of the most important design principles is that of matching, which was discussed in Chapter 6. It can often be seen as the basis of more specialised principles, as has just been shown in the case of cascade refrigerators. Here is a very elegant example which comes from the big end of a large marine diesel engine, that is, the crankshaft end of the connecting-rod which connects the piston to the crank so that the force on the piston can turn the crankshaft.

Figure 7.22(a) is a diagrammatic version of the junction between such a big end, a large ring-like hydrodynamic bearing which encircles the crank

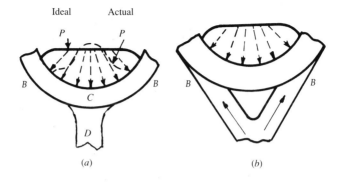

Fig. 7.22. Connecting-rods: junction of big end and shank.

pin, and the shank. Now the ring of steel surrounding the bearing is about 300 mm thick radially and commensurately long, quite rigid, one might imagine. But under the enormous loads to which it is subjected when the cylinder fires, it does deflect slightly, of the order of hundredths of a milli-metre. It might be thought that such a small deflection was of no account, but consider that this enormous load is transmitted to the big end by the sort of hydrodynamic oil film that was discussed in Section 7.2.2, which is very thin and whose very working depends upon the wedge action produced by the way that that tiny thickness varies.

In the fairy story of the princess and the pea, a queen placed a pea under the mattress of a prospective daughter-in-law. In the morning the soi-disant princess complained of the uncomfortable night she had had, and another mattress was placed over the first. The next morning she still complained, and so it went on. Eventually, the queen was satisfied that anyone whose sleep was spoilt by a solitary pea under 13 mattresses must indeed be a true princess.

Now the oil film is the princess, and the heavy steel ring of the big end is the mattress. Imagine the oil film applying a huge distributed load to the big end, like the princess applying a small distributed load to the mattress, over the length BB in Fig. 7.22(a). Near the Bs the ring will give more, that is, it will be relatively soft. Nearer C it will be relatively hard because of the proximity of the direct support of the shank D, and at C there will be a really hard spot, like a brick rather than a pea under the mattress. Because of this hard spot, there will be a concentration of load there, as shown by the broken line P, which represents the local pressure in the oil film. Now too high a value of P will cause the material of the bearing to fail by fatigue, and so the hard spot limits the total load the bearing can carry, and hence the power that the engine can develop. If a more even pressure distribution could be achieved, such as that shown by the solid line in the figure, then the power of the engine could be increased.

The engineers at Wartsila Diesel saw all this, and what is more, they came up with the very pretty solution shown diagrammatically in Fig. 7.22(b). They removed the hard spot at C by forking the shank into two just above the big end. This new form does not produce hard spots at the points B, because the support comes into the ring tangentially, and is therefore soft, just as a young tree is soft if you push it sideways, so it bends, rather than pushing straight down the trunk, when it is hard. We have given our sensitive princess a hammock, instead of a mattress with a brick, rather than a pea, under it.

Wartsila calculated the effect of such a change, no easy matter but practicable nowadays, predicted that it would allow them to increase the peak pressure in the cylinder, and hence the power, by 40%, tried it with a real engine, and found it worked. That is how engineering should go – but, of course, it doesn't always.

An important design principle says that one should only support a thin-walled structure by tangential supports, and this is an example. However, it takes some effort of the imagination to recognise that in this context, a massive steel ring is effectively thin! Alternatively, the more general principle of matching could be invoked to the same end.

7.5 Some design 'philosophies'

7.5.1 The philosophy of wave energy systems

By the philosophy of a design most designers who would use the term probably mean some central idea that informs the whole, so that two approaches to the same problem using different philosophies would lead to very different designs.

As an example, the writer has invented a device for extracting power from sea-waves, and the philosophy of this device is very clear. It runs as follows.

Wave energy is a very diffuse, dilute form of energy. Also, the level is very variable, with an average of, perhaps, 50 kW per metre of wave front, rising to perhaps 50 times that figure in a storm. The problems of wave power, then, are a very spread-out source of energy and very severe overloads. Two functions required of any power extraction system will be a surface interacting with the sea and some means of gathering up or concentrating the energy expended by the waves on that surface.

Consider the collecting surface first. It must be very large, and so to make the capital cost acceptable, it must be very cheap per unit area. It must also be able to resist storm damage, a property which may be achieved broadly in one of two ways: the first is by making the surface strong and rigid, and the second is by making it tough and yielding. The former can be done with steel or reinforced concrete, but the latter might be done with rubberised cloth of the kind used for inflatable boats. The 'tough and yielding' solution seemed more promising to the writer, especially if the surface could be presented sideways on to the sea, so the waves ran alongside it, rather than smacking flat up against it.

Now let us look at the problem of the concentrating mechanism. This could be a solid mechanism, of levers, pivots, shafts, and so forth, or it could be fluid. Given the decision to use a cloth collecting surface, the choice of a fluid needs no explanation. The practical choice lies between air and water, and there are strong reasons for preferring air, with its low density. In Chapter 2 the importance of 'quickness' in bows was explained: because the speed of travel of waves is determined by the inertia of the water, it can be seen without making any calculations that water is an unsuitable concentrating medium because it is not 'quick' in this connection – indeed, it is exactly as 'slow' as the waves themselves. On the other hand, air is about 30 times as quick as water in this application. Because it is so much less dense, in the same circumstances it moves about 30 times as fast under the same pressure differences (inversely as the square root of the density).

The scheme resulting from this philosophy is shown in Fig. 7.23. A long concrete beam floats in the sea head-on to the waves, carrying air-filled bags along the sides. The beam contains two longitudinal ducts A and B, running its whole length, and communicating with the bags by non-return valves, such that air can only flow from duct A into the bag, and from the bag into duct B. As a wave crest travels along the beam, as at C, it compresses the cells, squeezing air into duct B. As a trough passes, the pressure of seawater on the bags is reduced, and air flows into them from duct A. By this means

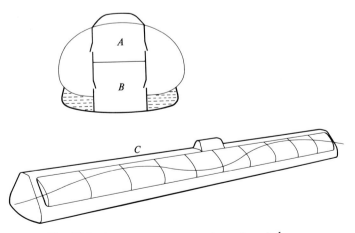

Fig. 7.23. A wave energy converter using air bags.

duct B can be maintained at a higher pressure than A while air is continuously pumped from A to B by the waves working the cells like bellows or
those simple concertina pumps used for inflating air-beds: the air returns
from B to A via a turbine, thus completing a closed circuit and providing
power from the sea. Which type of turbine will be suitable is easily determined with a few simple calculations and reference to Fig. 7.16.

Once the original decisions are taken, that this shall be a tough, yielding
device with the waves sliding peacefully along its smooth flanks, using air
to concentrate the power and deliver it to a small, fast turbine (a water
turbine would be bigger in diameter by about $\sqrt{30}$ times, because its characteristic velocity would be about 30 times less, because water is 30 times less
quick) then most of the rest flows from these central ideas, not by any
mathematical process, but by its consistency, its coherence, its consonance
with the theme, the 'philosophy', of the design.

7.5.2   A different philosophy

This invention, the Lancaster Flexible Bag, appeared at the time to be very
promising, but although some still believe in it enough to continue work
on a derivative of it elsewhere, the inventor, the writer, now regards it as
hopeless. It is a case of a failure of insight leading to an inappropriate philosophy. The besetting problem of energy from sea-waves is the large size of
the devices, leading to excessive costs: a better philosophy must be aimed
at small size. Let us see where this thought leads us.

A long device can at best collect the energy incident upon its length, with
perhaps a little more at the ends. Thus a device 200 m long parallel to the
waves can only collect the energy from about 250 m of wave front. However, a small device can draw energy into itself from either side, so that one
only 20 m wide can collect the energy from 50 m of wave front. Moreover,
the long device will be subject to severe bending, so that it has to be mostly
strong structure to survive, and so mostly expensive, whereas the small
device is not subject to severe bending and can be mostly cheap ballast.

Now power is force times velocity (Chapter 2): any wave power device
has a working surface on which the sea exerts a force, so we aim to make
that surface as large as possible relative to the device in order to make the
force large. We also aim to make it move as rapidly as possible, which in a
wave motion means with as big an amplitude as possible. One thought is
to make the entire body a working surface, with a paddle form to give a
large area for a given volume.

Much of this new philosophy was first put forward by two Norwegians,
J. Falnes and the late K. Budal, who drew attention to the virtues of a large
amplitude, which was to be obtained by maintaining the device in resonance
with the sea, like the glass resonating to the soprano's voice in Section 3.2.7.
Unfortunately, in this analogy, the sea is a very quavery singer, so the glass

has to be very smart to keep in resonance with it. Budal and Falnes showed how to do this in their device, which was a buoy (see Fig. 7.24(*a*)) sliding up and down ('heaving') on a rod pivoted to the sea bed; power was taken off by resisting the sliding, and resonance was maintained by changing the resistance at the right times in the cycle.

At Lancaster University we took aboard the Norwegians' philosophy, which seemed to us eminently sound. After working with buoys we came to prefer the paddle form, and we also had misgivings about the attachment to the sea bed. The wave energy converter designer has the problem expressed by Archimedes, when he said, 'Give me a fulcrum and I will move the earth'. Throw an object in the sea, and the sea will move it around; to extract power, this motion must be resisted. We need to produce a reaction to oppose the wave forces. There are three possibilities:

(a) attachments to the sea bed
(b) very large devices, so that the wave force at one point reacts against that at another, or
(c) an inertial mass.

Now the idea of reacting against some massive object does not at first seem attractive, but both (a) and (b) must be very expensive. Sherlock Holmes in several of the stories says something to the effect, 'It is an old

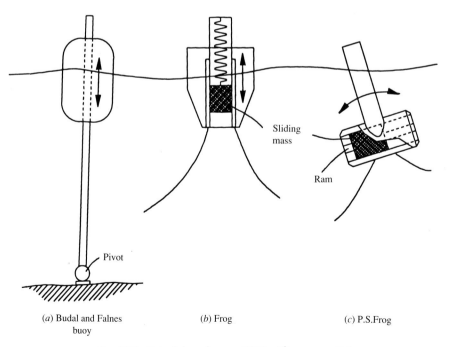

(*a*) Budal and Falnes buoy   (*b*) Frog   (*c*) P.S.Frog

Fig. 7.24. *Principles of some WECs (diagrammatic).*

maxim of mine, Watson, that when you have excluded the impossible, whatever remains, however improbable, must be the truth'. Change 'truth' to 'only hope' and the advice is applicable to design. Accordingly, we looked at what was left, the inertial mass, and we thought it could be made workable. Figure 7.24(b) shows the essence of the idea; a mass hangs on a spring in a buoy. Suppose the buoy to be lifted by a wave; the spring will be stretched and the motion upwards of the buoy will be resisted. Eventually the mass begins to rise, but now the buoy is starting its fall and the spring starts to resist that fall. If the waves are regular and of the same frequency as the natural vibration of the mass on the spring, the motion will be resonant. In an experiment in the wave tank we have had the amplitude of the buoy 20 times that of the wave, and the amplitude of the mass four times higher still, with very small waves, naturally. With larger waves we can extract power from the moving mass, so that its amplitude will not grow too large. With irregular waves, we can modify the stiffness of the spring four times a cycle and maintain resonance and a high energy capture.

However, the paddle form was more attractive, so we abandoned the buoy and produced the device shown in Fig. 7.24(c). It is rather like a huge table tennis racket with a cylinder containing a sliding mass in place of the handle. It floats in the sea facing the oncoming waves, batting them back whence they came, and moving in combined surge and pitch as shown. The sliding mass is restrained by hydraulic rams, which work most of the time as pumps, some of the time just hold the mass still and occasionally push the mass to accelerate it, all under the control of a microprocessor to maximise the energy captured.

There is one other important element in the design philosophy which has led to this device. Besides the doctrine of smallness, the maximisation of the working surface by making the whole thing a paddle which is all working surface and the use of a control system that maintains a state of resonance (or quasi-resonance, if you will not have resonance used of the response to an irregular input from the sea) there is also the hermetic philosophy, to have no moving parts outside the hull. The whole thing is a sealed box on a compliant mooring – one which does not restrain it stiffly, but allows it to move just like the moorings of ships. The sea pushes the box around and the box generates electricity entirely by internal workings. There are no problems with barnacles or seaweed or leaking of lubricants.

The progress since the air-bag device is enormous. Against about 60 W of average electrical output per tonne of displacement, we now believe we could produce 600 W per tonne. What is more, most of the air-bag device was expensive structure, while over 80% of the mass of P. S. Frog, as we call it, is cheap ballast. So far we have only worked with small models, but there is much experience to suggest that in this field models give reliable estimates of full-scale behaviour. In a review for the UK Department of

Trade and Industry by the Energy Technology Support Unit published in December 1992, they gave P. S. Frog the lowest cost estimate of all the devices studied, at 6p (say 9c) per kilowatt-hour; since then we have made an improvement which should reduce the cost on the same basis to 4.2p/kWh. There is hope for sea-waves as a source of renewable energy.

7.5.3 Suspended roofs

An example of design philosophy in structures is the suspended stadium roof shown in Fig. 1.1. Here a large area is to be covered in by a light, economical structure, and the best structural members for such a purpose are ties, as was explained in Chapter 5. However, as was also seen, most structures require a finite and irreducible volume C of material acting in compression. Because of the problem of buckling, compression members are not suited to airy structures, being at their best and most economical when short and fat, so that it makes sense to concentrate all the volume C in a few stout and relatively short members. The tension members, being at their best when thin, are very numerous and fine, and serve to diffuse the point supporting forces provided by the struts over the huge area of the roof – a good solution to the load diffusion problem. A ridge or bell tent shows the same philosophy, though in a less developed form.

7.5.4 Dams

Dams are among the earliest structures built by man: more than 3000 years ago the Ancient Egyptians built one which still survives and functions. It is a simple gravity dam, which opposes the thrust of the water by sheer weight. But many dams, particularly high ones, conduct the force to solid ground by means of relatively small quantities of strong material used in economical ways, which vary according to the nature of the site.

Figure 7.25 shows two different locations for dams. Imagine that you are looking at them from the up-stream side, so that the water would

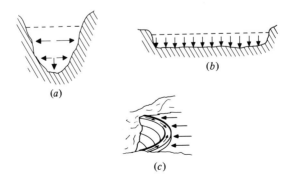

Fig. 7.25. Dam sites.

be pushing the dam away from you with a very large force – perhaps hundreds of thousands of tons. Then in the narrow gorge (a) it is most economical to lead the force sideways to the walls of the gorge, and for a designer working in a material which is strong in compression (like concrete) the logical form of structure is the same as that for a bridge in such material, i.e., an arch. Because the water load on the face of the dam is horizontal, however, the arch must lie in a horizontal plane, as in Fig. 7.25(c).

In the location shown in Fig. 7.25(b), the shortest way by which to lead the water force on the face of the dam to solid earth is downward. By tilting the dam backwards, the force can be inclined downwards, as at P in Fig. 7.26(a), which helps a great deal. However, this simple design is very extravagant in material, and most of the material behind the face can be removed, leaving only isolated buttresses, as in Figs. 7.26(b) and (c). But the slab between the buttresses is subject to bending, which severely limits the distance or pitch p that can be spanned. Fewer, more widely-spaced buttresses are possible by using arches, a more efficient structure, to span the gaps between them, as in Fig. 7.26(d). In such an arrangement, each buttress is subjected to abutment thrust from the arches, which balance one another, and forces in its plane, with regard to which it is very strong. It can therefore be made very thin, so thin that struts may be needed to prevent it from buckling sideways under the compression load P in it. The structural philosophy of the buttress dam is to carry the forces sideways, as shown by the smaller

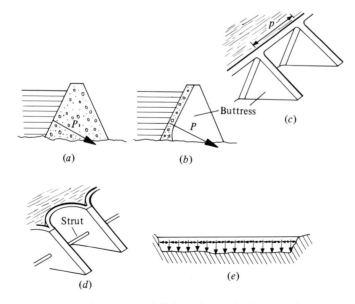

Fig. 7.26. The structural philosophy of the buttress dam.

arrows in Fig. 7.26(*e*), thus concentrating the load on a number of thin deep buttresses. Alternatively, it might be thought of as starting with a number of buttresses which divide the width to be dammed into narrow gaps, which are suited to that known means of damming narrow valleys, the arch dam. The arch dam, though, usually has a nearly vertical face, and relies on strong natural walls of rock to take the loads at its springings. The buttress dam relies on the downward component of the force *P* to keep it pressed to its foundations. Most dam failures, incidentally, have been due either to faults in the rock (in arch dams) or to water seeping under them, rising in pressure and lifting them from the bed.

7.5.5 The gas-cooled nuclear reactor

All the reactor in a nuclear power station does is to supply heat to the boilers of a steam power cycle: the nuclear fission processes in it develop heat, which is carried away by streams of water or gas. In the gas-cooled reactor, the streams of gas are passed through boilers to which they give up heat, returning via compressors which act as pumps to keep the circulation going. Figure 7.27(*a*) shows an early arrangement of the kind, with six boilers and six compressors grouped round one reactor in six circuits, each of reactor–boiler–compressor. These gas circuits have to operate at a high pressure, perhaps 30 times atmospheric, and the components are large. Large steel vessels to take high pressures at high temperatures and built to the strictest standards are expensive, and become rapidly more so with size, because of the problems of forming and welding the thick plates of which they must be made. The arrangement of Fig. 7.27(*a*) is thus a sensible one, in that it houses the reactor and six boilers in seven separate pressure vessels, the biggest of which is very much smaller than a vessel to contain them all, not to mention the ducts connecting them and the compressors.

The difficulty with this arrangement is that the reactor, six boilers and six compressors are all separately supported but joined by ducting subject to large temperature changes, and hence to thermal expansions which must be accommodated in some way. This was done by means of three hinges in the length of each duct, basically of the kind shown in Fig. 7.27(*b*), with a metal bellows or concertina-piece to provide a gas-tight connection between two rigid pieces of duct hinged together. Figure 7.27(*c*) shows the action of these hinges (*A*, *B* and *C*) in the upper duct leading from the reactor to the boiler; the duct has been shown as a single line, for simplicity, and the dotted lines show the changed position after the boiler has expanded straight upwards, the duct has expanded, and the reactor vessel at the attachment of the duct has expanded upwards and to the left. The movements have been exaggerated for clarity. It is simple to work out that three hinges per duct

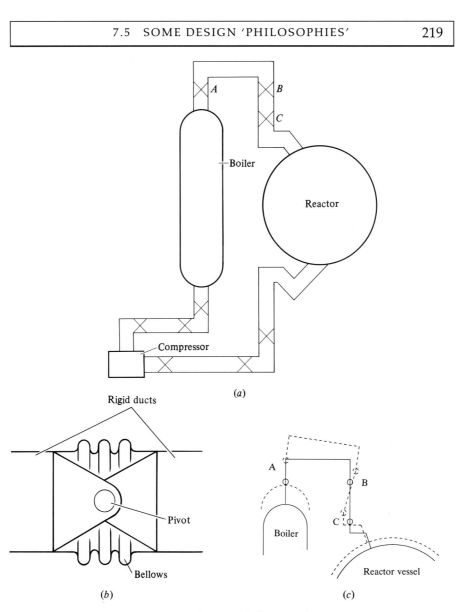

Fig. 7.27. Early gas-cooled reactor design.

is the least that will do, which means three each in three ducts in each of six gas loops, or 54 hinges in all.

The philosophy of these early British nuclear power reactors (the Magnox generation) was thus one of small separate steel pressure vessels running hot (they are insulated on the outside) and joined by articulated ducting.*

To improve the efficiency, it was desired to raise the working temperature (Chapter 2 explained how higher temperatures are associated with higher

* Booker, P. J. (1962). Principles and precepts in engineering design. *The Engineering Designer*, Sept.

efficiency) but this would have aggravated the pressure vessel problem, demanding special, very expensive alloy steels. This Gordian knot could be cut, however, by putting thermal insulation on the *inside* of the vessels. This would be difficult to do with the arrangement of Fig. 7.27(a), however. An alternative would be to put the reactor, the boilers, the compressors and the gas passages all inside one large pressure vessel insulated on the inside, and that is what was done in the next generation of advanced gas-cooled reactors (AGRs). The gas circuit is surrounded by insulation consisting of many thin layers of stainless steel spaced apart, and then by a thin layer of steel lining the prestressed, reinforced-concrete pressure vessel. This lining acts as a gas seal and has a tracery of cooling water pipes attached to it to supplement the action of the insulation in keeping the pressure vessel itself cool. The pressure on the lining is borne by the concrete behind it, which gathers up this diffuse load and transmits it to the 'string bag' of very strong and relatively cheap steel tendons which hold it all together.

Because the whole is wrapped in one giant pressure vessel, there is no need for hinges in ducting. There are internal differential expansion problems to be coped with – and often the principle of least constraint will apply here – but they are not complicated by involving the pressure containment and the sealing-in of the gas. The large diameter at which the radiation shield must begin means that it must contain a much greater mass of material, but the concrete itself supplies part of this function.

These are the central ideas behind two generations of gas-cooled reactors, which can be traced through innumerable details of the two sets of designs. Each has its own rationale, its own characteristics, almost its own style.

7.5.6 'Philosophies' of living designs

Some readers may object to the use of the word 'philosophy' in this context, perhaps because they do not understand what deep and rich thought is involved. Nevertheless, it seems the most suitable word in English, as the translators of that most remarkable book, Torroja's *Razon y ser de los tipos estructurales* must have thought when they rendered it as *The Philosophy of Structures*. Perhaps the Spanish architect and engineer's choice of title, literally, 'reason and being' rounds out the meaning as well as may be in a few words.

The designs of living organisms have their philosophies; they are often daring and always subtle: Blake must surely have had some appreciation of the kind when he wrote *The Tiger*. Perhaps more amazing than a tiger is a bird of prey, an eagle, say. It was seen in Chapter 2 that it is only just possible for living machinery to achieve flight at all at the all-up-weight of the eagle. Then it has to capture prey, an operation depending on refined information-gathering and processing and advanced control systems, with a very narrow margin between missing altogether and dashing itself against the ground. It scours a huge area for its prey, and constitutes with its mate a very rare, thin, intense and energetic presence in a large territory

At another extreme we might consider a ubiquitous organism like grass, covering great stretches of the earth, outweighing the eagles which fly over it and which, indirectly, it feeds, by perhaps a million times, deceptively simple but rich in resources to defend itself against the hazards to which it is exposed – drought, frost, disease, competition by other plants, damage by grazing animals – and to survive.

8

Economy, form and beauty

8.1 Economy, form and beauty

The forms of living organisms have evolved purely to perform vital functions economically. Most of our ideas of visual beauty derive from these strictly functional organisms, trees, flowers, animals, birds and the human body. The natural forms we have come to regard as beautiful are simply very effective and practical designs.

Consider the human arm. We have seen that any limb should have its mass concentrated towards the end attached to the body, so that the inertia to be overcome in movement is kept down: this is particularly important in fast-running animals, where the foot becomes little more than a minimal bony strut and is relatively long, but it is still a consideration in an arm. We have seen how both the minimisation of inertia and the maximisation of reach, or the volume of space which can be swept by the hand or foot, make it desirable to put a three-degree-of-freedom joint at the shoulder and a one-degree-of-freedom joint at the elbow. It is also possible to produce strong, but not rigorous, arguments for three principal joints (shoulder, elbow and wrist) dividing the arm into three segments of which the first two (upper arm and forearm) are roughly equal in length and the third, the hand, much shorter. It is fairly clear, too, that when standing relaxed with the arms hanging loosely the palm of the hand should be towards the body, rather than facing in any other direction.

We have already seen why the fingers should have simple hinge joints in their length, and for grasping it is clear that two joints are rather few and more than three unnecessary. On the number and arrangement of digits, however, we are on less sure ground. Would three fingers be virtually as good as four, but simpler and lighter for the same strength? More radically, would *two* opposed thumbs be useful? Clearly, for some tasks they would, but it seems likely that on balance they would not be an advantage, even though many birds have such an arrangement, with two digits opposed to two digits.

These considerations of inertia, of reach, of economy in degrees of freedom, of high versatility without sacrificing strength or agility, more or less dictate the bone structure and its articulation: only the fourth finger appears of doubtful justification. If now the muscles are added to operate all these joints, with sizes apportioned to their functions, and the whole is clothed with skin, the outside appearance of the arm is complete. We find it beautiful, but its form is dictated by its function as a powered mechanism, versatile and useful.

By contrast, the mechanical robot arm on the production line is a crude design cosmetically packaged. It has only about a quarter as many degrees of freedom, and its muscles, stronger than animal muscles, are not wrapped in a close elastic skin but concealed in loose-fitting sheet metal boxes. Its hoses and wires, corresponding to blood vessels and nerves, are not tucked neatly away between its muscles, but are all exposed lying over them when the casings are removed, or all the time at the joints. Compared with the human arm, it is simple, rough and not very beautiful, but it is precise and it is tireless.

8.2 The form of plants

To appreciate some of the niceties of the form of plants, it helps to consider a very simple problem of supplying some fluid – it could be water, or blood, or sap – to two points from a third. Consider Fig. 8.1(a), in which points A and B are both to be supplied from C, with twice as great a flow going to A as to B. Now there is a cost associated with any flow: there is the cost of the pipes and, in the case of water or blood, the cost of the pump and the pumping. Moreover, this cost will depend on the distance the flow travels and on the quantity flowing. However, while cost will be proportional to distance, it will be less than proportional to the flow: doubling the flow will not double the cost, but increase it by rather less, perhaps by 50%.

In scheme (a) of Fig. 8.1 a very simple plan has been adopted, of a pipe straight from C to A with a branch coming off at right-angles at D and leading to B. Suppose now we move the branch point to E, as in scheme (b). Then the section D to E now carries only the flow to A, and so it costs less than the corresponding section in scheme (a). On the other hand, the branch pipe EB is slightly longer than DB in scheme (a). For a small movement of the branch point towards C, i.e., where ED is small, there is a saving on balance, so scheme (b) is more economical.

Can we improve the scheme further? In scheme (c), the pipe from C to A has been deflected towards B. This makes the line AFC a little longer, and so a little more expensive, but it shortens BE to BF, and overall there is a net economy. The very cheapest version has F a little off the line BE for the most economical choice of E in scheme (b). The savings which can be made over the simple scheme (a) are not large. The schemes in Fig. 8.1 have been optimised for the relative costs shown in circles by the branches, e.g., the line carrying the whole flow from C costs 1.9 times as much per unit length as the line arriving at B, which carries only one-third of the

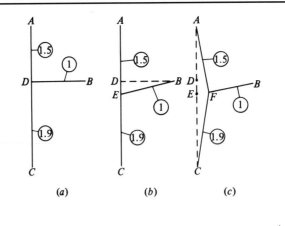

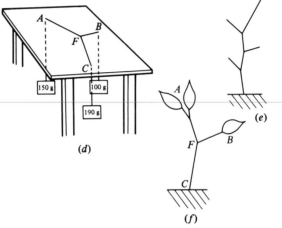

Fig. 8.1. *Economy of branching systems.*

flow: the saving of scheme (b) over scheme (a) is 1.8%, and the saving of scheme (c) over scheme (a) is 4.5%. Nevertheless, savings of this order are crucial in the fiercely competitive world of nature (*pace* Kropotkin).

One way of looking at scheme (b) is that the most costly pipe has been shortened, and in scheme (c) we can think of the small pipe drawing the main pipe off its straight line. Indeed, if we let the junction point F find its own position under pulls from the three pipes proportionate to their costs per unit length, 1.5, 1 and 1.9 from A, B and C, respectively, then the optimum layout will result. This suggests an easy way of making a self-optimising model. If we plot A, B and C to scale on a horizontal board, marking them by drilled holes, tie three strings together, the knot represent-ing F, and then drop the ends of the three strings through the three holes, we have the basis of the model. If we attach weights proportional to the costs to the three strings, say, 150 grams to the string through A, 100 grams to the string through B and 190 grams to the string through C, the model is complete (Fig. 8.1(d)) and the knot F will find its way to the optimum position. In order to reduce friction in any actual experiment, it would naturally be advisable to fit pulleys for the strings to run over at A, B and C.

Supply systems with more delivery points will show similar features when optimised (and can be represented by the same kind of analogue, too). Figure 8.1(e) shows such a branching system, with the 'side-shoots' drawing the main 'stem' out of line, and the resemblance to natural plant forms is clear. In fact, the principles of economy which dictate the branching forms of trees and shrubs are much more complicated, but give similar results. Structural needs, for example, give different relationships from those of flow, but the results of optimisation will look much the same.

One simple aesthetic manifestation of the economy of branching systems is balance. Suppose that Fig. 8.1(f) (derived from Fig. 8.1(c)) represents a sprig of a plant, with two leaves at A and one at B. Then the weights at A and B will roughly balance about F, and if the wind blows hard, the forces at A and B will exert roughly equal but opposite twisting moments on the stem CF. There is an appearance of balance about Fig. 8.1(f), like that of an artist's composition.

8.2.1 The form of trees

Real plants are branching systems more complex than those we have considered in the last section. Not only do they have many branch points, but they are not confined to a single plane and may, for instance, spiral round a central stem. In shrubs and trees, these central stems may themselves be branches of larger stems, often showing a similar pattern at each level of the structure, which is effectively a tiered one like the feather and the floor considered in Chapter 5 (although, plant growth being less systematic than animal growth or engineering design, it is difficult to say to which tier every stem belongs, the boundaries between twigs, small branches, branches and major branches being vague).

The principal function of all this structure is to support a pattern of leaves in such a way as to capture as much of the sun's energy as possible for the minimum cost in living material, a function which is complicated by the change in direction of the sun throughout the day and competition with other plants. The structural cost per unit area of leaf is least if the plant is compact, but then, at any particular time of day, a large proportion of the total area will be in the shadow of other parts. A straggling, open habit of growth avoids self-shading, but imposes a relatively high structural cost. Height is expensive structurally, but can be a winning advantage in competition with other plants. To grow in a forest, a species of plant must either be a tall tree to fight it out with the other tall trees for the light up high, or a climber than can use the structure of other plants to save growing its

own, or an aerial plant, with no roots in the earth, like mistletoe, or it must
be adapted to use the weak level of the sun's rays struggling through to the
forest floor – or else it must be a saprophyte or a parasite, living on the
material photosynthesised by other plants, dead or alive (see Fig. 8.2): to go
with each of these possibilities, the tall tree, the climber, and so on, there
is a variety of appropriate forms, each having its own advantages and disadvantages, but all being good designs in their own way.

Not only does the extended surface, as the engineer would call it, of the
leaves serve to capture the sun's rays, it also collects food from the air, in
the form of carbon dioxide. From carbon dioxide and water, using the high-
grade heat energy of the sun, it synthesises the carbohydrates from which
it mostly builds itself. The gathering of this essential food requires a large
surface because the concentration of carbon dioxide in the atmosphere is
only about 0.03%, or much less where plants are closely spaced. It also
gathers oxygen, a relatively simple task since oxygen is present in the air
in about 500 times as high a proportion as carbon dioxide. As with animals,
the life and growth of plants requires a continuous supply of energy, most
of which they derive from the combination of oxygen with fuel in the form
of carbohydrates (as with animals, there are stand-by anaerobic energy-
producing processes, too).

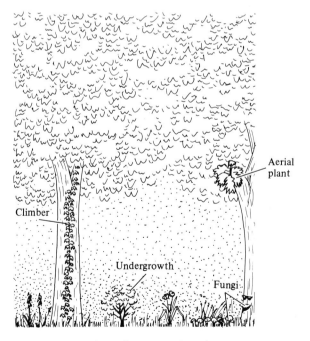

Fig. 8.2. Light and structure in a forest.

8.2.2 A tree regarded as an engineering design

Trees are remarkable designs, and it is interesting to see the extent to which they fit the categories used by the engineering designer in his work.

The chief divisions within engineering science, the fundamental areas of knowledge upon which the creative thinking of the designer is based, are usually labelled after some such scheme as the following:

(a) Mechanics, which deals with the forces between and the movements of solid bodies, part of which is covered in Chapter 4
(b) Structures, which we have seen in Chapter 5
(c) Fluid mechanics, which deals with the flow of fluids
(d) Heat and mass transfer, the diffusion of heat and mass through systems
(e) Materials, which we have seen in Chapter 3
(f) Thermodynamics, which we touched on in Chapter 2
(g) Systems and control (Chapter 6)
(h) Electricity and electronics

Trees offer elegant object lessons in design in all these areas except the first and the last. The leaf system is subtly designed to gather radiation and carbon dioxide, the last an exercise in mass transfer. However, the very system of millions of tiny pores (stomata) in the leaves which collect the carbon dioxide would lose water in dry weather at a rate which would soon kill the tree, were it not for one of the very many control systems it has, which closes them when necessary. Many more refined devices are also used in plants to prevent dehydration and help to account for the great variety of forms of leaf. Heat transfer is also important in plants, which have to keep cool in hot weather, and leaf forms are partly dictated by this consideration also. Mass transfer processes take place where the roots extract water and vital minerals from the soil, and throughout the tree itself.*

Fluid mechanics is perhaps not so impressively exhibited in a tree, but there is one very intriguing aspect which is still not thoroughly understood. Some trees grow to a height of about 120 metres. An engineer faced with the problem of moving sap from the roots to the leaves would be able to offer two solutions, broadly speaking; the first would be to use pumps in the roots capable of pumping to 120 metres; the second would be to provide a succession of pumps all the way up. In the second solution, the lowest pump could not be more than 10 metres up, because 10 metres is the maximum height to which water can be sucked. What we mean by sucking is creating a reduced pressure in, say, a drinking straw so that the atmospheric pressure outside forces liquid up, and so the maximum height to which

---

* Zimmermann, M. H. and Brown, C. L. (1971). *Trees: Structure and Function.* New York: Springer.

water can be sucked is that at which the pressure at the bottom of the water column due to the weight of the water itself is equal to that of the atmosphere, i.e., about 10 metres.

However, the tall tree adopts neither of these solutions: it appears to pull the water up from above, incredible as it may be, rather as a cotton thread might be pulled up through a tube from above. The free-hanging length (see Section 3.2.5) of water under these circumstances appears to be about 300 metres, which is not very grand, but the idea of a liquid in tension is a difficult one anyway. Under these conditions a very small bubble would cause the thread to break, since it would expand very rapidly to fill the tube, and then the liquid would fall away from below it.

Thermodynamically, the tree is a very elegant design. It uses a cascade pattern of chemical reactions driven by the sun's radiation to build up carbohydrates. This is at the other extreme to the crude irreversibility (see Chapter 2) of the solar heating panel. Because the surface of the sun is very hot, at about 7000 °C, the radiation from it is of high-grade and ideally convertible almost entirely into any other form of energy. Absorbing it in a solar panel at about 70 °C immediately reduces it to low-grade heat energy, of which only about 20%, even ideally, can be converted into other forms.

There are also great subtleties in the use of materials and in the structural aspects of trees. Their basic strength is provided by cellulose, a carbohydrate, helped out by substances called lignins, laid down originally in fibres lying along the walls of long thin cells in the layer immediately below the bark, the cambium, which is where growth takes place. This gives rise to the grainy, tubular structure of wood which is well-adapted to serve as a medium for the transmission of sap and is very strong in tension but is not so good in compression. This defect is partly overcome by the fibres being laid down in tension, throwing the middle of the trunk into compression. As it becomes more and more overlaid with new wood laid down in tension, the stress in any particular layer gradually decreases, until it moves at last into compression. In this way, at all stages of its growth, the trunk of a tree will be in tension at the outside, and in compression in the middle. The tree is most likely to break because of the wind bending it. In Chapter 5 we saw that the material on the inside of the bend, that is, the side away from the wind, will be thrown into compression by the bending, and it is here that collapse may occur (Fig. 8.3). But because of the stress built into the tree by its pattern of growth the outer layers of wood are everywhere in tension to start with. The effect is to reduce the compressive bending stress which is the dangerous one, and to increase the tensile bending stress, which the wood is much better able to resist.

This is very like what the engineer does in a prestressed concrete beam, except that there it is compression which the concrete can stand, and tension is the thing to be avoided. Also, the bending of the beam is usually in one

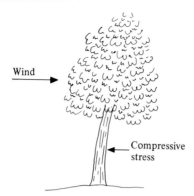

Fig. 8.3. Stress in tree trunk.

direction only, so that an asymmetrical and more economical system of prestressing can be used.

Thus the structural design of the trunk regarded as a beam is very good, but an engineer might be disposed to question whether the overall design concept is so sound, and put forward a proposal like that in Fig. 8.4. This design would unquestionably save a great deal of vital material, and the interesting questions of why this form of tree does not exist, and why it is unpleasing at first sight, probably more so than might be attributed to mere unfamiliarity, will be discussed later.

8.3   The form of animals

In plants, the overall form, pattern and texture are dictated by functional design considerations deriving from all the major non-electrical areas studied by the engineering designer except mechanics. In animals, mech-

Fig. 8.4. 'Structural' tree.

anics are added to all the others, and fluid mechanics, not very conspicuous in plants, becomes of first importance in birds and fish.

Simple mechanisms, like those of steam engines, have considerble charm. The constituent motions of rotation and reciprocation are dull in themselves, and it is their interrelation, their speed and the power they can develop that gives them their fascination. The movements of animals are much more complex – they have dozens of degrees of freedom to the engine's one – their rhythms are less smooth and insipid.

Both engine and animal are designed to reduce the mass of reciprocating parts, because they tend to shake the engine and they waste energy in both engine and animal. We have seen how this happens in the bird (Chapter 2) where flapping kinetic energy is lost, and how the insect overcomes this loss (the loss in the engine is of a different kind, and smaller). This leads to the use of elegant and graceful components for such purposes in both animal and engine. A particular effect is that in a limb mass at the end is four times as serious as mass half-way down, where its velocity is one-half and so its kinetic energy is only one-quarter, and mass close to the body is least serious of all. Also, a tendon is much thinner than the muscle it connects to the bone, so it is desirable to keep muscles close to the body and use tendons to carry the forces they exert to the bones they act on. This in turn leads to the charateristic tapered form of limbs, so marked in animals that run fast.

The more difficult and numerous the design problems presented by an animal are, the more complex it will be and the more elegant the solutions will appear. Thus a fish, being effectively weightless, or nearly so, can be almost any shape if it is not required to move fast: some fish have almost rigid bodies, with just a few fins to waft them around attached here and there. A swift fish, however, must have a narrow streamlined form and a flexible body composed mostly of propulsive muscle, arranged, as we saw in Chapter 6, so that every fibre can contribute fully.

The design problems presented by the bird are very exacting upon power/weight grounds alone, and only just within the scope of muscles to solve, as we saw in Chapter 2. The thinness of air, which causes these difficulties, also means that large changes in geometry are needed to produce the forces required in manoeuvres, e.g., braking forces on landing. A bird at rest has a smooth and compact outline which is completely transformed in fast flight, with the wings and tail spread and legs tucked up. When landing its form is transformed yet again, with wings arched and legs outspread and the tail bent down and forward and splayed to the utmost to create drag while maintaining lift at a low speed (Fig. 8.5). For a long time, aircraft designers were not driven to similar extremes because they commanded much higher power/weight ratios and their designs did not have to land in such confined spaces. However, increases in flying speed could not

Fig. 8.5. Bird alighting.

be met with equal increases in landing and take-off speeds, so that aircraft like Concorde or the swing-wing fighter display very considerable changes in form (Fig. 8.6).

Quadrupeds have different but very severe design problems. As we shall see in the next chapter, the wheel would present serious difficulties of production in nature, but it is also not at all clear if it would be useful in living organisms, which have to cope with an irregular terrain. The motion of a horse strikes most people as singularly beautiful, and yet it is no more than an elegant solution to a very practical design problem dictated by two powerful constraints, the need to conserve energy and the need to travel fast over rough ground without mishap. The first is apparent in the kinematics and the subtle rhythms of the animal's movement, but the second is not so obvious: it is an example of a very advanced control system, the eyes of the horse gathering information which is processed in the brain and modifies slightly the impulses sent to the muscles so the movement of the limbs is adjusted to fit the surface beneath the hoofs. Man has recently built rather simpler control systems of this type, for example, in hydrofoil craft which travel just above the surface of the sea, supported by short hydrofoils or wings of aerofoil form flying in the sea, so as to produce lift. In some

Fig. 8.6. Concorde.

such craft the surface of the sea just ahead is continuously measured and the angles of the hydrofoils slightly adjusted so that the hull travels smoothly and remains level in spite of waves.

However, the most refined mechanism produced by nature is man. The upright two-footed locomotion of the human being seems almost a recipe for disaster in itself, and demands a remarkable control to make it practicable. On top of this there is the sheer versatility of the human hand and arm and the astonishing feats of co-ordination and control displayed in playing some games or the piano.

8.4   Regular and organic curves

It has been remarked that the motions of the components in a steam engine are inherently rather dull or insipid, while the motions of animals are more complex and interesting.

Fig. 8.7. Regular and organic curves.

A motion is only a shape described in time, and in the same way very smooth or regular shapes, straight lines, circles, even ellipses and parabolas, are rather dull and insipid and produce an interesting effect only in patterns or where numbers are seen together. On the other hand, irregular curves can be much more interesting, and in living organisms it is usually irregular curves which are found.

A circle can be defined by one quantity, its radius. An ellipse can be defined by two, say, the lengths of its major and minor axes (see Fig. 8.7). An irregular curve such as the profile of a rather smooth-outlined leaf might require six or seven quantities to describe it tolerably satisfactorily, and a curve with even more character, like the line of an arm from point of shoulder to ball of thumb, might take 20 or more, and still the shape drawn to that specification might not pass the critical eye of a skilled draughtsman. Put simply, there is much more in the irregular curve, much more capacity to convey either information or feeling.

The engineer makes simple shapes most of the time: his machine-tools generate planes and cylinders which will suffice to define most of his components, and there is little incentive to use less regular shapes. From a functional point of view, however, the ideal shapes are often much less regular. In very large or very critical constructions, the shapes often look much freer, because there the saving resulting from using the ideal, less regular form is greater than the extra cost involved in making it. As an example, consider the dam profiles shown in Fig. 8.8. These are vertical sections through arch dams and it will be seen they approach that degree of purposeful irregularity characteristic of organic forms. The extreme variety

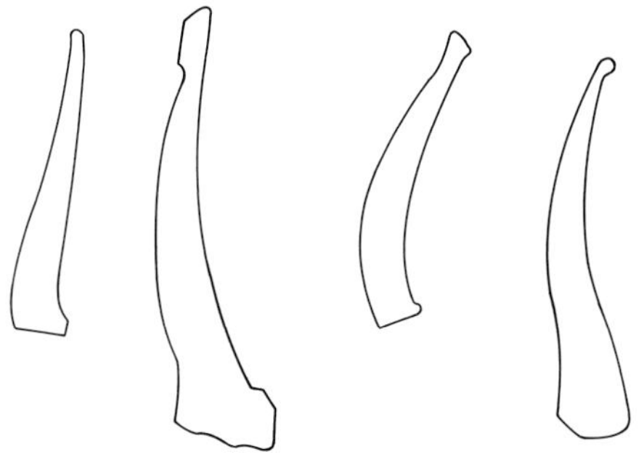

Fig. 8.8. Dam profiles.

is partly attributable to variations in the design problem, partly to variations in the design philosophy: at the time most of these examples were built, several schools of dam design flourished.

The word 'regular' will do very well for most of the forms made by engineers, but the less regular but, nevertheless, far from random shapes of living organisms need some less wide description than irregular. For all the importance of the subject and the great amount of writing on art and aesthetics, there does not seem to be a widely-accepted term for such curves, and perhaps 'organic' is as good as any.

Form in design in nature and by man is dictated by function and modified by production methods. The shapes of living organisms have been adapted to their function by evolutionary pressures, and thus might be expected to be 'optimised' or perfected. It may be that some changes that would be improvements lie outside the range that evolution can span, as we shall see in the next chapter, but within that range we might expect the design to be the best possible, and we usually assume that the production process does not limit the forms that can be produced.

The forms drawn by the engineering designer, however, are strongly influenced by the production methods to be used. Only where function and shape are intimately connected, as in the outer surfaces of ships and aircraft, will organic forms appear, and even those will be slightly compromised in the interests of production. For example, the middle length of a merchant ship generally has a section rather like Fig. 8.9, consisting of straight lines joined by arcs of circles at the corners (or, more expertly, the 'turn of the bilge'). This shape combines a large hold capacity of convenient form for stowing cargo, which tends to come in brick-shaped pieces, with low resistance to motion. It also enables flat plates to be used, except for a few at the turn of the bilge, and the stiffening ribs need only be bent over a short portion of their length, and there to a fixed radius. The more 'organic'

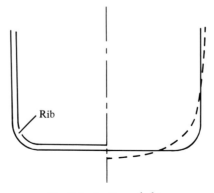

Fig. 8.9. Section of ship.

section shown dotted on the right-hand side would probably produce very slightly less drag, but would require all the plates to be bent, and the ribs to be bent throughout their length to a radius varying from point to point.

There are also many cases where the function requires regular forms: a ball bearing with egg-shaped balls and non-circular tracks would not be very satisfactory. Nature uses regular shapes whenever they are the most suitable: the stems of grasses and bamboos approximate closely to cylinders, and eggs are sensibly circular in cross-section. Often the departure from a regular form – say, the imperfect sphericity of a pea – may seem purely fortuitous and of no great consequence one way or the other. But sometimes, as with the departures from a sphere of the ball-joint surfaces in hip and shoulder, we understand, to some extent at least, that the differences are important and represent improvements over the simple regular forms.

8.5 Function and beauty

Whatever the nature or purpose of the aesthetic sense that humans undoubtedly possess, its primary nourishment has come, and still comes, from the observation of nature, and mostly living or 'designed' nature at that. So far as we know, every manifestation of evolution is strictly functional, and this view is continually being reinforced with the growth of understanding, as in the case of the structure of fish muscles discussed in Chapter 6. Thus whether we believe that beauty lies in the eye of the beholder or that there is absolute beauty, or even both, we must accept that function has an important role, if not a dominant one, because it furnishes us with much of the stock of images which we invest with, or in which are inherent our aesthetic values. The chief other source of such images lies also in the designed world, that of objects designed by man, but these also have been derived largely either from nature or from functional considerations.

One way of tracing this dependence of aesthetic values on origins in function is to consider some of the terms commonly used in the discussion of works of art, such as form, composition, balance, symmetry, unity, rhythm, pattern, texture, etc. All these have strong roots in function. Form has already been considered.

Discussions of composition in paintings, say, often refer to ideas of a functional nature, such as the stability of a triangular grouping with a side of the triangle at the bottom (see Fig. 8.10). It is interesting to speculate whether this association with stability is purely statical, deriving from the difficulty of overturning a pyramidal object with its large base and low centre of gravity, or whether it owes something also to the elementary structural principle of the triangle. However that may be, when critics talk

Fig. 8.10. Triangular composition (based on Leonardo's Mary, St Anne *and the Infant Jesus).*

of the stability and strength of such compositions they invoke either a conscious or a subconscious appreciation of the functional virtues of such arrangements.

Balance (and lack of balance) are more clearly functional in their origins. As particular aspects of composition, they carry overtones of movement or the potential for movement, as well as statical values. Symmetry is very nearly confined to the designed world, only the discs of sun and moon, the rings of rainbows and rare crystals possessing this appearance in the inanimate world known to early man. It is interesting to reflect on the contrasting preponderance of symmetries in the designed world, and some of the few exceptions to it.

Nearly all locomotive systems are symmetrical, but it is interesting that three lopsided kinds of boat (at least) have been developed. One is the well-known Venetian gondola, which is built with one side much deeper and less strongly curved than the other, so that it may be worked efficiently with the single oar on one side. Another is the traditional crooked stern junk of Fou Chou for which no reason is known, if we disallow the local tale that none of its inhabitants could see straight, so sunk in depravity were they!*

Finally, the proa of the Ladrone Islands was asymmetrical about its length but symmetrical about a tranverse axis. To change from one tack to the other, it changed bow to stern like a tram, and went the other way, so

---

* I have since come across an explanation of this crooked stern, which is that it enables an extra steering oar to be mounted to assist in descending dangerous rapids, a regular requirement for these vessels. Worcester, G. R. C., *Sail and Sweep in China*, HMSO, 1966.

keeping its single outrigger on the leeward side to prevent capsizing (Fig. 8.11).

Some crustaceans have asymmetrical claws, adapted to different purposes, a sensible functional arrangement. Also some internal organs are asymmetrically placed, and the shells of molluscs are often asymmetrical. But, on the whole, the animal kingdom is symmetrical.

Unity is a more difficult concept, and may have more to do with design for production than with any other function (if we consider 'producibility' to be a function, which seems a useful generalisation). Thus the hypothetical stayed tree of Fig. 8.4 lacks unity because the stays are not in keeping with the rest of the structure and not suited to the production method of plant growth. An impression of unity may also spring from a consistent design philosophy, as in most living organisms (indeed, perhaps in all living organisms viewed by a sufficiently knowledgeable and sensitive critic).

8.5.1   Unity and chimeras

A 'chimera' meant a creature with a lion's head, a goat's body and a serpent's tail, but the use has since been generalised to any hybrid organism of the kind. In *The Revolt of the Angels*, Anatole France made the angel Arcade speak somewhat as follows of such fancies:

> You could also say that a being with both arms and wings is a monster and belongs to teratology. In Paradise we have cherubins or keroubs in the shape of winged bulls; but they are the clumsy invention of a God who is no artist. It is true, however, that the Victories of the temple of Athene Nike, on the Acropolis of Athens, are beautiful with their arms and wings; it is true that the Victory of Brescia is beautiful, with arms extended and long wings falling on powerful loins. It is one of the miracles of the Greek genius, to have understood how to create harmonious monsters.

He does not mention that breathtaking fragment of honey-coloured stone in the Louvre (Fig. 8.12), which is perhaps the more convincing because it

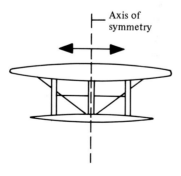

Fig. 8.11. Proa.

Fig. 8.12. Winged Victory.

has lost its arms, the *Winged Victory of Samothrace*. We can appreciate the mastery with which the Greeks scarfed together incongruous parts to make a congruous whole, even though we know what the Greeks did not, that no such tiny wings could ever lift a human body by muscular action. But a plausible simulacrum of a functioning system is there, and it achieves beauty. It may be said that the statues are meant to represent gods, who are not subject to physical laws, but that argument lacks weight because the rest of the body, wings apart, has a proportion and form strictly adapted to functioning in the mammalian flesh on the planet Earth. A part which is too obviously not of this kind will destroy the unity of the whole.

It is an old joke that a camel is a horse designed by a committee, a joke which does grave injustice to a splendid creature and altogether too much honour to the creative power of committees. For a camel is no chimera, no odd collection of bits, but an elegant design of the tightest unity. So far as we can judge, every part is contrived to suit the difficult role of the whole, a large herbivorous animal to live in harsh climates with much soft going, sparse vegetation and very sparse water. The specification for a camel, if it were even written down, would be a tough one in terms of range, fuel economy and adaptation to difficult terrains and extreme temperatures, and we must not be surprised that the design that meets it appears extreme. Nevertheless, every feature of the camel is of a piece: the large feet to diffuse load, the knobbly knees that derive from some of the design principles of Chapter 7, the hump for storing food and the characteristic profile of the lips have a congruity that derives from function and invests the whole creation with a feeling of style and a certain bizarre elegance, borne out by the beautiful rhythms of its action in a gallop (Fig. 8.13).

Fig. 8.13. Camel.

In contrast with the camel, the monsters of legend and heraldry are for the most part clearly the product of essentially unimaginative minds, mere crude and inconsistent assemblages of known components, 'undesigns' of the feeblest sort, of the same kind as those creatures dismissed in the past, by harsh critics of the hack science fiction they populated, as 'bug-eyed monsters'.

8.5.2 Pattern, rhythm and texture

Ideas of pattern derive from the natural world, and most abundantly and richly from the living world – waves, snowflakes, leaves, scales, feathers, branches, flowers – and are present in the earliest technologies – flint tools, weaving and plaiting, brickwork. Rhythms derive from animal rhythms – heart beat, walking and running, the complicated gaits of quadrupeds (the walk, trot, canter and gallop of a horse all have different patterns, with gradations between as well). The concept of texture comes to us heavily loaded with functional associations (including those of production) – stone, brick, fur, the smooth lining and rough exterior of shells, cloth, glossy and matt leaf surfaces, the elastic smoothness of the skin. All these properties have their place in human experience because of their functional virtues (where function includes producibility): it follows that the aesthetic qualities attributed to them will be related to function.

8.5.3 Aesthetics and the experience of design

Most of the terms used to describe the properties of works of art have strong roots in, or associations with, functional design. This might lead us to expect some sort of correspondence between beauty and function, particularly since the strictly functional designs of the living world are so often said to be beautiful. Of course, the stems and branches which grow from roots are not the same as the roots themselves, and even if the origins of our aesthetic sense lie in the appreciation of purely functional qualities, it does not have to stop there, as is shown by the existence of non-functional objects of purely aesthetic intention. But there are certain very interesting questions to ask. In particular, is there any truth in any of the following set of four propositions (not always recognised to be all different):

- Proposition 1: If it looks right, it is right
- Proposition 2: If it is right, it will look pleasing
- Proposition 3: If it looks wrong, it is wrong
- Proposition 4: If it is wrong, it will look uncouth

The odd propositions bear on the extent to which our aesthetic (or some related, rather subjective) sense may help with functional design. The even propositions concern the extent to which functional design may take care of aesthetic considerations.

I believe that there is a great deal of truth in all four, subject to certain qualifications which reduce their usefulness rather substantially. For example, the first proposition is useful in helping us judge the functional quality of a new design, say, of a boat or of a hand-tool, only when our aesthetic sense, or rather, the particular aspect of it that we exercise in looking at a boat or a hand-tool, is sufficiently well-educated in the first place. Suppose we take the water-nut pliers of Fig. 8.14 as an example. The proportions are everywhere well-suited to the function: both the jaws and the handles, which all act as cantilevers, are ideally tapered for that role; the flattened portions adjacent to the pivot, where the bending moment is a maximum and the section is weakened, are made very deep, and the thickest portions are the jaws where they grip the softer material of the nut, to increase the torque which can be applied without causing damage. The shapes are 'organic', the proportions and balance are palpably excellent, the textures and material are appropriate and pleasing. They look right and feel right before ever they are used, and that is a view which has often been expressed in this particular case.

Hand-tools, however, are a field in which many of us have a well-educated aesthetic sense. Although most people might find the scythe of Fig. 1.2 beautiful, how many would really appreciate its function, how many would simply 'feel that it looked right' and how many would merely be conforming to the prejudice that all ancient implements are beautiful? The scythe is really rather a strange device, a sort of camel of the tool world, a thought improbable and bizarre. It is almost easier to believe it is function-ally efficient because it is so strange, strange in a convincing way, just as the camel is convincing where the chimera is not.

Fig. 8.14. Water-nut pliers.

The case of a boat is of yet another kind, where even a well-educated eye can be deceived into thinking that a hull looks right when it is not. Nor is Proposition 3 of sure application here, for it has been recorded that a new and successful form has been scorned by some of the experts when first introduced. Nevertheless, it is probably true that good lines in a hull look good, and bad ones look bad, to the majority of educated eyes, most of the time.

When a successful new departure is made in such a field, it is not long before the educated eye has added to its education, and the new shape is seen as beautiful. To what extent is this an enlargement of the aesthetic taste and to what extent is it simply the justification of technical hindsight? Almost certainly, both are involved. Also, I believe that it is difficult to separate the two effects, not just because they are both subjective, *but because they are so nearly the same thing anyway*. The difficulty of distinguishing between an aesthetic and an intuitive technical judgement must surely be known to all enthusiasts for cars, aircraft, horses and musical instruments.

In sum, I offer the tentative conclusion that Proposition 1 is generally true, but is likely to fail in the most crucial cases, so that it is a help provided too much reliance is not placed on it, while Proposition 3 may be just slightly more reliable.

Turning to the even propositions, hardly anyone is likely to agree with No. 2 (if it is right it will look pleasing). They might quote against it gas holders, electricity pylons and truss bridges, all of which are sound functional designs which few find aesthetically pleasing. But the first two have unloved associations, so that any gas holder or any pylon is likely to be found unpleasing, whatever its design. For a fair test, we have to restrict our enquiry to whether a good functional design of gas holder or pylon is pleasing as gas holders or pylons go. Now a gas holder is a very extreme structure, very large and subject to relatively low loadings. Moreover, it moves. The traditional Victorian design, with its spidery guiding columns and bracing wires enclosing the great featureless smooth cylinders of the holder proper, seems to me to have considerable beauty, once set aside the prejudices excited by its associations. The variety of heights of the holder and the shadows of the tracery of guides on its curved surface add interest to its appearance (Fig. 8.15).

Truss bridges (Fig. 8.16) are not generally attractive, but neither are they usually very well-designed: this reflects to a certain extent the principle that it is uneconomical to design very well anything of which the total value is small, that is, where the product of individual cost and number to be made is small. Commonly, only one bridge would be built to a particular design, and the value would not be great. In particular, detail is treated in

Fig. 8.15. Gas holder.

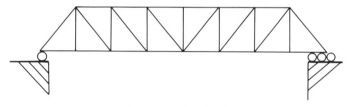

Fig. 8.16. Truss bridge.

the most expeditious way, and the appearance of any frame structure tends to be sensitive to detail. But very large frame structures lovingly executed, such as the Forth rail bridge, or very complex frames made from a limited number of standard components, like certain space-frame roofs, can be very attractive (Fig. 8.17).

Because large numbers of a single pattern of electricity pylon are used, the design can be very carefully done. Nevertheless, I feel they are not as pleasing as they might be, setting aside the prejudice they raise, and so they must count as a failure of Proposition 2. On the whole, though, I believe that it is difficult not to see beauty in functional excellence if you understand it, and even more difficult to see beauty in what you recognise to be functionally unsound (Proposition 4). The whole question of the link between beauty and function rests on the level of understanding of the beholder, and the extent to which an intuitive recognition of rightness for function is possible.

Fig. 8.17. Space-frame roof.

8.5.4 Insight and intuition

The Concise Oxford Dictionary gives as the meaning of intuition 'immediate apprehension by the mind without reasoning; immediate apprehension by sense; immediate insight', which will do very well for its use here. For example, the appropriateness of the tapered form of the handles of the water-nut pliers can be gasped without reasoning, although it is a simple matter to produce a reasoned argument to back up the intuitive view. The combination of suspension bridge and arch used in the Saltash bridge (Fig. 5.12) can be understood easily by people who do not understand a triangle of forces, because their experience has given them an intuitive understanding of the way in which the loaded chain pulls inward and the loaded arch pushes outward. The word 'insight' means much the same as intuition, except that a pedant might argue that since intuition derives from Latin roots meaning 'untaught', what is based on what has been taught (other than by experience) can only be insight, and not intuition. The important aspect of both is the immediacy of understanding, without the need to go through a series of steps of reasoning, often without the ability to perform those steps. Ideally, the designer should have a well-developed insight into all the aspects of the field in which he is working, based upon a thorough aprpeciation of the science.

What should we say of those critics who possess no understanding of function but offer opinions on functional design? For example, the reader of this book who has been patient with the complexities of Chapter 5 will know that a prestressed concrete bridge, such as that in Fig. 5.36, is much more difficult to understand than the Saltash bridge of Fig. 5.12, and, above all, that vital elements of its structure, the tendons and the hinges, are either totally invisible or as inconspicuous as a snake's ears. Nevertheless, an art critic has expressed the view that such a bridge has a 'visual meaning' and the Saltash bridge, because its 'structural principles are not clearly expressed', does not.

The appreciation of the beauty or otherwise of functional designs must be strongly influenced by the degree of intuition or insight possessed by the viewer: if we believe beauty is an entirely subjective phenomenon then we must accept that beauty is in the eye of the beholder and depends, among other things, on the education of that eye. Once we admit that that education may contain objective principles, as it certainly does in the case of functional design, then we are well on the way to admitting an objective content in aesthetics.

The case of the Waterloo road bridge over the Thames in London (Fig. 8.18) illustrates this question. Greatly praised when first built, it is rarely mentioned now. It is a set of beams made to look like a series of arches from the outside by a non-load-bearing masonry cladding, and rather like a

Fig. 8.18. Waterloo Bridge.

padded bosom, its external appearance is a fraud unrelated to the underlying structure. It is tempting to those of us who find it an insipid design to think that this is not simply because it is a fraud but because it is an unconvincing fraud. A true arch bridge would be less flat underneath, with deeper haunches, a point which does not require an education in engineering to recognise. We may be prepared to accept a fraud, but not a fraud so transparent as to insult our intelligence. It is tempting to believe in the underlying absolute values of functional beauty, making themselves felt through the educated eye of the engineer and the intuitive eye of the practical man in the face of the mannerisms of architects and artists. And perhaps these values do, after a generation or so, come to shape opinion at last.

An example may be found in the splendid roof for Paddington station, for which the ornamental details were done by the famous architect, Digby Wyatt, while Brunel himself designed the structure. Brunel wrote to Digby Wyatt in early 1851, 'Are you willing to enter upon the work professionally in the subordinate capacity (I put in the least attractive form at first) of my Assistant for the ornamental details'.* So writes the arrogant philistine engineer to the true artist, with much else about his own ability. Nothing but an aesthetic disaster could be expected from such a subordination of the architect to the engineer. But the great roof is admired today as it should be, for all that much of its original splendour is dimmed now, and the case of Digby Wyatt and Brunel is cited as a good example of co-operation between the two professions.

8.6 Style and the means of production

One important influence on the appearance of man-made products is the means of manufacture. In particular, there are *styles* which arise from different means associated often with particular products. Thus the style of eighteenth-century English furniture runs through nineteenth-century coaches, the early rolling-stock on the railways, early cars and even early aircraft, dictated not just, or even chiefly, by fashion but by traditional and appropriate methods of construction in strong, expensive hardwoods and a minimum of metal at points of load concentration and load transfer. These materials favour straight members of constant or gently-tapered rectangular cross-section. Decorations like stringing, fluting and moulding follow the lines of force, because it is so easy to produce these forms with the grain, which is also the direction in which the wood is strong. When furniture took to heavier and more bulbous forms, made by veneering soft-wood carcasses, this style was not followed in coachwork, with its requirements of supple strength and lightness. Where decoration was painted, and so free

---

* Rolt, L. T. C. (1957). *Isambard Kingdom Brunel*, pp. 231–2. London: Longmans.

from the constraints of grain, it still followed the same general pattern. Altogether, the themes of cabinet-work, coach, railway carriage and early car are straight beams jointed together and panelled in with substantially flat panels, with longitudinal decoration of the beams and peripheral decoration of the panels (Figs. 8.19, 8.20 and 8.21).

Modern cars are not built from beams and panels. Usually they are made from steel pressings welded together, and this gives rise to a quite different style. Large flat areas in pressings of thin sheet are weak, so that curved forms are commonly used, called 'shells' by the engineer, and having indeed much in common with sea-shells structurally.

Early aircraft belonged to the furniture and coachwork tradition, but modern ones relate, not to modern cars, but to ships. Wooden ships were

Fig. 8.19. Writing table.

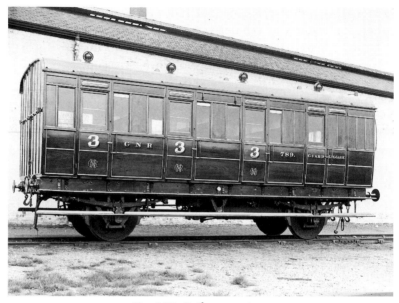

Fig. 8.20. Railway carriage.

Fig. 8.21. Vintage car.

built of transverse frames or ribs planked across, a style of construction which continued into iron and steel ships and finally into modern aircraft, although in both cases with a much increased emphasis on longitudinal members or stringers to take bending, as in the feather. But the means of production of aircraft nowadays have little bearing on their appearance, which would be much the same however they were made.

It is interesting to see how a good style evolves for each method of construction. Just as eighteenth-century furniture developed in beauty with the passage of time and with improvements in skill and technique, so in a much shorter time modern means such as the die-casting, the steel pressing and the plastic moulding have been given style, by the efforts of both industrial designers and engineers. This style often comes as much from functional considerations as from strictly aesthetic ones: there is perhaps a tendency to be too free in form with some of these processes, but rationalisation tends to remove the variety and produce a restraint and discipline which can often be pleasing, just as the beauty of the sonnet is heightened for most of us by the more or less strict form it takes, or a restricted but well-chosen range of colours on his palette may help to give character to the work of a moderate artist. This progress towards style is speeded by the industrial designer, whose different training and lesser preoccupation with function often enable him to work in a most fruitful partnership with his engineering colleague.

8.7 The aesthetics of mass-production

In the recent past it was often suggested that mass-production and beauty were in some way irreconcilable. In the face of the care put into the design for appearance of many mass-produced articles, this view is probably much less common than it was. Indeed, it is easy to see now that very large-scale production makes possible inexpensive goods with a very high finish and carefully-modelled forms – and of the highest quality, too, like the pick-up cartridge in Chapter 3. But because the costs of design and tooling of such products are high, they must be spread over a large number of articles to be acceptable.

Early mass-produced articles often had a large number of simple hand-finishing processes applied to them, and are very pleasing; some sewing machines from the first quarter of this century fall into this category. But many products were left rough, and the unimportant parts of castings, for example, were given little attention and were often crude in appearance: this was true of some parts of the Trojan engine described in Chapter 1, which did not have the elegance of form to match the elegance of its principles. Nowadays, however, it is usual for care to be taken with almost every detail of a design for mass-production. A pair of bolts may be prevented from working loose, locked, as an engineer would say, by a small piece of sheet metal with two projecting tabs bent up against the bolt-heads, called a tab-washer (see Fig. 8.22). Quite a crude shape of washer would be effective but in mass-production it is not merely no more expensive, but likely to prove cheaper, to form a carefully thought-out, neat shape. It is attention to details like these and the careful use of colours and finishes, for practical reasons as well as for appearance, which gives even strictly utilitarian products a beauty of their own. The engine and its surroundings in a new car present a neat complexity, attractively patterned by the tendrils of colour-coded electrical connections, black hoses with contrasting clips, thin pearly tubes for washer fluid, bright cadmium plate on pressings, contrastingly-painted castings, and so on. Electronic circuitry, too, often has a simple visual charm, with its variegated rows of resistors with their coloured rings and the Lilliputian detail of small components.

8.8   The aesthetics of functional design

So far, this chapter has been concerned with the visual aspects of aesthetics. But aesthetic experiences or impressions are not confined to those derived through the senses. We speak of an elegant proof in mathematics or an elegant theory in physics, and we are expressing the same general kind of admiration and acknowledging pleasant and satisfying feelings similar to those we derive from fine music or painting or architecture. Functional design can have the same kind of elegance, apart from any question of appearance: the Trojan engine, given as an example in Chapter 1, was indeed rather uncouth in execution and scarcely beautiful to look at; its aesthetic quality is confined to its elegance of principle so that an understanding of the engineering aspects is necessary before it can be appreciated. In Chapter 2, we saw the beautiful economy of the bow, whereby in the last stages of the discharge of an arrow the kinematics of the system feed the kinetic energy of the tips of the bow itself via the string into the arrow. The urban car of the future, with its energy store and regenerative braking, may well

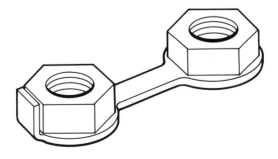

Fig. 8.22. Tab-washer.

do much the same, but it will lack the charm given the bow by its economy of means, the extreme simplicity of form combined with refinement of function.

8.9 The aesthetic basis of design

Besides the aesthetic properties of designed objects – and here as throughout most of this book, design means design for function – the very thought processes of the designer may hang upon decisions made on a basis which may best be descibed as aesthetic.

To understand this controversial statement, it is necessary to anticipate a later chapter a little. A large part of design consists in choosing from a very large number of possible ways of performing the various functions required of the thing designed, a single way for each function. The choice is complicated by the interactions of one function with another and by the many different ways in which the functions may be combined. Such a choice can rarely be made by an exhaustive analysis, although the attempt is sometimes made. What generally happens is that the designer chooses on a basis which is best described as one of aesthetic appeal: he thinks about the problem and the possible choices until some pivotal idea seizes his imagination, some particular way or ways of performing one or two of the more crucial functions which appear particularly satisfying or elegant, which have unlooked-for advantages or combine in a peculiarly helpful or apt or economical fashion. Then he tries to add all the other functions, to complete the outline of the whole solution; it may be he will come up against insuperable difficulties, and have to seek for another promising starting point, again, by the use of his aesthetic judgement. It is aesthetic in the sense that the emergence of the pivotal idea is accompanied by the same kind of intellectual pleasure as is excited by an inspired development in a great piece of music or a striking painting, and it may be recognised by that circumstance. It is a beautiful idea.

A very similar state of affairs occurs in mathematics. Poincaré pointed out that the possible combinations of mathematical relations are far too many to be studied, and that the idea of some entirely mechanical 'sieve' for selecting the fruitful ones is absurd. The sieve which does perform the task is the aesthetic sensibility of the mathematician, because, 'the useful combinations are precisely the most beautiful, I mean those that can most charm that special sensibility that all mathematicians know, but of which laymen are so ignorant that they are often tempted to smile at it'.

Another observation of Poincaré's will ring true to the engineering designer. He wrote

> When a sudden illumination invades the mathematician's mind, it most frequently happens that it does not mislead him. But it also happens sometimes,

as I have said, that it will not stand the test of verification. Well, it is to be observed almost always that this false idea, if it had been correct, would have flattered our natural instincts for mathematical elegance.

Most designers could quote cases where some idea turned out to have a fatal flaw but, nevertheless, was 'beautiful' or 'pretty' or 'elegant'.

The mathematician G. H. Hardy wrote an excellent little book (*A Mathematician's Apology*) to try to convey to laymen the aesthetic aspect of mathematics, for he recognised how difficult it was. He used just two examples. I have tried to convey the aesthetic appeal of functional design with two examples, the Trojan engine in Chapter 1 and the steam catapult in Chapter 10, but I recognise the task is nearly as difficult as Hardy's. Nevertheless, the role of the designer's special aesthetic sense is paramount in functional design.

9

Production, reproduction, evolution and design

9.1 Production methods

In both nature and the works of man, methods of production exercise a powerful influence on design. Consider first two manufactured articles – a piece of furniture and a glass bottle. The first is made from pieces of wood, cut to shape in a large number of operations, fitted and glued together. If it is made by a craftsman working from drawings, the time taken will be very long, but he can easily make one piece of furniture a little wider, and the next a little taller. The production process is slow, but very flexible. Partial automation could be introduced, so that a leg, say, is put into a machine and the mortise for a tenon is cut as quickly as the words can be read, but then both the location of the mortise and its size will be fixed. Thus the manufacture can be made much faster, and it will be cheaper if enough pieces are made, but flexibility is lost.

The glass bottle is made in relatively few operations. A gobblet of hot, soft glass is roughly shaped and blown up to fill a mould, which dictates the outside shape, even to a manufacturer's name in relief. In fact, the outside shape of the bottle is a replica of the inside of the mould, for which reason this production method will be called 'replicative'. The ultimate form of the piece of furniture, however, is nowhere to be seen in the tools which make it. The most piecemeal unautomated methods of manufacture, consisting of a large number of elementary operations, will be called 'knife-and-fork' methods as they often are in industry.

With some exceptions, knife-and-fork production is generally most economical when only one or two of a design are to be made; replicative production is usually cheapest when large numbers are required. However, particularly when the material is metal, the accuracy of cutting or other removal processes (most of which come in the knife-and-fork category) is much higher, so a combination of methods may be used, mainly replicative but with just a few crucial areas to be machined away to a precise dimension. Thus, for decades, car engine cylinder heads have been cast in metal which is left in the rough 'as cast' state except for the under surface which seals

253

the cylinders, the seats and guides for the valves and sparking plugs, and so on. Until recently, the inlet passages leading to the cylinders were also left as cast but now it is common to machine them also to improve the flow of the air and petrol mixture and to distribute it more evenly between cylinders.

For some jobs, knife-and-fork methods have achieved a new economy by the use of machine-tools under the control of a computer instead of a human operator, and some rather standardised types of part may be made very simply now. Also, by the use of clever computer programs, very 'organic' shapes can be made from relatively simple drawings which do not fully define the surfaces: the computer 'fairs' the rest of the surfaces into the defined parts (see Chapter 1 for the explanation of fairing in ship design). However, these programs are generally used to make, not the object itself, but tools or dies to be used in its manufacture by replicative processes. Thus the form of a bottle may be 'drawn' incompletely by the designer, using a computer. The computer then defines the shape more fully, pictures this complete form for the designer to improve and furnishes the instructions for another computer which will control the machine-tool which makes the moulds.

The important replicative methods are casting or moulding, and printing, which may be regarded as typical three- and two-dimensional processes, respectively, although methods based on printing and allied techniques are used very effectively to produce three-dimensional devices. Characteristically, these methods require a considerable initial investment in patterns, dies, blocks or negatives (the first cost) after which the additional cost to produce each item (the marginal cost, in the language of the economist) is relatively low. Thus the dies for a die-casting may cost £5000, but the additional cost of making one casting, the marginal cost, may be only £2. The total cost of making x items is then £$(5000 + 2x)$, which works out at £$(5000/x + 2)$ each. Figure 9.1 shows the average cost per item against x, the number made. To make a single item by knife-and-fork methods would be much cheaper than making the dies, which have to be more accurately

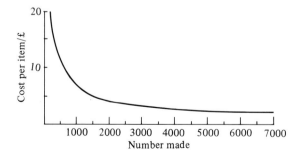

Fig. 9.1. *Cost per item/number made, replicative process.*

made, with a high finish, in tougher material, and must consist of pieces accurately articulated together so that they can be opened up to remove the casting.

9.2 Replicative methods: casting and moulding

The best known and probably the oldest replicative method of manufacture used by man is casting, the pouring of a liquid material into a cavity in the shape of the required article, where it is allowed to set. A familiar example is a jelly, which is in a liquid form when it is poured into a jelly-mould, in which it sets and from which it can then be turned out on to a plate. The cavity may be formed in sand or similar material which is broken up to extract the hardened casting (sand-casting), or it may be formed from re-usable pieces which can be separated (die-casting). In the case of sand-castings, the cavity is formed by moulding the sand round a 'pattern', a model of the final product, which is slightly bigger than is required to allow for shrinkage on setting. Split planes in the sand (which has additives to make it strong enough) enable the mould so formed to be separated to remove the pattern (see Fig. 9.2). Alternatively, the pattern may be a casting itself, in wax, which can be melted out of the mould (lost-wax process).

The material in which the casting is made may be a solid which is temporarily melted by heating (e.g., cast iron, aluminium, some plastics) or a liquid mixture of a resin (plastic) and a hardener, which sets by polymerisation when heated, or wet concrete, which hardens by the water entering into combination with cement. The material may be a ceramic slurry, a creamy mix of water, clay and other substances, which is made into a weak solid by the water being drawn out by the absorbent material of the mould. The weak solid is then fired in a furnace to partially fuse it into a hard solid; earthenware is often made in this way (slip-forming).

The material may not be liquid enough to flow under gravity, as in the case of the glass bottle, where air-pressure is used to stretch the toffee-like glass to fit the mould, which has no solid body (or 'core', technically speaking) to shape the inside surface, or with many plastics, which are forced into the mould under high pressure (injection moulding). Considerable pressure is often used in the die-casting of metal, because the cooled die tends to chill and so solidify the surface of the casting and prevent flow into corners, and pressure helps to ensure sound material free of holes. Sometimes, especially for shapes like pipes, the mould is rotated fast during casting. This not only forces the metal into the mould, but increases the tendency of bubbles of gas to move to the surface and escape, giving sounder castings (centrifugal casting).

Where the material is toffee-like, or 'plastic', rather than a liquid flowing under gravity, the process is generally called 'moulding'. An extreme case,

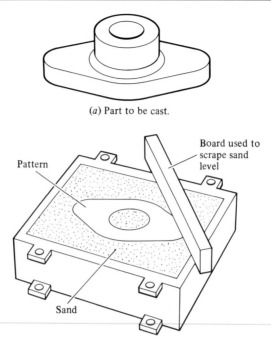

(a) Part to be cast.

(b) Pattern buried in sand.

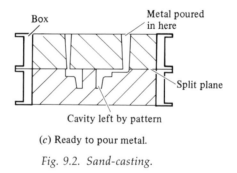

(c) Ready to pour metal.

Fig. 9.2. Sand-casting.

where the material is like a very stiff toffee, is drop-forging, where hot metal lying on one die carrying part of the final shape, say, a spanner, is struck very hard by the die bearing the other half, which is attached to the face of a drop-hammer or forging press. By the use of very high pressure, metals like mild steel can even be moulded cold ('cold-forming').

9.2.1 Replicative processes: blanking, pressing and drawing

These are processes applied to material in sheet form. In blanking, shapes are cut out of the sheet rather as a pastry cutter is used to cut shapes from rolled-out dough. In pressing, a sheet is pressed down all over a die to bend it to the required shape, as in forming the larger pieces of a car-body. Drawing is a rather more extreme form of pressing, requiring more plastic flow in the material, as, for example, when a saucepan is made by pushing a sheet through a hole by means of a die shaped like the inside of the saucepan (see Fig. 9.3). Many of these processes are carried out cold.

9.2.2 Planar replicative processes: printing, photography and microelectronics

Two of the most profound influences of engineering on human history have depended on replicative processes of production in which the thing produced is simply a pattern in a plane – like a picture, or a printed page or a photograph. The first was the introduction of printed books in the middle of the fifteenth century, and the second was the production of small cheap powerful computers in the second half of this one.

Consider etching. A copper plate is coated with a thin film of wax. The artist uses a pointed tool to cut a drawing in the wax, stripping it away along the lines he has drawn to bare the copper. The plate is immersed in acid, which attacks and dissolves part of the copper where it is exposed, but is resisted by the wax 'resist', so that the copper beneath is left unaffected. The plate is carefully cleaned of acid and wax and covered with ink. The ink is then wiped off the surface of the plate, but remains in the grooves cut by the acid. A sheet is paper is placed on the plate, and paper and plate are squeezed together with great force, so that the ink in the grooves is transferred to the paper and a print has been made. What is important, however, is that the plate can be used again and again, to make a series of prints all the same, until it wears out.

There are other printing processes in which the ink does not penetrate the hollows and sticks on the top surfaces, as in the rubber stamp, the linocut and the woodcut (as against the wood engraving).

The most important application of these kinds of 'planar replicative' processes has been the printed word. Consider a newspaper, perhaps man's nearest approach yet to the extravagant production processes of nature. Feeble as it is by comparison, it is, nevertheless, impressive; a million copies a day and a million characters per copy, perhaps, and all more or less correct and with a message, however trivial, for someone.

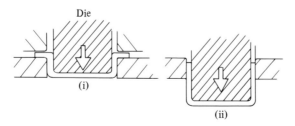

Fig. 9.3. Drawing a saucepan body from sheet.

Photography has led to some very important planar replicative processes. Using a wafer of silicon instead of a copper plate, replace the wax in the etching process by a photographic emulsion and then project an image on it. Now develop the emulsion by treating it with a chemical which removes either the part which has been exposed to light or the part which has not (according to the nature of the original emulsion or 'photoresist'). We can now etch away the bare material, or change its chemical nature, for instance, by oxidation from silicon to the oxide of silicon, or add impurities by allowing them to diffuse in from a hot gas, the foreign atoms penetrating the surface and slghtly alloying or 'doping' it. We can lay down a metal layer on the wafer by condensing the metal vapour on it, and other kinds of layer can be produced by chemical deposition from vapour.

We can etch away these superimposed layers in different patterns imposed by photographic means, and so build up simple three-dimensional structures, of silicon with various properties (according to its doping) of the oxide or nitride of silicon, and of metal; these simple structures are the individual components of the 'microchip', and one slice may incorporate thousands of them.

Figure 9.4 shows a few stages in the making of a very simple device of the kind. First, the slice of silicon is oxidised over its entire surface, and then a film of photoresist is laid down on it. A mask with transparent and opaque patches in the required pattern is clamped on top and the whole is exposed to light (Fig. 9.4(a)). In the subsequent development, the resist is removed where it has not been exposed to light.

The oxide is then etched away except where the resist protects it, exposing the underlying silicon wherever the first mask was opaque. The wafer is then heated in a gas containing boron, which diffuses into the silicon to modify its properties locally. Figure 9.4(b) shows the wafer at this stage.

There follow three rounds of processing, each in this sequence;

- add layer all over (oxide, oxide, metal)
- apply resist
- apply mask (second, third and fourth)
- expose to light
- etch away latest layer to expose the one below (silicon, silicon, oxide)
- remove resist

Figures 9.4(c)–(e) show the cross-section at the etching stage of each of these cycles. It will be noted that of the metal patches in the final stage, two reach through to the silicon but the central one does not, because all of them correspond to opaque patches in the second mask but only the outside ones correspond to opaque patches in the third mask.

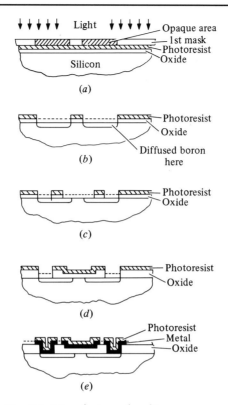

Fig. 9.4. *Manufacture of a chip.*

Yet further processes are needed, and a fifth mask, to provide another partial layer of metal to connect the patches together, but what we have in this part of the chip now is a MOSFET or metal–oxide–silicon (or semi-conductor) field-effect transistor. The manufacture usually involves many more processes and some devices may require 15 masks.

In the manufacture of a very large-scale integration (VLSI) chip thousands of such devices, in several sorts, are made simultaneously on one wafer, together with the 'wiring' to connect them. These chips are remarkably complex and yet cheap, so that in them you may be getting hundreds of transistors for a penny.

Two things are particularly worthy of note about the chip: firstly, that such a thing can be designed at all, and secondly, that once it is designed, it can be made.

The very complicated task of design is made possible by the very thing of which the chip will often form part, the computer. Without it, it would be almost impracticable to undertake the design and the preparation of the masks, with their masses of detail, all of which must be right. But that has been part of the pattern of engineering for centuries; good tools today make better parts for better tools tomorrow. Furthermore, the computer provides the means to test the design (by simulating its behaviour) and to test the product when it has been made.

The manufacture of VLSI chips is another triumph of design and technology which stretches to the utmost the resources of optical and mechanical engineering. First, the structures are so small in some cases that light can no longer be regarded as fine-grained and travelling only in straight lines – it diffracts round the edges of the images in the masks to give fuzzy effects, so that ultra-violet light is used for its short wavelength. Just as these same limitations of light in microscopy led to the use of electrons in its place, so electron beams are now often used in chip manufacture instead of light.

Consider now the problem of register. The chip surface has to have projected on it several very finely-detailed images, one to each mask, and all these details must align correctly between one image and the next; they must register, as the printer would say. In colour printing, where four superimposed images in different colours are laid down one over another, any imperfections of register give a fuzzy mess. In the manufacture of chips this problem of register is made much harder by the fineness of the detail and the processes in between the application of one mask and the next, which may involve heating the wafer to 1000 °C. Under such heating, for example, 'doped' areas may expand differentially, bending the wafer, which does not return completely to its original shape.

For this and other reasons it is common to build up the image from a number of separate images, by a 'step and repeat' method. To do this, the chip must be moved and aligned very accurately many times, so that to keep up the throughput of chips the movements must be rapid. The two requirements together, extreme precision and rapid stepping, constitute a very difficult specification for the designer to meet.

The manufacture of VLSI chips is a triumph of mechanical and other kinds of engineering. With thousands of individual sub-systems in each chip, failure in any one of which renders the device useless, it is remarkable that so many of them work. Once made, they are much more reliable than most engineering products, and they enable us to contemplate with confidence such achievements as a crane which will pluck a hovering aircraft from the sky.

9.3 Some other interesting manufacturing methods

There are very many interesting manufacturing processes, and only a few will be described here. The justification for the great variety used lies not only in the different shapes made, but also in the different properties of the materials they are made from. Solids which behave plastically, like dough and most useful metals, are particularly easy to shape. For example, flaky

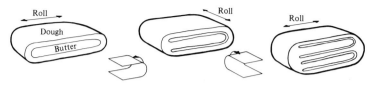

Fig. 9.5. Making flaky pastry.

pastry is made by wrapping a slab of butter in a sheet of dough (see Fig. 9.5). The whole is then rolled out flat, folded over (so there are now two layers of butter between three layers of dough), rolled out again, folded over, and so on. Each cycle doubles the number of layers of butter, which, like the layers of dough between, become thinner and thinner, the same process as was often used in making swords, where the 'butter' was the steel and the 'dough' was the oxidised scale on the outside (Chapter 3). In both cases, reactions between the layers during cooking are important, but with the steel, 'cooking' followed every operation of folding and hammering.

An other production process to be seen in the kitchen is *extrusion*, where a paste of water and icing sugar is squeezed from a bag or a syringe through a nozzle having in it a hole of attractive shape, say, a star (Fig. 9.6). This process is often used with metals or polymers (plastics) to make useful sections, e.g., for curtain rails. Figure 9.7 shows an aluminium alloy extrusion used in making a central heating radiator. The tubular centre carries hot water, and the lengths of extrusion run vertically side by side. The 'nozzle' is called a die, a rather over-worked word but one which usually denotes a tool carrying some elements of the shape of the product it makes, and to that extent replicative.

An interesting variant is *hydrostatic extrusion*, which is where the whole process is conducted at a high pressure, as if you were squeezing out icing or curtain rails at the bottom of a deep ocean. The advantage is that some materials which are not plastic enough to extrude into the air can be extruded from a very high-pressure region into a high-pressure region. This

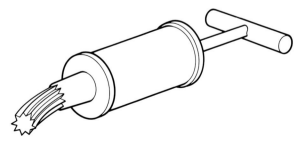

Fig. 9.6. Extrusion of icing.

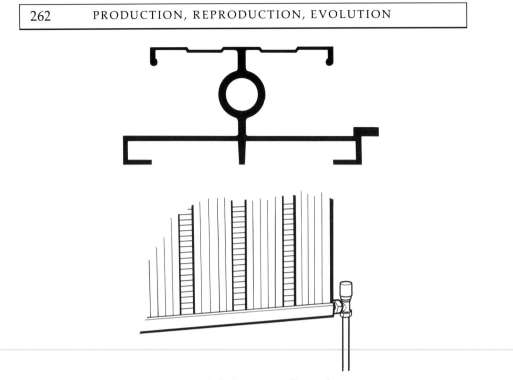

Fig. 9.7. Extruded aluminium alloy radiator.

is true, for example, of the rather brittle metals used in superconducting wires.

At very low temperatures the resistance of these wires to the passage of electricity disappears, offering the possibility of thin power cables of very high current-carrying capacity or electric generators and motors of smaller size and higher efficiency.

Another interesting process is that by which seaside rock is made, the kind which has the name of the resort all the way through (see Fig. 9.8). A short, very fat piece of rock is built up from red and white sweet (by knife-and-fork techniques!) and then rolled out to the required size, much as

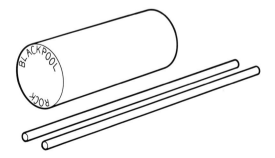

Fig. 9.8. Making seaside rock.

plasticine can be rolled between the hands to make a long thin cylinder. If the reduction in diameter is to one-twentieth, say, then since the volume of sweet remains unchanged, the length must increase $20 \times 20 = 400$ times. Besides rolling, the lengthening process could be done by extrusion through a die, or, alternatively, by pulling through a die (or rather, a series of dies, if large reduction is required: wire is made by this method). The extrusion of a composite cylinder like this is seen in those toothpastes which have coloured stripes in them.

The superconducting cables that may be used for the generators and motors of the future may be made on the seaside rock principle, by hydrostatic extrusion, with a large number of fine wires embedded in a larger wire of a different metal. But a more interesting example is the image intensifier shown in Fig. 9.9. This device is used to convert a faint optical image into a much brighter one: it consists of a disc pierced by about three million very small, but relatively long, holes. It starts as a short fat cylinder of glass surrounded by a layer of a different glass. It is drawn out into a long thin cylinder which is then cut into lengths which are formed into a bundle, which is fused together and in turn drawn out, so that now there is a long thin cylinder of the outer kind of glass containing many longitudinal filaments of the other kind. This is repeated several times, ending with a short fat cylinder with the required number of filaments, which is cut across the grain, into thin circular slices: the glass forming the filaments is then chemically leached away, the different nature of the matrix glass protecting it from attack.

An interesting process is that of explosive forming, where, for example, explosive charges of carefully-calculated size and distribution are used to press a sheet into a die. This method is sometimes used because no large

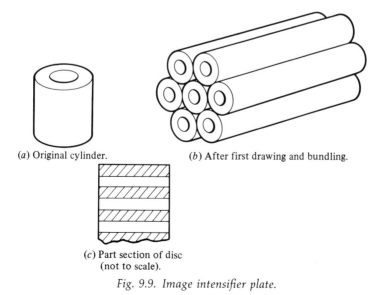

(a) Original cylinder.   (b) After first drawing and bundling.

(c) Part section of disc
(not to scale).

Fig. 9.9. Image intensifier plate.

and expensive machine is needed, because the die can be much cheaper and also because the material often forms much better in a very rapid process. In tubes or rings of metals which are good electrical conductors effects very like those of an explosion can sometimes be produced by electromagnetic means.

It would be possible to fill a large book with brief descriptions of manufacturing methods, so many are there, and another with methods of joining pieces together – the many kinds of welding, screws, nails, glue, bolts, studs, staples, sewing, and so on. For the purposes of this book, it is enough to know that this great variety exists, and that the choice of method depends in intimate and complicated ways on the materials used, the shapes to be made and the work which they are called on to do.

9.4   The relation between manufacture and design

Both because the cost of a product is often very largely the cost of manufacture, and because the feasibility of using a less expensive material often hinges on devising a production method to suit it, designers must study these aspects of their work with great attention. Usually the prototype of a design is made by knife-and-fork methods to avoid the costs of all the tooling, moulds, dies, patterns and special production machinery which will be needed eventually if the design is to go into full-scale manufacture. Nevertheless, the good designer will have carefully thought through all the means of manufacture to be used, in conjunction with the production engineers, and done his best to ensure that the final cost is low, in line with that celebrated definition of an engineer as one 'who can do for one dollar what any fool can do for two!'

9.5   Production in nature: reproduction and growth

The designs of engineers are recorded in drawings, reproduced in prints that used to be in white on blue and hence called blueprints. Other designers then make drawings for the jigs and tools required to make the parts, and craftsmen make the jigs, tools, patterns, templates, dies, and so forth.

In nature the role of blueprints, jigs and tools is assumed by remarkably complex molecules of substances called as a whole deoxyribonucleic acid (DNA) and ribonucleic acid (RNA). Each living cell contains the blueprint of the entire organism of which it forms part in the form of very long molecules of DNA, the famous double helixes. These helixes are composed, rather as letter-press is composed from type, from a sequence of four building blocks, the bases adenine, cytosine, guanine and thymine, which are conveniently represented by the letters A C G T, strung together with sugar and phosphate. Thus the formula for DNA may be largely represented by a sequence of these four letters, say

...A C C G T G...

giving the order of these bases as they occur along one of the helixes. The formula for the other helix is easily derived from that of the first, because opposite an adenine base there is almost always a thymine one, and opposite a guanine base there is almost always a cytosine one. Thus the short length of DNA given above would have the complete formula

...A C C G T G... (1st helix)

...T G G C A C... (2nd helix)

Now growth in multi-cellular organisms takes place by cell division, each cell of the resulting pair having its full complement of DNA, its complete blueprint of the complete product, the animal or plant. Thus division of one cell into two requires the duplication of the DNA, which happens by the two helixes separating and each building its own complementary helix from the available building blocks in the substance of the cell. Thus the first helix, the top line in our formula, would latch on to itself a thymine base opposite the first adenine base shown, a G base opposite each C base, and so on, building up its new complementary second helix, while its old second helix was building a new first helix. The end result is two identical sets of DNA, which separate to form two separate cells about them; the DNA has replicated itself.

In sexual reproduction, the fertilised egg is at first a single cell which has received half its DNA from one parent and half from the other. However, in the replication of DNA from one parent, errors may occur in the ordering of fairly large sections, called genes, or groups of such genes, as if in the printing of a book pages had been put in the wrong order. To pursue the analogy, sometimes pages or several pages in a run are left out, and sometimes pages or chapters are duplicated.

The DNA of the cells of plants and animals occupies bodies called chromosomes, which occur in pairs: thus man has 23 pairs of chromosomes. In the book analogy, a pair of chromosomes is a pair of books giving the instructions for the same parts of the organism. In the process of sexual reproduction, one book of each pair comes from the mother and one from the father. If the one from the mother calls for blue eyes and that from father for brown eyes, then the man (or woman) will have blue eyes, because blue eyes are what is called a 'dominant' trait: where genes for both blue and brown eyes are carried, blue eyes will develop.

In spite of the rapid progress of molecular genetics, we still know very little of how these remarkable great molecules are effective in assembling living substance from small molecules, according to patterns built into them, although we do know that RNA, another variety of long-coded molecule, is important and that the whole business hinges on processes in which one molecule, acting as a template, assembles another on itself. Still less do we understand how the DNA determines the development of the entire organism, although in many cases we can locate the actual pages in the particular pair of books that determine a particular feature, without yet following the mechanism by which the determination operates. How do the 'coded instructions' of the DNA (although to regard them strictly as instructions would clearly be wrong, since they represent the tools as well as the drawing), how do these 'instructions' present in all the cells ensure that two groups of them in the right place become arms, and that in these groups smaller groups become bone, nerve, muscle, and so forth, all with the right and very complex interrelationships? Most of this fascinating process still remains to be unravelled.

9.5.1   The genetic inheritance

The amount of DNA in every human cell is about 5000 million pairs of bases (i.e., AT or CG pairs). Reading along one helix through all the chromosomes we thus have about 5000 million characters in a four-letter alphabet; to convey the same amount of information in a 26-letter alphabet would require only about 2000 million characters, or, using the same unit as we used in Chapter 6 as a measure of information, about 400 bibles.

Now language is highly redundant, in the sense that many more characters are used than are absolutely necessary to convey the information contained. If we used every possible two-letter combination as a word, regardless of whether it could be pronounced or not, then, with a 26-letter alphabet we should have $26 \times 26 = 676$ 'words', such as PA, QX, and so on, which is enough for a very basic vocabulary. The potential number of unpronounceable three-letter words would be about 18 000, enough for a fairly rich vocabulary, whereas the average length of words is about five letters, or perhaps twice as long as is strictly necessary to convey a written message. Moreover, many words are redundant, too, and not only in political speeches. Just as redundancy in structures has its uses, so has redundancy in the transmission of information. Any mistake in a non-redundant message makes it incomprehensible, or, worse, misleading.

The genetic code appears to be redundant, to a greater extent than just having two sets of 'books' or chromosomes. Many sections of DNA are repeated elsewhere, and some parts have no apparent function. The value of redundancy in improving the reliability of systems has been discussed in Chapter 6, and it may have a similar role in the scheme of genetic design. If a particular gene governing some feature is missing or defective in the

DNA inheritance from one parent, it may be present in that from the other, and the organism may not be badly affected. A few characteristics are not inherited from both sides in the case of males, because 'maleness' is itself the result of the deficiency of a particular chromosome (no doubt some women readers have long suspected something of the kind). One 'chapter' of this kind is the one on colour vision, which daughters have from both their father and their mother; as long as one of these inheritances is sound, a daughter will have unimpaired colour vision, but if one is unsound, she will give half her sons and half her daughters the deficient or damaged gene. Her daughters will all have colour vision, provided her husband has, since they will inherit a sound 'chapter' from their father. But her sons inherit no chapter on colour vision at all, sound or unsound, from their father, and so half of them (on average) will have only their mother's defective gene and will be colour blind. The lucky sons who have drawn their mother's sound gene for colour vision will have normal sight.

Colour blindness is not often fatal in its effects, but the inability to produce a particular enzyme may well cause death, so that redundancy of a higher order than that provided by the double inheritance may have evolved in the past, just as high redundancy is provided for some of the most vital functions in aircraft.

The inheritance by an individual of genetic material from two parents is the function of sex, nature's first great invention, in hastening evolution. It provides for the continual reassortment of characteristics in a *particulate* way. If a few members of a tribe have kinked hair while all the others have straight, then their genetic legacy will appear, not as a slight kinkiness in everyone's hair some generations later, but a continuing small proportion of people with hair just as kinked.

9.5.2 Mutations

The sexual mechanism ensures that a great variety of combinations of the genetic variations of a kind of animal or plant, a species, will arise, and that favourable modifications can be transmitted: it does not explain how new characteristics can enter the genetic stock. This happens by mutation – accidents which occur to the genetic material. As an example, when the DNA is replicating, the right partner bases are tagged on to the bases in the separating helixes. However, right bases do not fit into the right places with quite the exclusivity with which keys are supposed to fit locks. For instance, adenine (A) normally pairs with thymine (T), but there is a very rare form of adenine which will pair with cytosine (C), which normally pairs with guanine (G). Thus, at the instant of replication it is possible for a C base to find one of these rare A bases to hand, and write it into the new strand where there should be a G. Here the analogy with the book breaks down, because a single wrong letter in the genetic code makes it 'read' entirely differently, a problem that rarely arises with written languages.*

Mutations, or spontaneous changes in the DNA, are much more commonly harmful, or even fatal, than advantageous to the indivual that possesses them. But the few favourable ones determine the direction of evolution of the species, under the driving force of natural selection.

9.6 Design in nature: evolution

In anthropomorphic terms, regarding nature as a designer, DNA is the working drawing (and also to a large extent the tooling); mutations are an entirely random way of introducing modifications, mainly small and mainly harmful; sexual reproduction provides for the trying of these blind changes, when they are not lethal too early, in various combinations; natural selection provides the process of evaluation which determines the direction (or directions) of evolution. This system of 'designing' is very powerful, but exceedingly slow, with 'lead times' of millions of years for major new departures, and very wasteful, with millions or billions of scrapped models. If life began without sex, as seems probable, if it all began with chance occurrences in some primordial and inorganic 'soup' full of the chemical building blocks of life, then it must have been slower still to begin with.

In sexual reproduction, evolution produced a way of speeding up evolution, probably by many times. Since then, there has probably been very little improvement until the advent of man.

It is difficult to accept that all the complexity of the living world can have evolved by natural selection, an automatic design process in which small variations are made in every generation, the favourable ones accumulating by reason of the higher survival rates they confer on the individuals possessing them. Some of the arguments have been dealt with elsewhere, to the small extent that is appropriate in a book whose proper concern with nature is with the design to be seen there, not with how that design has occurred. But one objection to the idea of functional considerations as a determinant of evolution ('survival of the fittest', or better, fitter) lies in those living things whch seem purely aesthetic in inspiration – butterflies, birds and flowers. Some readers may well have wanted to raise this objection in Chapter 8, but it seemed more appropriate to deal with it here. In many ways, flowers are the most striking and familiar example, and so I shall discuss them.

9.6.1 The pollination of flowers

Not being able to move about, the higher plants must depend on other agencies to assist in their sexual reproduction and in the spreading of their

* Stent, G. S. (1971). *Molecular Genetics*. San Francisco: Freeman.

seed. It is not surprising, therefore, that they spend a great deal of their precious living substance in overcoming the limitations of their fixed position. Many important plants, trees and grasses particularly, rely on the wind to carry their pollen to other members of the same species. But many rely on insects (or more rarely, birds) which they attract by means of conspicuous flowers and scents. Nevertheless, it is difficult at first encounter with the idea to believe that the great range of form, colour and markings seen in flowers has all evolved for such a simple end.

Firstly, the bright colours of flowers have evolved to attract the insects which will carry their pollen to others of the same species and bring them pollen to fertilise their ovules and produce seed. Some flowers that seem inconspicuous to us are strongly 'coloured' in ultra-violet light, which we cannot see but some insects can. Secondly, the markings of the flower are often honey-guides (as was recognised almost 200 years ago) that show where the nectar is that rewards the insect for its services. Thirdly, the form of the flower is adapted to the shape and actions of the particular animal that pollinates it, even to the extent of mimicking a female insect so as to attract the male. Fourthly, the form is adapted to avoid self-pollination, as in the design principle called heterostyle. The need in pollination is to transfer pollen from the anthers of one plant to the stigma of another. In the primrose, the nectar is at the bottom of the tubular central part of the flower, the corolla; the plants are of two kinds, pin-eyed, in which the stigma is near the entrance to the corolla and the anthers well down inside it, and thrum-eyed, in which the anthers are at the entrance and the stigma is well down inside (see Fig. 9.10). Thus the region of a bee's body which comes in contact with the stamens of a pin-eyed flower, and so picks up their pollen, is likely to touch the stigma of a thrum-eyed flower, and vice versa. Since each plant has only one kind of flower (this being coded in its DNA) pin-eyed plants will generally fertilise thrum-eyed, and vice versa. Because the

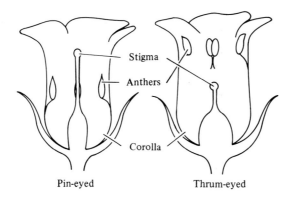

Fig. 9.10. Primrose.

thrum-eyed gene is dominant (i.e., plants with thrum-coding from one parent and pin-coding from the other will be thrum-eyed), there are roughly three times as many thrum-eyed as pin-eyed primroses. Some plants come in three kinds, with flowers whose geometry is such that one sort will fertilise either of the other two readily, but its own kind only rarely.

An interesting design for ensuring cross-pollination is found in the early purple orchid, a fairly common British plant. When a bee visits the flower, two small club-like bodies, called pollinia, become stuck to its head by a rapidly-setting gummy secretion, and stand out like two horns. After about half a minute, these 'horns' bend at the base through about 90°, so that they are directed forward instead of upward, and in this position they will deposit their pollen on the stigma of a flower. However, after a lapse of half a minute, the bee is unlikely to be still busy in the flowers of the same plant, but will have flown to another.

In passing, it is quite simple to demonstrate the action of this orchid flower by simulating the entry of a bee's head by inserting the point of a pencil. If the flower has not yet been visited, you will see the two pollinia attached to the pencil point, and in about 30 seconds they will change position (see Fig. 9.11).

There are very many other interesting functional aspects of the design of flowers which could be quoted, and perhaps even more that we do not understand. Certainly, there seems no need to invoke causes other than natural selection to explain the wealth of colours, markings and forms that wild flowers show.

9.6.2 Niches and nichewise evolution

An important idea in ecology, the science which deals with the mode of life of organisms and their relationship to their environment, is that of a *niche*, which might be called an opportunity for a life-form. For example, in the forest of Fig. 8.2, among the niches which exist is one for plant forms which

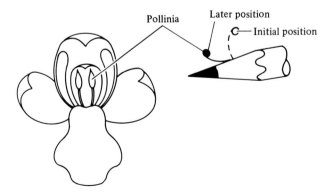

Fig. 9.11. Pollinia of the early purple orchid.

do not grow up from the soil, but establish themselves in the branches of the trees; to define this niche completely, we must identify in it all the needs of the organism – in the case of the plant, light, air, carbon dioxide, water, minerals, means of pollination, means of seed-dispersal, etc. For the aerial plant, water and seed-dispersal present special difficulties. Mistletoe gets its water and minerals from the tree it grows on by a system of root-like organs invading the branch – it is a parasite to that extent. Birds eat its 'berries', in which the seed is embedded in a sticky pulp (tenacious enough to have been used to catch birds). Some seeds become stuck to their beaks and the birds remove these encumbrances by scraping them off on the branches of trees.

In tropical rain forests it is possible for plants (epiphytes) to grow on trees on which they depend only for support, although they may need special provision to retain and store the water which comes their way. Some do this by forming from some of their leaves a tank or cistern. This tank itself provides another ecological niche, which frogs use for breeding and which is the only habitat of some waterweeds.

Just as 'Nature abhors a vacuum', so she seems to move in to fill all empty ecological niches, provided they last long enough for her very slow evolutionary design process to produce the appropriate organisms.

While progress may be difficult to define, most would agree that there is an evolutionary progress, from single-celled to multi-cellular organisms, from lower to higher, from fish to amphibian to reptile to mammal, and so on. But sometimes a line degenerates, and this is still evolution – an organism evolves to fill a niche in which it no longer needs all the apparatus of the life-style it came from, and it becomes simpler. For progressive evolution, there must be a series of niches demanding new features in the design. Recent studies of the descent of man have brought out its 'nichewise' nature very strongly. Thus, though we are no longer very good tree climbers (the undoubted talents of small boys notwithstanding), life in the trees was a vital stage in developing the characteristics that enabled us to reach our present state. The arboreal life, with its need to judge distance for jumps, demands binocular vision – two eyes looking ahead, with a large overlap in what they see, and a stereoscopic effect arising from the slight differences in the two images (binocular vision is also valuable to hunting animals, but their prey may be better off with the wider field of vision that goes with eyes on the sides of the head). For clinging to trees, sharp claws will do as squirrels demonstrate, but grasping hands and feet work too. Once the hands and feet are capable of grasping, claws can be abandoned in favour of flat nails (still needed to support the friction pads on the finger-tips) and greater versatility.

Life in the trees was also an essential preliminary to walking on the hind legs, a method of locomotion only fully developed in man. A creature with binocular vision has difficulty in looking behind unless it adopts an upright posture, which is very practical in the arboreal life. If it takes to the ground again, it will naturally adopt a largely two-legged form of locomotion.

Now the possibility and advantages of an erect terrestrial biped with binocular vision, a thumb capable of working against the fingers in a grasping action and a well-developed brain must have existed long before man evolved: the niche we humans occupied in primitive times was there long before we were, but none of the great wealth of terrestrial forms was able to evolve to fill it. It was filled only by this long circuitous route via life in the trees, which favoured some of the attributes required.

Imagine a man who stands facing a high hill which he has determined to climb, *moving always upwards*. It may be that this restriction he has laid upon himself will involve him in a very circuitous route, round the head of a valley, say, so that he can avoid ever going downwards. This is perhaps how it is with evolution: some potentially successful designs, like man himself, can only be achieved by a circuitous route, through a series of niches in which the attributes required in the ultimate design are evolved one by one.

Another example is the large fish-like creature which might have been postulated in the time before the fishes began to colonise the land. This design would have been warm-blooded and air-breathing, and the largest creature that had ever lived. The ecological niche for it existed, but the creature to fill it, the whale, did not evolve from fish directly. Instead it came through a long line of most unfish-like, four-legged forms on land, and took about 200 million years to do so. In between, there were air-breathing pelagic dinosaurs of fish-like form, but they did not survive.

Why did no great fish evolve to fill that ecological niche? What advantages did its circuitous evolutionary history confer on the whale? Warm-bloodedness would have helped, but would not have been enough by itself. Breathing air confers advantages over extracting dissolved oxygen from the sea, but in the case of the whale the need to return frequently to the surface for air must be a handicap that might be expected to outweigh any gains.

One suggestion is that the crucial advantage conferred on the whale by its ancestry lay in greater intelligence, and this is supported by the fact that after the whales first appeared the size of their brains increased considerably. But why was there no evolutionary pressure on fishes to develop in the same way? Or was a quantum leap required that could only come about via life on land? Another suggestion is that the crucial advantage lay in the nurturing of the young, developed by the mammals on land, and almost unknown among fish. The fact that the largest fish, the big sharks, are viviparous and produce few or only one large young at a time supports this explanation.

Whatever the answer may be, the whales are an example of the roundabout routes that evolution has taken in filling ecological niches.

Two weaknesses should be noted in the analogy between nichewise evolution and a man climbing hills without ever descending. The first is to do with degrees of freedom (Chapter 4); the man in the hills can only decide his direction and his decision can be expressed by one quantity, e.g., 'bearing 120°'. But in changing a design, there are very many degrees of freedom – longer or shorter legs, thinner or thicker tail, less or greater slope between chin and forehead, etc. The second is that the hills remain the same during the man's climb, while the evolutionary 'hills' are moving and changing with the climate and the rise and fall of other life-forms. But there is another respect in which the analogy may be helpful; a life-form which has reached the top of a local hillock, as it were, will become virtually fixed and remain unchanged, since all directions are downwards. Plants and animals are known which have remained apparently almost unchanged for hundreds of millions of years.

9.6.3 The severity of the design specifications of living creatures

The extreme difficulty of the design problems faced by nature has been touched on before, in the case of the bird in Chapter 2. The difficulties of a hostile environment are added to by those of growth; many organisms must fend for themselves from an early stage; for example, fish may be all on their own when only a centimetre or so long, though they may eventually reach a metre in length. The caterpillar–pupa–butterfly and tadpole–frog metamorphoses are familiar. If the caterpillar is eaten there will be no butterfly: each stage must be viable. Insects and other arthropoda, such as crabs and shrimps, have a hard outer skin which cannot grow with them, and must be moulted periodically as it becomes too tight; until the soft new armour hardens, they are relatively defenceless. Molluscs, like snails and limpets, overcome this difficulty by having a *gnomonic* design of shell – that is to say, one that remains the same shape while growing at an edge. The simplest such shape is a hollow cone (see Fig. 9.12); when added to round the rim it grows larger without change of shape. Coiled snail shells are just coiled cones, and change in shape with growth round the rim of the opening only in that the number of coils increases.

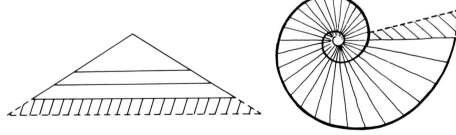

Fig. 9.12. Gnomonic shell forms: growth at the edge only (shaded) leaves shape virtually unchanged.

One of the devices used to cope with the extreme severity of the design problem of living creatures is a very high production rate, so that out of millions of embryos a handful may survive. But some less wasteful approaches have appeared in the course of evolution, principally, the protection of the young by adults among the higher animals and termites, ants and bees.

9.6.4 A speculation on hypothetical limitations of the genetic code

The section which follows is mostly pure speculation by an unqualified amateur, and so should not be taken very seriously. It is concerned with questions which may be resolved in a very few years concerning the degree to which the genetic code written in the DNA of the living organism is able to represent any creature, or only certain creatures, amongst all the designs which might otherwise be possible.

Firstly, let us consider what may be called the Djinn's dilemma, or the Preponderance of Fortuity. Imagine Aladdin rubbing his magic lamp to summon the Djinn and demanding a palace, just a modest little palace, and let us suppose, not being familiar with the creation of palaces from nothing, that a Djinn merely has to specify anything in his mind for it to appear. Naturally, this means that he has to decide the general plan, and the dimensions and the numbers of windows, and the materials and precise form of the window sills, say. And so let us suppose the window sills are to be of a fine limestone, that in reality would be built over millions of years from the inorganic remains of living creatures and then compacted and heated for some hundreds of millions of years, so that it would be full of the vestigial patterns of myriads of tiny shells and skeletons. Suppose we focus upon some tiny corner of one of these sills, and the size and orientation of the crystals in a minute spicule in one of these skeletons. Unless these details are specified, how are they to be determined? and how much information would be needed to specify the palace *completely*, leaving nothing to chance?

The answer is roughly this: to specify the exact composition, state, position and orientation of every molecule of the palace at one particular instant (for they would not remain unchanged for a fraction of a second) would require a volume of written material, in the concisest form of expression and in type sizes and thicknesses of paper of the smallest and thinnest, exceeding the volume of the entire earth by several thousands of times. It follows that in all designs we can readily conceive, the vast preponderance of the relationships between the parts have been determined by chance, and only a few 'leading dimensions' and a rough description of the general nature of the materials have been specified by the designer. The history of the ex-spicule in the limestone has involved a chain of accidents extending over billions of sunrises and sunsets, and the Djinn's task is a prodigious one indeed, if we assume that every detail in a creation from nothing must be determined by his will, lacking this ancient chain of fortuitous events to decide it.

Now this point is only of philosophical interest, but the living organism presents a refinement of the puzzle. Here is a bewilderingly complex design built from instructions running into, perhaps, 400 bible-equivalents in the case of the highest animals. Is it enough? This problem has another aspect. Is the information specific enough to determine all characters separately? Are some of the features of living organisms 'accidents' arising from the limitations of the genetic code?

To take a celebrated example, the giant elk, a creature rather like the North American moose, flourished in Britain until relatively recently on the evolutionary scale. Through the ages its whole body grew larger, but its antlers grew disproportionately to the body, so that some have conjectured that they must have been a handicap, and that they may have contributed to its extinction. There are other examples of a growth in size of the whole animal being accompanied by a proportionately much greater growth in some organ or attribute. This phenomenon is called allometry, and it is believed to be caused by the mode of action of the genes controlling growth.

This view of the cause of the extinction of the giant elk is not supported now, and the real cause may have been hunting by man or some other development. However, while this allometric effect may not have been harmful, it is not clear that it had any evolutionary advantage either.

Now it is easy to understand why the elk should evolve to a larger size in the cold climate of ice-age Europe, for larger size reduces the relative amount of energy needed to keep warm (see Chapter 2). It is possible that in the short term the genetic material could not code a larger elk without also coding proportionately even larger antlers. As yet, very little is known about the way in which the DNA controls the development of the organism, and there may be complicated interrelationships between characters which limit the directions open to the evolutionary process.*

If these speculations are correct, then for a species of organism to exist, it must meet a number of criteria:

- it must *work* mechanically and chemically
- it must be capable of being coded
- it must be able to survive, at all stages of growth
- its design must be able to be evolved through a series of forms from some other form, each form being viable in its own niche of this nichewise progression.

---

* Since this was written, I have found related arguments in respect of butterflies in Ho, M.-W, Saunders, P. and Fox, S. (1986). A new paradigm for evolution. *New Scientist*, No. 1497 (27 Feb.), pp. 41–3.

9.7 Design by man

With the evolution of man, nature produced an organism that could extend or modify its own powers as required, rather as perhaps had already happened in the very limited case of antibodies. Here was a creature that could make armour for itself, or machines to enable it to fly or to live outside the atmosphere which was natural to it. Moreover, it produced these things not by evolutionary design, not by random variation followed by suck-it-and-see selection processes, but by *directed* design, making great imaginative jumps, visualising, calculating and shaping with increasing confidence. The culmination to date of this process has probably been the Apollo flights to the moon; in this tremendous enterprise, the truly critical stages had to work first time – and they did.

But even in design by intelligence, there is evolution of a kind. The original plans for reaching the moon were very different from those finally used, because of the evolution of the design concepts during the programme. Originally, a staging post was envisaged which was to be a space station orbiting the earth. But because of the improved structural design of rockets it was possible to increase the fraction of the all-up-weight which was fuel: as was briefly indicated in Chapter 2, this fraction has a critical effect on range. Some of this stuctural improvement, incidentally, came from stiffening the tubular shell of the rocket by an internal pressure, just as we saw nature giving structural strength to organisms such as small plants and worms in Chapter 5. As a consequence, it proved possible to dispense with a staging post and fly direct to orbit round the moon.

9.8 Nichewise evolution in the history of design

Until very recently, the evolutionary or step-by-step aspect of design by man was predominant, as was illustrated in Chapter 1, by the example of sailing ships in particular. As was explained there, advances in design became truly deliberate in the nineteenth century, so that progress accelerated vastly. Nevertheless, some characteristic features of natural evolution persist in engineering design.

Just as a life-form must have an ecological niche, so an engineering invention or design, if it is to be more than a mere toy or curiosity, must have a niche, that is to say, an area of need which it fills better than any available alternative. In 1712, when Newcomen produced the first real steam engine, there was a niche available for it as a source of power for pumping water from mines, and that was probably the only niche in which its grave disadvantages would allow it to gain a foothold. Fifty-seven years later, when Watt produced his improved steam engine with a separate condenser and, consequently, a much reduced appetite for coal, the industrial world had developed and there were more and larger niches open to this more advanced design.

Watt's improvements to the steam engine also illustrate the influence of parallel evolution in the means of production. Developments had just been made in the method of boring out large cannon, and these were available for machining the cylinder bores of the new engine to a higher accuracy than was possible in Newcomen's day, so reducing the difficulty of sealing the piston. Watt recorded his satisfaction that the cylinder could be bored accurately within the 'thickness of an old sixpence'.

Boulton, Watt's partner, wrote to him to enquire whether he thought their engine was threatened by some early work then being done on steam turbines. With remarkable insight, Watt replied:*

> In short without god makes it possible for things to move 1000 feet per second
> it can not do much harm.

This was a case of a niche which required other developments elsewhere before the 'organism' to fill it could appear, much as the whale required other developments on the land before it could appear. The steam turbine finally arrived in the twentieth century, after considerable advances in materials, manufacturing methods and our understanding of the dynamics of machines, advances which took place mostly in quite different fields of engineering and enabled 'things to move at 1000 feet per second'.

Until very recently, this nichewise progress was most important in methods of production, in particular, in the production of accurate work at low cost. One of the first ventures into mass-production was that by Marc Brunel, father of the famous Isambard. Around 1806 he built special machinery for making parts for the huge numbers of pulley blocks needed by the Royal Navy, till then made by hand. Later, typewriters were among products which were important in the development of the capacity to make large quantities of good, high-precision parts at low prices. Bicycles were perhaps less demanding in accuracy, but needed strong and light components. The knowledge and experience gained in such manufactures paved the way for the cheap internal combustion engine, the car and the aeroplane.

In the latter half of the twentieth century, however, the nichewise effect has become less obvious. Engineers have become more calculating, longer-sighted, more capable of judging what can be done, and so less dependent on experience. In the future major developments will hinge less on the accidents of what has gone before.

9.8.1 Evolution in the design of particular products

Even if the nichewise effect in the evolution of engineering design has become small, there remain points of resemblance with natural evolution.

* Dickinson, H. W. (1939). *A Short History of the Steam Engine*, p. 188. Cambridge University Press.

Of these the most striking is the evolution of single products from the early forms to more advanced later versions. Often the first stages of the process are marked by a greater variety and more extreme forms, as in the early days of the car when, for a time, electrical, steam and internal combustion prime movers seemed all to be in the running. Later the range of forms narrows down to a few dominant ones, often with long struggles between closely-matched types. At the moment, for example, in some cars the engine drives the front-wheels, while others have rear-wheel drive. For decades, however, the front-wheel drive car was a relative rarity.

For most of the history of the railways, the steam locomotive has dominated the field. Its first serious challenge came from the electric locomotive, with the diesel engine appearing later as a contender. But steam seems now to have had its day, and electric motors will probably come to dominate all busy lines, with diesels retaining a niche on the less heavily-used routes. For very high speeds, though, a newcomer is gaining a footing, the gas turbine.

In engineering design, just as in living organisms, certain arrangements prove apparently unshakeable winners. Just as nearly all large life-forms have a backbone (the squids and octopuses are an exception), so nearly all engines of the displacement variety (see Chapter 7) use the piston-in-cylinder geometry. Any of the many other displacement forms of pump could be used as the basis of a steam or petrol engine instead, and much ingenuity, time and money has been spent in the quest for an engine to run more smoothly and with fewer working parts than the familiar piston engine, which is rough-running and complicated. A recent and major attempt, and the one to have had the most success, is the Wankel engine (see Fig. 9.13). On the face of it, the Wankel has winning advantages over the piston engine: its parts rotate smoothly, without causing much vibration, unlike pistons and connecting-rods; it has very many fewer moving parts; it is smaller. But it does have difficulties with sealing the space corresponding to the cylinder, so the gases do useful work rather than escaping. In the piston engine, this sealing function is done by the piston rings, which themselves were a considerable source of difficulty in the early days and are now a very highly-developed design of great subtlety.

It may be that developments in design and materials will make the Wankel design a successful challenger one day or that one of the older designs of engines with smoothly-rotating parts will be revived and improved and will bite into the empire of the piston engines. It may be that some completely new 'rotating-piston' engine (as designs like the Wankel are sometimes called) will appear and prove successful. Or it may be that the piston engine will go on unchanged and unchallenged, until perhaps engineering has no need of such crude devices at all.

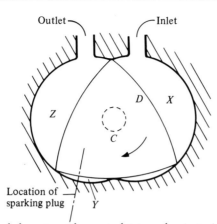

Outlet   Inlet

Z   D   X

C

Location of
sparking plug   Y

Fig. 9.13. Wankel engine. The central triangular 'rotor' rotates in the 'oval' casing, while its centre C moves round the circular orbit shown just three times as fast in the same direction. During one revolution of C in its orbit, i.e., one-third of a turn of the rotor, the space X between side D of the rotor and the casing will expand slightly, contract greatly to the size of Y at the instant shown in the figure and then expand again to the form of Z. During the next orbit it will repeat the process. These two expansions and contractions can be made to correspond to the four 'strokes' of the Otto cycle, by judicious siting of inlet and outlet (which are simple ports as in the two-strokes of Fig. 1.10) and sparking plug. The orbiting motion is arranged to drive the output shaft, and the relation of rotations to revolutions is ensured by gearing.

One interesting aspect of evolution in engineering design is the wealth of new versions that appears when some long-established device is threatened by extinction. In the late nineteenth century the steam engine (that is, the kind with pistons) had reached a high degree of refinement. The steam pressure and temperature had been raised greatly above those used in the time of Watt, and the steam was expanded through several cylinders in succession to make possible the highest extraction of energy from it. But around the turn of the century two rivals emerged to challenge it – the steam turbine and the internal combustion engine. It was soon swept from most of its niches by one or other of these rivals, but in two fields (besides the locomotive) it struggled on.

One was in the provision of power for factory workshops, textile mills and similar places, where a new type of steam engine with a single cylinder proved the most able to resist, by using a very large expansion ratio – that is, the steam was admitted at very high pressure to a tiny space, the 'clearance volume', between the piston and the cylinder head when the piston was furthest into the cylinder (see Fig. 9.14). Thus by the time the piston reached the other end of its travel, the volume of the steam had increased

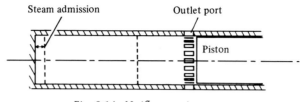

Fig. 9.14. Uniflow engine.

many times, so extracting energy that otherwise would require two or three cylinders in series. At the end of the stroke exhaust valves in the cylinder wall were uncovered, allowing the steam to escape: by this arrangement, the steam flowed always in the same direction through the cylinder from the hot inlet end to the cool outlet end (hence the name 'uniflow') which improved the efficiency for reasons related to the second law of thermodynamics and the regenerative principle discussed in Chapter 7.

These large, slow, single-cylinder uniflow engines were very simple and reliable, and managed to keep a foothold for a long time. They did have one difficult design aspect; steam had to be admitted only for a very short time when the piston was at or very near its extreme leftward position in the figure, or else the full efficiency was not obtained. To achieve this, very advanced designs of inlet valve were used, which opened widely and shut again, all in a few thousandths of a second, and many different and very ingenious forms were developed. Anyone wanting a steam engine then could have his choice of many different versions of the uniflow engine, but within a few years they ceased to be made.*

However, the complex steam engine had survived in one niche, the propulsion of ships, and the triple-expansion engine (so-called because the steam is expanded in three stages, through three cylinders, each larger than the last) enjoyed a triumphant revival in the 1939–45 war. Then hundreds of them were built to power, not only the famous Liberty merchant ships, but naval vessels as well. The reasons were that steam turbines could not be built without special production equipment, but piston engines could, and steam plants could use any fuel, while in those days internal combustion engines could burn only kinds of fuel urgently needed for other purposes.

9.8.2 Aircraft propulsion

Sometimes evolution in engineering design seems to turn back on itself. The first aircraft had propellers driven by piston engines, which began to be displaced in military aircraft after 40 years, by jet engines, gas turbines whose power output is in the exhaust jet. Shortly after, civil aircraft and

---

* For an account of the complexity and elaboration of late steam engines, see Allen, T. (1931). *Uniflow, Back-Pressure & Steam Extraction Engines*. London: Pitman.

military transports had propellers driven by gas turbines (turboprops) but later jet engines were used. But now civil airliners are driven by ducted fan engines, which are gas turbines, most of the power output of which goes to drive a fan which is really a propeller in a tube with some stationary blades to improve the efficiency. Only a small fraction of the thrust comes from the exhaust itself in a ducted fan turbine, which is really a turboprop engine adapted to high speeds. Now the ducted fan may be replaced by a true propeller of novel form, with many high-speed swept-back blades.

9.8.3 Typewriters

The typewriter also exhibits this recursive effect. Early typewriters used a disc with all the type round the rim and later versions had the type in rows round a solid block. But these designs were swept away by the triumphant formula of a type block for each character, each on the end of its own lever connected to its own key in a piano-like mechanism. They survived only in the child's toy which had rubber type moulded on the rim of a disc.

Eventually, however, the 'golf-ball' tyepwriter appeared, with all the type on a single sphere, and more recently the 'daisy-wheel' typewriter or printer, harking back to even earlier forms.

Incidentally, typewriters also illustrate the nesting of degrees of freedom noted in Chapter 4. To type the next letter in the line, shall we move the type-head to the right, or the paper to the left?

9.9 Genetic engineering

In recent years it has become possible to alter the genetic material, the DNA, of organisms. It is possible to break the sequence of bases at certain special points and graft in other sections of DNA from other organisms. The first applications have been to the development of modified forms of lowly organisms, such as the bacterium *Escherichia coli*, which can produce useful drugs, such as insulin or interferon. It is not nearly as easy a matter as it might sound, and great ingenuity is required to design a sequence of treatments and operations which will break the chain at the right place and to obtain a form of the desired section of DNA which will be accepted into the gap.

At last man is creating new organisms, not by breeding, which is a kind of artificial and accelerated evolution to a conscious end, but by fabricating new combinations of genetic blueprint. These new organisms are designed, to the extent that they are contrived with a useful end in view – the production of valuable drugs.

Admittedly, these first few steps are only joining up bits of natural DNA, without any great certainty of success. But 10 000 years ago man was just chipping flints and scraping hides, while today he is growing perfect crystals and infusing them with traces of other substances in delicate patterns indiscernible to the naked eye. It is not conceivable that something like the same increase of mastery will occur in genetic engineering, not in 10 millennia perhaps, but in 10 decades rather?

9.10 Non-evolutionary design

The idea of deliberate innovative design has been clear for over a century, yet evolutionary aspects remain, for a variety of reasons. One of these is the difficulty of predicting exactly how a design will perform, which, strangely enough, is easier for a spacecraft in flight than for a car on a road. Many designs have to proceed stepwise, because of the learning which has to be done before the ultimate product can be designed. Another reason is that the market for advanced designs usually requires preparation by less advanced designs; it is no good producing brilliant products if people do not want them. Very often, the market has to be made more critical, more demanding of high standards, and often this requires a series of steps.

Nevertheless, the evolutionary aspects of design are declining, and jumps are becoming longer and are more confidently taken. Evolution itself is becoming extinct.

10

Designing and inventing

10.1 Designing and inventing

This chapter looks at the activity of functional design and inventing, the sort of people the designers and inventors are, and some aids to design and invention.

By this time, the reader should have a clear idea about what functional design is, and invention is nothing but a grander word for a particularly original or important step in design. Although invention is generally much more impressive than design, it often turns out to be the easiest part.

A recent example in which invention was rather obvious, even trivial, and all the difficulty lay in design, was the application of disc brakes to cars. The brakes on bicycles had been of this form for decades, but a great deal of ingenuity and hard work had to be put into producing a disc brake for cars that could compete with the highly-developed drum brake on favourable terms.

In affording legal protection to designers and inventors, the law makes a distinction between an invention, which can be patented, and a design, which can only be registered or copyrighted, forms of protection which can be evaded by relatively trivial changes. In general, invention is more fundamental and less specific in form and application than design, but no sharp line can be drawn between the two.

10.2 The character of inventors and designers

During the last 20 or so years there has been great interest among psychologists, educationalists and others in what they call 'creativity', and many quantitative studies have been made of creative people. Inventors and designers do not seem to have been included. However, a particular kind of designer, architects, have, and so have scientists, and it is reasonable to expect that any qualities present in the more original members of both these groups would be shared by the people in the bordering and overlapping group we are interested in. Also, some of these qualities are shown by yet other groups of creative people, such as writers, so that it seems even more likely that they will be shared by inventors and designers.

One striking finding is that these creative people combine high psycho-pathology ratings with high ego-strength, that is, compared with the average, they have both a higher tendency to psychological disorders and greater strength to overcome this tendency. This is a particularly interesting finding in that the *combination* of high psychopathology and high ego-strength is rare in the population at large.

Another pronounced trait of creative people is what is called psycholo-gical-mindedness, a responsiveness to other people's interests and motives, but they are not particularly sociable and are low on social conformity and little concerned to produce a good impression on others; these findings are what one might expect. They show a marked tendency to 'femininity' of interests, and high levels of self-assurance and self-confidence.

It has sometimes been reported that there is little or no correlation between creativity and intelligence, but such statements are misleading. Sometimes they refer to series of tests in which the samples of people involved, architects or school pupils, say, had average intelligence quotients (I.Q.s) well above the average in the population at large (in two cases, around 130), so that all they show is that above a certain high level of I.Q. there is very little relationship between creativity and score in intelligence tests. Moreover, the intelligence tests themselves are open to severe criti-cism, particularly on the grounds that they discriminate in favour of certain aspects of intelligence, e.g., the verbal rather than the non-verbal: in the cases of architects, and to a lesser extent scientists, some intelligence tests might be expected to be of rather limited relevance. Perhaps the best inter-pretation of these results is that they show the lack of correlation between I.Q. scores and intelligence rather than anything else.

Besides these systematic studies of creative people, there is a wealth of anecdote and history, much of which supports the finding of high psycho-pathology ('genius is next to madness') and high ego-strength among creat-ive people. It must be remembered, however, that pressures of convention are so great that it often requires considerable strength of personality to do something entirely reasonable. For instance, many awkward loads are most easily and safely carried on the head. This practice is rarely seen in London today, however. It may be that there are many potentially creative people who lack strong personalities and whose talents therefore never emerge.

10.2.1 Measures of creativity

Psychologists have tried hard to design measures and tests of creativity, but it is clearly a very difficult task and it is not surprising that the results are not very convincing.

One of the most celebrated of these tests is the one called 'Unusual Uses', due to Guilford. The subjects who are being tested are asked to list as many unusual uses as they can for a common object, say, a barrel, a blanket or a brick (examples given by Hudson). A subject's score is based on the number of 'original' uses he thinks of. As a way of measuring creativity, Unusual Uses is singularly unconvincing for a number of reasons. Firstly, creativity generally involves the reverse process, not devising many uses for one object, but picking one thing out of many that might be chosen for a particular purpose, the next phrase or modulation in a piece of music, or the configuration of an engine, say. Secondly, how are we to value a trite unusual use for a flowerpot like breaking a window, compared with a clever one, like wetting it and using it as a milk cooler? Thirdly, we shall see later that producing unusual uses may be stimulated by systematic processes, at least in so far as the properties of the object are concerned.

Hudson would call those people with high scores in such tests not 'creative' but rather 'divergent'. His division of people into 'divergers' and 'convergers' he sees as reflecting emotional rather than intellectual differences, the avoidance of, or commitment to, practical action. 'In a caricature', he writes, 'the converger takes refuge from people in things'.

It is worth mentioning these ideas because they have gained so much popular currency, which is the more unfortunate because of the particular terminology used. Essentially, it would make much better sense to call creative processes convergent. Consider the sculptor carving a form from a block of marble – he removes marble in such a way as to converge on a particular shape. The designer is choosing one form from many, narrowing down to a particular solution, and the quality of the final design depends on the quality of his selection, which is essentially a convergent process. What could be more convergent – sometimes surprisingly, but always convincingly, convergent – than the music of Mozart? It will be argued later that one of the distinguishing features of the good designer is his ability to *converge from a wide base* on a good choice.

Now this sort of convergence on a choice and 'divergence', as measured by tests such as Unusual Uses, may well have much in common, but they are not the same and it is not at all certain that they always go together. The trait that Hudson refers to as convergence might better be called 'conformity', the tendency to follow set procedures, rather than being given a title which suggests an element of lateral movement just as much as divergence does (only the one suggests purpose, and the other rather a lack of purpose).

By and large, the psychologists do not seem to have achieved much in the study of creativity, and indeed it may be unreasonable to expect that they should. They have indicated that personal factors may be as, or more, important than intellectual ones. Their choice of terminology may have been unfortunate in other cases besides that of convergence and divergence, and this perhaps stems from an imperfect view of thinking processes which is not limited to psychology.

10.2.2 Analysis and synthesis

As a simple illustration of the last point, consider the use of the words
'analysis' and 'analytical', which are often associated with rigour, single-
minded and invariable procedures, impersonality, objectivity, and so on.
Now analytical papers written after the analysis has been done tend to read
in that way, and so they should: the argument should not be confused by
an account of the mistakes, the blind alleys traced and retraced, the flashes
of insight, the half-baked ideas that sometimes resurrect themselves fully-
baked, the fancies and the humour that sometimes turn out to have helped
on the work. Lord Baker writes of the physicist Bernal, standing at the
centre of a shallow bomb crater in the 1939–45 war, 'he wrapped his arms
about him and was clearly a bomb, thinking what he would do'.*

Just as analysis is in reality no mechanical progression neither is synthesis a matter of sheer inspiration. Creativity involves reasoning and calculation, too, as I shall try to show later.

10.2.3 Mathematicians, physicists and others

Perhaps the most important evidence about the nature of invention has come
from mathematicians and physicists, and it supports two principal ideas. The
first is that the essential process is one of choice, and that the choice is made on
principles which are best described as aesthetic. The mathematician Poincaré
wrote of aesthetic sensibility as 'the delicate sieve' with which the most useful
combinations of ideas are separated from all the others which might be
studied. (There is an interesting parallel here with the crude sieve, natural
selection, with which nature separates the most useful combinations of genes
from all the others which sexual reproduction generates.)

The second is that the process of invention usually falls into stages, of
which the first three may be described as preparation, incubation and illumination. The later stages are not clearly distinguished, but may be
described as verification and making precise (Hadamard), development, elaboration, embodiment, working out, etc., according to the nature of the
invention and the preference of the inventor. It is the first three which are
of most interest, however. A common finding is that after working hard on
a problem for some time (preparation) and after dropping it for other work
or some form of relaxation (during which incubation is supposed to take
place) there comes, out of the blue, as it were, illumination.

The physicist Heisenberg expressed the nature of this kind of illumination very well in a letter to Einstein, when he wrote:

> 'You must have felt this too: the almost frightening simplicity and wholeness
> of the relationship which nature suddenly spreads out before us . . .'

* Baker, J. F. (1978). *Enterprise versus Bureaucracy*, p. 13. Oxford: Pergamon.

The experience of illumination is often shared to some extent by those who later study the work of the inventor, particularly when they begin to see the development of the first idea ahead of their reading of his paper, as when in an exciting story we anticipate some ingenious turn of the plot.

It is not just in the fields of mathematics and theoretical physics that invention is associated with the phenomenon of illumination. Archimedes' cry of 'Eureka' was prompted by the very practical invention of a way of measuring the density of a crown, and Watt relates how he was suddenly struck with the idea of the separate condenser for steam engines while out for a walk on a Sunday. In the more innovative forms of design, illumination of a kind often marks the emergence of a design philosophy.

It is interesting to note the discrepancy between the ideas of Poincaré and others, of invention as the choice of combinations based on aesthetic principles, and psychological tests such as Unusual Uses and terminology that associates creativity with 'divergence'.

The inventive man is marked out by the range of elements he considers combining, and the skill and judgement with which he selects or converges upon his preferred combination. When good and experienced designers tackle a problem together, the speed with which one recognises the fitness of the element the other proposes and guesses the way in which he would use it provides the most satisfying evidence of their quality. Their thinking is most often characterised by a rapid and confident convergence from a wide base of possible components or methods.

10.2.4 Who are the designers and inventors?

Almost anybody may be an inventor or a designer. Many people are seized, once or more in their lifetimes, by an idea which they think is brilliant and will make their fortunes: alas, though they may often be right about the brilliance of their scheme, brilliance is not enough to make a fortune; most inventions do not represent a real improvement on what is done already, and so are doomed to failure.

A great many designs and inventions fail in practice, in spite of being very ingenious in concept and thoroughly worked out in detail. It is hard on the originators that they achieve neither fame nor fortune, in spite of their very notable achievements. Perhaps one day a wiser society will reward such near misses, in recognition that they are great manifestations of human intellect, as much so as many others which are equally 'useless' but where usefulness is not expected.

However, a common fault of inventors and designers is to be too blindly devoted to their brainchildren, especially to a first and only brainchild. It is sad when a clever man spends his lifetime or his savings persisting with an idea long after its faults have become manifest, when he might have made other inventions, not so flawed, had he dismissed the first. Of course, it is very difficult to judge when a weakness is fatal and inescapable and when it is tolerable, or susceptible of elimination or amelioration. There are many cases where success has come only after many years of hard work and struggles to convince others of the virtues of an idea.

It is commonplace that many inventors originally worked in fields and remote from that of their invention. For example, Watt first made the acquaintance of the steam engine in which he was to effect such striking improvements when he was working as an instrument maker for the University of Glasgow. He undertook to repair the little Newcomen engine used for demonstrations by the professor of natural science, and about two years later hit on the idea of the separate condenser. Two of the several inventors and developers of the screw propeller were an Austrian forestry officer and an English farmer. Many other such cases could be cited.

Although anyone may be an inventor and designer, over an increasing part of the field it becomes more and more necessary to have a thorough physical insight that takes years of education and experience to acquire, so that valuable innovations by laymen become less and less likely. Nevertheless, trained engineers and scientists of an original turn of mind remain able to contribute over a very wide range.

An interesting characteristic of innovation in engineering, as in science and the arts, is the importance of relatively small groups of gifted people coming together in areas of rapid development and later dispersing into others. A famous case is that of Henry Maudslay, who recognised some of the essential needs of the new era of precision engineering then dawning, such as accurate measurement and accurate screw threads, and his assistants, of whom several later made important contributions to the development of production methods, notably James Nasmyth, inventor of the steam hammer, and Joseph Whitworth. In the twentieth century new developments such as airships and the jet engine have attracted young men of outstanding ability who have subsequently dispersed to other fields in which they have made important advances.

While the fine arts and the sciences show similar phenomena, there are some significant differences. In the fine arts, the individual master of the atelier is the unifying influence, whereas in the great endeavours of engineering design the object to be created generally has the central role. The sciences come somewhere between, for the seminal areas will be studied in many places, but a great school will only emerge where there are one or more outstanding workers.

Nowadays, the practice of science runs in narrower grooves than in the past, so the influence of a great research team is not likely to spread much beyond its own field. On the other hand, the designer of today is as versatile as the scientist of the nineteenth or earlier centuries, so the effect of the dispersed design team is likely to be felt over a wide range. Thus emigrants from early jet engine teams have later influenced design in many other fields, including nuclear power plant and diesel engines.

10.3 Design in the schools (I)

There has been a great deal of interest recently in the schools in teaching related to design, I believe, largely for two wrong reasons. Some advocate such teaching for a vocational purpose, the beginning of the training of tomorrow's designers. Some regard it as desirable as training for tomorrow's consumers, so they will appreciate good design and purchase wisely from the great variety of goods, cheap and dear, shoddy and sound, appropriate and inappropriate to their needs, that are on offer to them in the free world.

The real justification for teaching something of design in the schools is its sheer educational value, in the broadest sense. Schools should try to develop the powers of thought in children as much as possible, but at present they do so in a very imbalanced fashion. They concentrate on verbal modes of thought and neglect, and even despise, the very important visual mode. But it is at least arguable that visual thinking is of a fundamentally higher order than verbal thinking, although both are essential to our culture.

Unfortunately, in British schools, the new subject called 'technology', which includes design, is developing in a dull fashion, lacking depth and intellectual grittiness and is likely to reinforce old snobberies rather than dispel them. For reasons of economy, it has to be done with a legacy of teachers originally selected and trained for other subjects, it is excessively verbal and largely innumerate, and there may be an unacknowledged desire to make it a subject with which the less gifted pupil will not find difficulties.

10.4 Verbal and visual thought

At times, some have felt that man's most distinctive characteristic, setting him off from the animals, is his use of tools and his making of things – he is homo faber, man the maker. It is true that animals do make things, especially social insects like the bees, wasps, ants and termites, but these are structures they are programmed to produce. But a more distinctive feature still is language, which is the principal means by which our culture is passed on from generation to generation. This role in education and society has greatly advanced the prestige of verbal skills. Since it is my present purpose to let down this rather overblown prestige, let us call the characteristic verbalising man, homo jabber.

Now homo jabber has had everything his own way for a long time – it is not possible to read a newspaper without realising his sovereign sway in human affairs, and a fine mess he has made of them. But it is at least a tenable argument that the highest forms of thought are not verbal at all, or at least, only insignificantly verbal.

In the first place, visual thinking is parallel, rather than serial, and so it is capable of handling information much faster. In Chapter 6 we saw the prodigious flow rate of information involved in television, something of the order of a bible every two seconds. Now, only a small proportion of that information impresses itself on the consciousness, perhaps one-thousandth penetrating the abstracting networks in the brain, but that is still equivalent to a bible in half an hour.

Consider how little time you need to look at someone before you can recognise them again, and how long it would take to describe them sufficiently in words for another to recognise them, if indeed you could do it at all. When you use a street map to go between two points in a city, you can immediately visualise alternative routes, compare the distances and decide which way to go. Given the same information in verbal form, what a lot of it there would be, and what a task it would be to work out the way to go!

Visually, we can take in information more rapidly and reliably, and in a very selective way – think how we look at once at the area we are interested in on a weather map on television. Moreover, our vision has powerful abstracting systems, like those that enable us to read handwriting or recognise an object at whatever angle it is presented to us and even when it is largely obscured. These great powers come from the parallel processing of information which vision can perform, with thousands of channels, whereas hearing has only one, though we may be able to listen to two voices at once, with difficulty.

Vision is also very closely linked with the sense of touch and the motor nerves which operate the muscles and which feed back information on the resulting position of the body and limbs to the brain (the proprioceptive, or 'self-perceiving' system). This whole system, which was discussed in Chapter 8 in the case of a horse galloping over rough ground, is the seat of our power to imagine something happening before it does. The hunting of large animals by primitive man must have depended on his ability to plan the encounter, essentially by visual thinking-through, and then to communicate the plan by speech.

Much scientific and most design thinking must be primarily visual. Einstein had this to say on the subject:

> The psychical entities which seem to serve as elements in thought are certain signs and more or less clear images which can be 'voluntarily' reproduced and combined.
>
>   . . . But taken from a psychological viewpoint this combinatory play seems to be the essential feature in productive thought – before there is any connec-

tion with logical construction in words or other kinds of signs which can be communicated to others.

The above-mentioned elements are, in my case, of visual, and some of muscular, type.*

Einstein was working with concepts of physical science as remote from the familiar experience of our senses as may well be, but designers of all kinds are working 'at home', as it were, so that their visual and tactile imagination need be stretched much less.

In some areas it may be possible to distinguish two schools of thought, in mathematics, for example.

There is the algebraic tradition – verbal, serial, short on insight and imagination, long on rules and rigour – and, on the other hand, the geometrical tradition – visual, relying largely on insight and inspiration, and not very systematic. The most undisciplined and creative part of mathematics in the schools was the riders in Euclidean geometry, now banished for its lack of rigour and its old-fashionedness (and, perhaps, its intellectual demands?). The tendency of modern mathematics is towards the blind and rigorous operation of rules: as Bondi has said, mathematicians are not particularly good at thinking, they are good rather at avoiding thought.

It is interesting to speculate to what extent even literature depends on the visual imagination. Do most authors first see characters and their actions in the mind's eye, or do they construct them from verbalised attributes?

Suppose we consider the action of even a very simple mechanism, like a striking clock, and try to describe it without diagrams, without visual props. As with the street map of the city, or the human face, the description becomes prodigiously lengthy and extremely difficult to follow, especially if we exclude special words which conjure up very special shapes, like 'escapement'. It is not just that words are an inappropriate medium for the job: the action of the clock is much more complicated than most of the things we usually describe in words. Even if we take a subject like history, it is much easier to understand a family tree which is drawn out as a diagram than a series of relationships described in words. If we try to trace the relationship of Richard II of England to Richard III from written material it will take time and probably the making of notes, but from a diagram it can be seen 'at a glance'.

The advantages of words lie in their convenience for recording and communication and their abstraction, i.e., their power to describe attributes in isolation, or classes of objects rather than particular ones. It is quicker to say or write 'dog' than draw one, and it impossible to draw a dog, plain and unadorned. If it is recognisably a dog, it will have peculiarities not common to all dogs – its ears will be pointed or floppy, its legs long or short, and so on – it will be a particular, not a general, dog. Still less can

* Hadamard, J. (1944). *The Psychology of Invention in the Mathematical Field*, Appendix 2. New York: Dover.

we render by drawing a property like generosity or unity. In the family tree, we use words to indicate the persons and a diagram to indicate the relationships, and this is often the way with drawings used for thinking, like those of engineers (Leonardo's are a good example). Form and structure are drawn, but parts are labelled and additional information is provided by words. Both verbal and visual thinking (short for visual-cum-muscular-cum-proprioceptive-cum-tactile thinking) are important, but there is no case for saying that verbal thinking is of a higher order. Nevertheless, homo jabber, the pedagogue, the politician, the preacher, the bureaucrat, the grammatician, the lawyer, thinks he is no end of a fellow, and looks down on humble homo faber, the engineer and the designer. Unfortunately, this snobbery is rampant in the schools and so is reflected in the attitudes of pupils to the subjects taught there.

10.5   Design in the schools (2)

Such design as is taught in the schools should form part of a co-ordinated development of the visual thinking powers of the pupils, the aims of which should be primarily educational in the broadest sense, and only secondarily vocational or a preparation for adult life.

The subjects already taught in schools and chiefly concerned with visual (and related) thinking are:

(a) art (drawing, painting, modelling, etc.)
(b) craft (work in wood, metal, cloth, etc., involving the knowledgeable and skilful use of tools and materials).

To those should be added two subjects which are new or largely new:

(c) design appreciation (the deep study of design, both aesthetic and functional, in nature and the works of man)
(d) drawing for thinking.

Important components of visual thinking come also from other subjects in the present curricula, notably from physics and some parts of mathematics (e.g., geometry, graphs and trigonometry) and probably games and physical education contribute substantially.

10.5.1   Design appreciation

Design appreciation should be a large new subject, covering much of what is now often called technology, as well as some biology and ecology in which interesting design is involved – in fact, much of the subject matter of this book more suitably treated. But it should also involve some design–build–test projects for the pupils, for these enlist their enthusiasm and provide the superior kind of insight that comes from doing rather than studying. Examples might be to design, build and test:

(a) a model steam-boat using a given steam engine, with a competition for the longest run using a given weight of fuel
or
(b) an aerial cableway for transporting money and papers across shop floors between counters and cash desks (a vanished system)
or
(c) from a limited supply of cartridge paper and carpet thread, a 'bridge' to span a gap (say 0.6 m) and support as great a load as possible at its centre (see Fig. 10.1 – Chapter 5 provides a useful basis for deciding how to do this).

Many teachers have been introducing this kind of project in the schools in recent years, and have shown what splendid results can be obtained.

In addition to design–build–test projects in engineering, this subject, design appreciation as I have called it, should include large numbers of very small exercises, experiments and experiences – as simple as breaking various materials in bending to see how strong they are and in what way they fail, or finding the characteristic (see Chapter 6) of a small electric motor, or assessing the performance of a simple jumping toy. The chief object should be to develop physical insight and visual thinking generally, but with a small quantitative element so that numeracy is involved.

10.5.2 Drawing for thinking

Just as we exercise and develop our powers of verbal thought by talking and writing, so we can educate ourselves visually by looking and drawing. The simple technique of orthographic projection (Section 1.8) can solve many problems, like the size of room of a given height we can build in a particular roof or whether a vehicle can negotiate an entrance off a narrow roadway. Rough drawings, part scaled, part free-hand, and simple calculations can be used to study the design of objects like nutcrackers (Fig. 10.2), karabiners (Fig. 10.3) or clothes-pegs (Section 10.9.3). Kinematics is involved in many insightful exercises, on subjects such as the

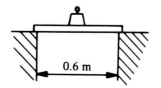

Fig. 10.1. Design–build–test problem.

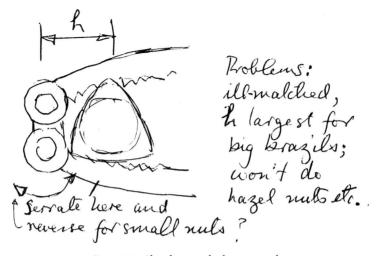

Fig. 10.2. Sketch to scale for nutcracker.

human knee and vehicle suspensions (Chapter 4). Planning the making of objects such as polyhedra out of flat sheet folded up, defining forms for boat hulls, designing wooden furniture and making drawings of plants to enable them to be identified are all good drawing-for-thinking activities. Using a contour map to find a route for a road not exceeding a particular gradient will help to prepare a student for the ideas of the calculus, among other things.

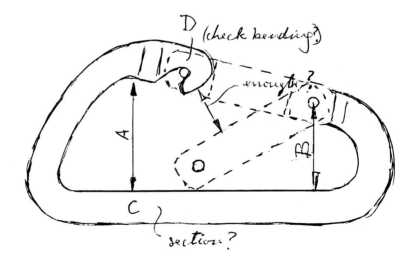

Fig. 10.3. Sketch to scale for karabiner.

10.6 The higher education of designers

The vocational or specific education of designers should not begin in earnest until they leave school; even then, it is probably too early to pick out the designers. Adequate provision is made for architects and industrial designers, and most engineering courses devote some time to design.* The serious weakness that does exist is that in the United Kingdom architects have five years of study at their university or college, but engineers have only three, a ratio which is clearly absurd. In Europe, the ratio is reversed or nearer unity, and the provision for educating high-grade engineering designers is generally far superior. The provision for industrial designers, who are concerned less with functional aspects and more with stylistic ones, is less inadequate in Britain.

The shortness of engineering courses in Britain has meant that the universities must concentrate primarily on engineering science, and this in turn has led to the neglect of design. This, again, means that there is a dearth of knowledge of design among engineering academics, except at a few universities which have paid particular attention to such aspects. It is high time that vigorous action was taken.†

10.7 Some aids to design

10.7.1 Ways of stimulating ideas: maieutics

Very often our reaction to a clever invention or an elegant design is 'how simple but effective', or 'why didn't I think of that?' or even 'how obvious – why hasn't it been done before?'. Most innovative ideas are easy to understand: the difficulty lies in hitting on them in the first place. There is therefore much interest in the thought processes that give rise to such ideas and ways of stimulating them.

The expression, 'hitting on an idea', is a vivid and convincing one, which conveys very well the apparent fortuity of brilliant designs and the mental impact of the onset of illumination.

Poincaré said of the choice of fruitful combinations of mathematical ideas that the aesthetic rules which must guide it

> are extremely fine and delicate. It is almost impossible to state them precisely; they are felt rather than formulated. Under these conditions, how can we imagine a sieve capable of applying them mechanically?‡

---

* French, M. J. (1981). Existing undergraduate courses in UK universities. In *Proceedings of Conference on Education of Tomorrow's Engineering Designers*, pp. 97–101. London: Institution of Mechanical Engineers.

† In 1838 Disraeli said, 'The English manufacturers unfortunately fancied that all they needed to do was to produce a cheap article, and the consequence was, that their cheap article was refused, while an article of a better kind, and *displaying more ingenuity in the manufacture*, the produce of other countries, was preferred.' (Hansard, March 15th: my italics.) This neglect of functional design has characterised much of British industry up to the present day. Important reports have recognised this, notably the Feilden Report (*Engineering Design*. London: Her Majesty's Stationery Office, 1963), but no significant action has followed.

‡ Poincaré, H. (1914). *Method and Science*. Translated by Maitland, F. London: Nelson.

Occasionally, a special-purpose sieve may be possible, designed for the solution of a single problem, but then the difficulty is simply transferred from the original problem to the problem of inventing a sieve.

In the majority of problems, all that can be offered are ways of increasing the odds on a hit, as it were. Such aids to hitting on ideas are sometimes called 'maieutics', a term which comes from the Greek for a midwife, because of their role in 'bringing forth' brainchildren. These maieutics, these midwives, vary from the broad and generally helpful, to the narrow, angular and powerful, and their relative usefulness depends on the nature of the problem in hand, as we shall see.

Two books between them cover most of the 'soft' maieutics, the generally helpful kind: they are *Synectics* by W. J. J. Gordon, firt published in 1961, and *Applied Imagination*, by A. F. Osborn, first published in 1953. Most other writings on such methods known to me add very little to what is in these two books.

It is convenient to take Gordon's approach first. One of his views, shared by Osborn, is that 'a properly operating group has advantages over an individual', i.e., that two heads are better than one, and three better still. He sees four interrelated psychological states as being involved in creative problem-solving:

(a) detachment, followed by involvement
(b) deferment of premature attempts at solution
(c) speculation
    and
(d) autonomy of the object

He believes that it is important not to attempt to arrive at a solution too soon (b). By autonomy of the object, he means a state where the 'idea seems to take over', to develop a life of its own, a state which will be recognised by many inventors and designers and which has been treated in Chapter 7, where it was called 'the emergence of a design philosophy'. He lays great stress on the composition of his 'synectics groups' who are to produce the ideas, and the surroundings and the circumstances in which they are to deliberate ('synectics' is a word he coined for 'the joining together of different and apparently irrelevant elements. Synectics theory applies to the integration of diverse individuals into a problem-stating, problem-solving group').

He proposes two basic operational mechanisms, which I would classify as maieutics: *making the strange familiar, and making the familiar strange*.

'Making the strange familiar' he sees as the function of analysis, which brings the problem within the scope of our understanding. 'Making the familiar strange' he proposes should be done by means of his chief maieutic, *analogy*.

Gordon distinguishes four kinds of analogy which may be used as maieutics:

- *personal analogy*, of which Bernal in the bomb crater was an example,
- *direct analogy*, of which there have been many examples in this book, such as the analogy between the fish muscle and a rope in Chapter 6,
- *symbolic analogy* – in which, according to Gordon, the innovator 'summons up an image which, though technologically inaccurate, is aesthetically satisfying' – and
- *fantasy analogy*.

While it is possible to question Gordon's classification, there can be no doubt that analogies of all kinds are among the most powerful maieutics and perhaps the most important.

Perhaps because of his emphasis on analogy, Gordon seems to overlook the full value of his own aphorism, to 'make the strange familiar'! It is of the utmost advantage to make familiar, or more precisely, to make *commonplace* (Gordon's word), the subject matter of the problem, because we can handle the commonplace with a certainty and insight which is not possible with the esoteric. Simple visualisations that can be thoroughly and completely grasped enable inventions to be made.

A splendid example of making a difficult subject commonplace is that of the concept of a magnetic line of force, a concept quite unjustifiable from the point of view of a purist, but of immense utility to the creative thinker because it is so easily visualised and so vividly apprehended. The attraction between a magnet and a piece of iron, say, is visualised as caused by large numbers of thin eleastic threads joining the two: these threads, the 'lines of force', have no substance but behave much as thin strands of highly-stretched rubber might, contracting if they have the chance and expanding in thickness as they do so. This analogy to the magnetic field is easily grasped and can be used to interpret many aspects of its behaviour into commonplace images, which is a great help in inventing or designing devices using magnetism. It is much easier to handle concepts which have been made familiar or commonplace, so that the chance of hitting on a good idea is greatly increased; this contention is supported (were support needed beyond that given by the common experience of us all) by the very simple and graphic accounts which inventors usually give of their ideas.

Osborn's book, *Applied Imagination*, is wider-ranging and more eclectic than Gordon's, and the maieutics it offers are even softer. He particularly advocates a technique called 'brainstorming', 'a conference technique by which a group attempts to find a solution for a specific problem by amassing all the ideas spontaneously contributed by its members'. Important aspects of brainstorming are that criticism is not allowed until afterwards – the *deferment of judgement* principle – and a high flow rate of ideas is aimed at, on the basis that quality will come with quantity (which often appears to be the case). He stressed that brainstorming is not a panacea that will produce painless solutions to all problems, but only one useful technique to be used in conjunction with others (a warning that applies to all maieutics).

Osborn listed many other approaches, of which perhaps the most important is that of deliberate variations, which he summarised under the headings, 'put to other uses, adapt, modify, magnify, minify, substitute, rearrange, reverse, combine'.

Another device which Osborn mentioned is attribute listing, due to Crawford. Here all the attributes of the thing to be designed or invented are listed and considered separately to decide whether they are susceptible of useful change – e.g., whether the shank of a screwdriver should be round (it might be made hexagonal in section, enabling a spanner to be used on it, a technique which is sometimes useful to loosen stubborn screws).

10.7.2 'Hard' maieutics

In contrast with the 'soft' maieutics of Gordon, Osborn and others, there are more positive and powerful idea-generating principles which are useful in specific cases.

In 1974 a competition in invention was run in which one of the set problems was, 'the direct production of electricity by pressure, as on a surface, or by heat without going through conventional engines. Existing methods are excluded'.

Now such a problem can be tackled by a 'hard' maieutic. It is a simple matter to write down all the known physical principles which might be used in such an invention. Indeed, a book summarising all known physical principles has been written specifically for this kind of use.*

Now any new method for generating electricity must be based on one of this limited set of principles or on some new principle, as yet undiscovered. It is then easily shown that no practical invention of the kind is possible unless some such new piece of fundamental physics is forthcoming. Naturally, no prize was awarded in that section.

Untypically exact is the maieutic which enables us to 'design' all possible regular convex polyhedra (solid bodies, with all the faces regular polygons of the same size and shape). The most familiar regular polyhedron is the cube, which has six faces, each of which is a square, meeting by pairs in 12 edges. The 12 edges meet by threes in eight vertices (see Fig. 10.4). If the three faces A, B and C which meet at the vertex Q are imagined to be opened up flat or 'developed', as in the figure, it is seen that an empty angle PQR has been formed between faces A and C. Now any vertex of a regular convex polyhedron must open up in this way when developed, otherwise it

---

* Hix, C. F., Jr and Alley, R. P. (1958). *Physical Laws and Effects*. New York: Wiley.

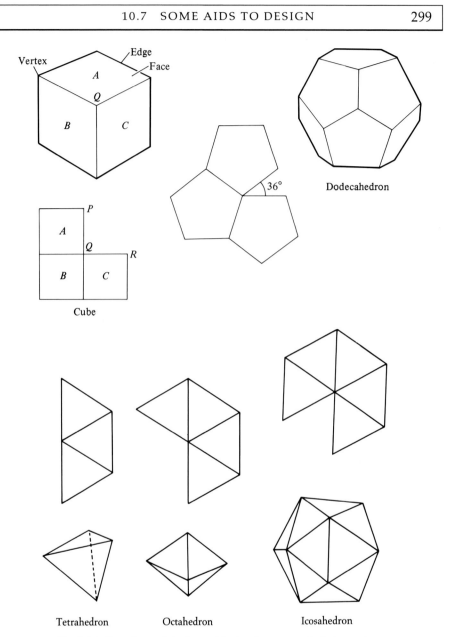

Fig. 10.4. Regular polyhedra, with developments of their vertices.

could not be folded up to a solid corner. It follows that the angles at the corners of the polygons meeting at any vertex of a polyhedron must add up to less than 360°. This shows at once that there cannot be a regular polyhedron with hexagonal faces, for the corner angle of a regular hexagon is 120°; since the least number of faces which can form a vertex is three, a solid vertex cannot be formed by regular hexagons, or any regular polygons with angles of 120° or more, i.e., with more than six sides. The search can thus be limited to polyhedra with triangular, square and pentagonal faces.

There cannot be a regular polyhedron with four or more square faces meeting at a vertex, since four times 90° is 360°. Similarly, the angle of a regular pentagon is 108°, so that only with three faces can a vertex be formed. With equilateral triangles, however, a solid vertex can be formed with three, four or five faces. It is thus plain that there can be at most five regular polyhedra, one with pentagonal faces meeting three at a vertex, one with square faces meeting three at a vertex and three with triangular faces meeting, respectively, five, four and three at a vertex. In fact, all five exist – they are the dodecahedron, the cube, the icosahedron, the octahedron and the tetrahedron (see Fig. 10.4).

A similar example, of a more functional nature, is the demonstration that there are only three kinds of moving joint with one degree of freedom (Chapter 4). More typical was the 're-invention' from fundamental principles of the spear-thrower and the sling in Chapter 2: there only two structural elements deemed likely to be available to a primitive culture, the strut and the tie, and two design principles were deduced, that the missile should move faster than the hand so the latter would not absorb too much of the available energy, and that the time taken by the hand should be as great as possible. The element and the principles once set out, the rest follows.

A hard maieutic gives a precise way of tackling a problem for which it is suitable, but its range of application is often rather limited. One 'hard' maieutic mentioned by Osborn is *morphological analysis*, which is particularly associated with Zwicky. A related idea is that of the analysis of function in which all the functions to be performed by the device to be designed are listed, together with the means by which each might be performed, including the elimination of the function itself.

10.7.3 Application of analysis of function to the invention of a wave energy converter

To see how the analysis of function works consider the functions involved in a device to collect the energy of sea-waves (a wave energy converter, or WEC) such as the one discussed in Chapter 7. First of all, there has to be a surface on the WEC which is pushed to and fro by the sea, where useful work is done by the waves, what may be called a *working face*. Now for a force to do work, it must be resisted somehow and somewhere by a *balan-*

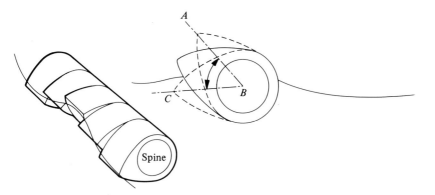

Fig. 10.5. Salter's 'nodding duck' WEC: cam-shaped bodies oscillate
through an angle ABC about a spine: the forces on one body are balanced
by those on another further along the spine, which is parallel to the
crests of the waves.

cing force. Then the work done by the sea on the working face must somehow be collected, a function that may be called *concentration*. There are
other functions to be performed such as converting the energy to a useful
form (e.g., electricity) but these three will do for a start. We can tackle the
design of a WEC by considering how these three functions, those of the
working face, the balancing force and the means of concentration, are to be
provided.

In the WEC considered in Chapter 7, the working face was provided by
the rubberised cloth forming the air-filled bags, but in some of the devices
now being developed it is a rigid surface, as in Salter's 'duck' (Fig. 10.5),
and in others it is simply an air surface (Fig. 10.6).

The balancing force can be provided in three different ways. Either the
WEC may be held rigidly, say, by building it into the sea-bed, or it may be
so massive as to resist the wave forces by sheer inertia. Finally, it is possible

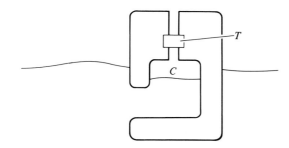

Fig. 10.6. National Engineering Laboratories' oscillating water-column
WEC: water rising and falling in the space C pumps air through the
turbine T.

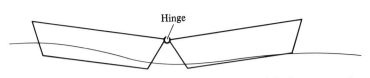

Fig. 10.7. Cockerell's contouring rafts WEC: bending of the hinge is used to pump fluid, so extracting energy from the waves.

to balance the wave force by another wave force somewhere else on the device, using a strong structure to connect the two.

In the flexible bag WEC of Chapter 7, the energy is concentrated from where it is extracted from the waves, all over the skin, by means of the pumped air which is collected up in a duct and delivered to the turbine: the means of concentration is air. In Cockerell's device (Fig. 10.7) the hinge between the rafts is flexed by the waves. This flexing is resisted by hydraulic rams which act as pumps, extracting energy from the flexing, so that in this case the means of concentration is hydraulic, as it is in Salter's 'duck'. In some other devices the means of concentration is seawater itself, used in a simple water turbine.

Each of the three functions of the WEC we have considered thus has at least three means of realisation, which we can set out as in Table 10.1.

A table of this kind is helpful in studying a design problem like that of a WEC, where there is a clean sheet to start with. But it is also useful in cases where, unlike the WEC, there are precedents in existing and successful designs, where what has gone before may blind us to other possibilities.

A good name for this kind of table is 'table of options', because it sets out in condensed form all the options the designer has, which are all the possible combinations of one means from each row. Table 10.1 has three means in each of three rows, giving 3^3 or 27 combinations. The Salter 'duck' has a solid working face, balancing by structure and concentration by hydraulics, as do the Cockerell rafts, but the oscillating water-column has a working face of air and concentration by air, while still using structure for balancing.

It is not practicable to go through all the possible combinations and produce a worked-out design corresponding to each one, and anyway, the same combination can lead to vastly different schemes, as the Salter

Table 10.1.

Function	Means of performing function
working face	flexible sheet, rigid solid, air surface
balancing	rigid fixing, inertia, structure
concentration	air, hydraulics, seawater

'duck' and the Cockerell rafts show. The approach, often called a morpho-logical chart, has been much favoured in German universities, but they tend to use many functions, often quite minor ones which confuse the issues, and elaborate tables. On the other hand, most experienced designers would think about which were the critical functions and select means for them without using a table at all, based on their experience and their judgement.

Nevertheless, the small or kernel table of options, including only the most important functions, has its uses, especially for new or unfamiliar types of problem, and particularly in the early stages. Once well-engaged upon the design, a return to this table is only likely to be useful when there seems no way forward, and fresh ideas are sought.

Progress in wave power at Lancaster has been briefly described in Section 7.5.1. It hinged on a number of sharp insights developed over the years, some leading to what was in effect a return to the table of options, a change in one of the critical means. The first of these insights was into the advantages of air over water for concentration of the power, and the second was into the high structural cost associated with large devices, leading us to a preference for small devices (called 'point absorbers'). The third was the recognition of the need to use advanced control techniques to maintain vigorous motion, or quasi-resonance. The fourth was the appreciation of the critical role of the balancing function and the recognition (using Sherlock Holmes's maxim) that inertia might be a practical means of balancing if it were that of a relatively small mass moving relatively fast. Important, too, in the development of each of the four major designs we have studied was the emergence of a design philosophy, as explained in Section 7.5.1.

In practice, the development of insights and the emergence of design philosophies are always likely to prove more important than the analysis of function, but it is not always in our power to bring them about. A designer struggling for insights is like a poet scratching for inspiration. Both can work away at bits, write or sketch outlines, rack their brains or go out for a brisk walk, but they cannot command the muse, as it were. On the other hand, any plodding fellow can sit down and draw up a simple table of options, with the major functions to be performed and the alternative means of performing them, and by thinking about it and its contents, perhaps develop useful ideas.

10.7.4 Classification

The table of functions suggests another maieutic; this involves producing a systematic classification of all possible designs, an arrangement of pigeon-holes in rows and columns into which they will all fit. This has been done for the WEC in a very simple form in Fig. 10.8, using the ways of per-

		Method of balancing wave force		
		Fixing to sea bed	Inertia	Large structure
Method of concentrating power	Air	6	4	3
	Hydraulics	3	1	4
	Sea water	4	?	??

Fig. 10.8. Simple classification of WECs.

forming the functions of balancing and concentration as a basis. The columns correspond to ways of balancing the wave force, and the rows to ways of concentrating the power. In this way, nine pigeon-holes are created. The WECs of Figs. 10.5 and 10.7 both balance the wave forces against one another by means of a relatively large structure, and so will fall in the last column: in both, hydraulic rams or motors are the most likely way of concentrating the power, so both fall in the middle pigeon-hole of the last column. The numbers in the pigeon-holes are the numbers of WECs of each kind known to me and recently being studied or developed. Question-marks indicate empty pigeon-holes, and two question-marks indicate an unlikely combination.

It is easy to make more elaborate systems of pigeon-holes, by including other functions or by using other kinds of subdivision. For example, WECs using structures for balancing wave forces may lie parallel to the wave crests (as with the 'duck' of Fig. 10.5) or at right-angles to them (as with the 'sausage' of Fig. 7.23). Wankel used a classification with 864 pigeon-holes in his search for an improved internal combustion engine: of these pigeon-holes, 252 corresponded to combinations of features which were impossible or very unfavourable, 149 corresponded to known inventions, and the remaining 463 were 'unoccupied'. Wankel and his colleagues spent years studying the possible contents of some of the most promising of these 'unoccupied' pigeon-holes and produced a remarkable series of rotary engine configurations, one of which was the basis of the Wankel engine (Fig. 9.13).

10.7.5 Breaking logical chains

In Section 7.5.1 it was explained how we came to look again at using inertia for balancing the wave forces in a wave energy converter, having at first judged it impractical on the grounds that the mass would be too large. The chain of the logic was then broken by examining this link, and perceiving this weakness in it: if the acceleration of the mass were high enough it would not have to be too large, since the reaction force is the product of the mass and the acceleration.

A sovereign method in design is to set up a logical chain that purports to lead inexorably to a conclusion, and then to test each link of that chain in the hope of breaking one. A link found to be unsound may open up a new direction. The maxim of Sherlock Holmes mentioned in Section 7.5.1 often reduces to the same device.

10.7.6 Combination and separation of functions

Two of the most useful steps in design are to combine functions in one component and its opposite, to divide functions previously performed by one component between two or more parts. On the whole, combination of functions is likely to prove valuable in simple or cheap devices, while separation of functions is often a vital step in advancing engineering design.

The 'cat's-eye' which marks the centre of a road consists of reflectors embedded in a rubber body whose principal function is structural: however, in addition, when squashed by a tyre, the rubber 'squeegees' the surfaces of the reflectors, keeping them free of mud. Thus one moulding supplies the functions of body and wiper.

In the original form in which Newcomen produced it, the steam engine used its working cylinder as a condenser as well: after the cylinder had filled with steam on the 'out'-stroke of the piston, water was injected to condense the steam. Unfortunately, this spray of water also cooled the cylinder, and so only about one-third of the incoming steam did useful work, the remainder being needed to heat the cylinder again. Watt's great invention was to condense the steam in a separate vessel, the condenser, which was always cool, while the cylinder remained hot all the time: this simple separation of the functions of working cylinder and condenser reduced the fuel consumption to less than half.

10.7.7 The gear-pump

Figure 10.9(a) shows a gear-pump, a device in which two gears enclosed in a casing fitting closely over the teeth pump oil from the high- to the low-pressure side. The oil in the spaces A between the teeth at the outsides of the gears is forced from the low-pressure to the high-pressure side, while only a small fraction returns through the mesh of the gears at B. The ends of the teeth are closed in by two end-plates.

When the pump is delivering at a high pressure, the end-plates are forced apart by the fluid, which leaks back between them and the ends of the gears. For a long time, this effect meant that gear-pumps were unsatisfactory for use at any but relatively low pressure. However, this situation was altered by separating two functions of the end-plates, those of *sealing*, preventing the fluid leaking back past the gears, and *retaining the pressure*, the struc-

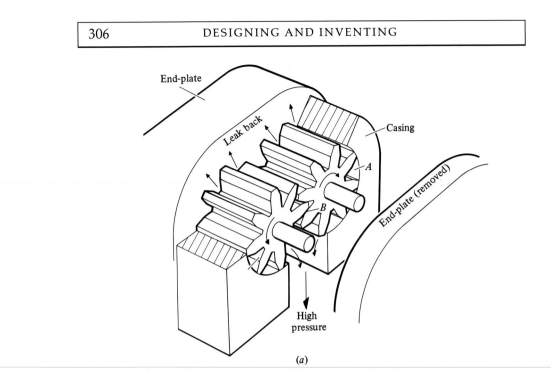

(a)

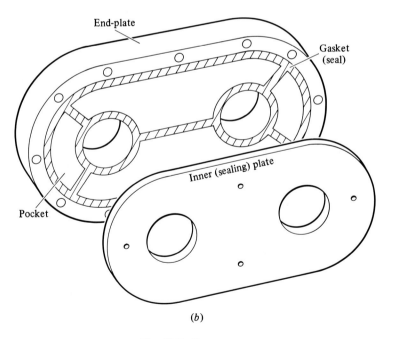

(b)

Fig. 10.9. Gear-pumps.

tural function. (If the distinction between sealing and retaining pressure is not clear at first, think of them as preventing leaking and bursting, respectively.) By providing two sets of end-plates, inner ones to do the sealing and an outer pair which are structural, gear-pumps can be made satisfactory for quite high pressures (see Fig. 10.9(b)). Pockets are provided between the inner and outer plates and holes through the inner plates admit fluid to these pockets: the holes are so positioned that the pressure distributions on the two sides of the inner or sealing plates nearly balance, the net effect being to squeeze them gently against the gears, so effectively sealing against leakage back.

Once the idea of using a separate inner plate for sealing is conceived, any competent engineer can design the pockets, holes, seals, etc., although it takes a very good engineer to obtain the best possible result in terms of low leakage and low friction. The difference between the simple end-plate of Fig. 10.9(a) and the double pocketed one of Fig. 10.9(b) is rather like the difference between the simple view of many organic mechanisms (like the knee) and the complex and subtle reality.

10.8 Abstraction as an aid to design

Functional analysis starts by adopting a very abstract view: for the end-plate, we substitute the abstract functions of sealing and retaining pressure, discarding as far as we can all our preconceptions, e.g., of a steel plate with holes in it. One or other kind of abstraction, of discarding attributes like 'slab-like form' or 'made from steel', is at the back of many maieutics. We adopt 'thing-language', as it were, so that relationships are stripped of associations as far as is profitable: we look for a leak-preventing 'thing' and a burst-preventing 'thing'. In the limit, abstraction becomes self-defeating, the whole problem melting away into complete vagueness: if we are designing a gear-pump, we had better keep the gears, even though we can change their shape, and certainly we must keep the pump.

Abstraction is essentially verbal thinking: we throw away hardware, and keep only the labels, 'seal', 'pressure casing' or whatever, and we then try to visualise all the forms which each label might be attached to, and choose the most suitable one. But once we start this convergent process of design, from the all-embracing abstract to the particular concrete, it is then that the difficulties begin, and it is then that the individual quality of mind of the designer becomes crucial. It must, for example, be well-stocked with ways of sealing and of resisting pressure, to be able to fill out the abstract shell with more realisable images, and it would help further if these ways were seen as part of a systematic and comprehensive view of sealing and resisting pressure, with all their special advantages and disadvantages clearly perceived.

10.8.1   The design repertoire

The mind of the designer should be like a rich, open soil, full of accessible resources, of which the chief is an ordered stock of ways and means, illustrated by many examples: the chapter of this book on structures (5) could be regarded as a small beginning for such a stock of ways and means in one important area. The sum total of all such stocks might be called the 'design repertoire'.

From the last section, it is clear that the repertoire should be organised on the most abstract lines practicable, so that any item in it is likely to be recalled in as many contexts as possible: too particular a label may cause it to be overlooked.

As an example, consider the problem illustrated in Fig. 10.10(a), where a part A is to be fixed to a part B by means of a screwed extension of A

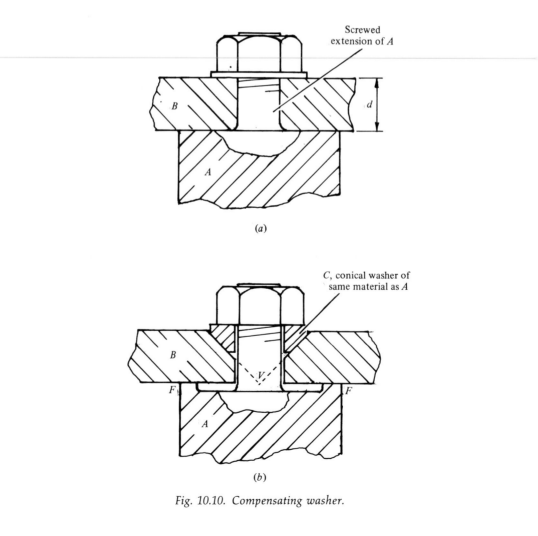

Fig. 10.10. Compensating washer.

passing through B and secured by a nut. A simple enough problem, you may say. But supposing the whole is subjected to great heat, and the best material for A expands more than the best material for B: then the distance d between the shoulder of A and the nut will increase more than the thickness of B, so that A will become loose on B.

What is to be done? It might be possible to use some other material for A or B, so that the difficulty did not arise, but the experienced designer with a good knowledge of the repertoire may suggest the elegant solution of Fig. 10.10(b), where a conical washer C is used, about which two points must be noted. Firstly, the vertices of both the cone of C and of the conical hole in B in which it fits are at V, in the plane of the interface FF of A and B. Second, C is made of the same material as A (preferably, also, the nut has about the same coefficient of expansion as A). When C and A are heated together, they tend to expand as a whole, growing in size without changing in shape, so that the vertex of the conical surface of C remains in the plane FF. Similarly, when B expands, the vertex of its hollow conical surface remains in the plane FF. Consequently, the fit between C and B remains unaltered because of the special kinematic property of cones, that they will still fit together if one expands uniformly relative to the other. This rare property of the cone, not possessed by, for example, cylinders or spheres, is often turned to advantage by the designer, and it should occupy the ready-access store of his memory.

Notice that if this item of the repertoire were recorded in the mind, say, as 'a way of fixing bolts where differential expansion tends to loosen them', it might not come to mind in some other context. Recorded among 'bits of geometry useful in problems of fit', it is likely to be recalled in a wider range of circumstances in which it might be useful.

The raw material for a comprehensive design repertoire has been published, and many sections or parts of it have been partially abstracted or condensed. Some of these abstractions are more or less comprehensive, but none seems to me sufficiently confined to the essentials or organised on a basis which is fundamental enough. However, the designer seeking to furnish his mind will find no shortage of excellent reading.

10.8.2   Standard problems

One advantage of an abstract approach is that many of the problems begin to look the same. Chapter 6 discussed matching, and the way this standard design problem manifests itself in bicycles, fountains, man-powered flight, and so on. Even more common is the problem I call 'disposition', which occurs where a lot has to be squeezed into a small space, or some commodity which it is difficult or expensive to provide has to be put to different purposes to the greatest effect. For example, the division of the floor area of a house into living rooms, bedrooms, hall, etc., is a disposition problem. Often it is not just the amount of the commodity, area in this case, given to each function that matters, but how its share is distributed or shaped. The areas of the various rooms in a house might be satisfactory, while their shapes and the way they communicated with one another were not.

The helicopter tail-shaft in Fig. 3.12 presented a simple disposition problem. Three structural functions had to be supplied, stiffening against bending, transmitting torque to drive the tail-rotor and preventing circumferential buckling, each requiring layers reinforced in different directions. For all three purposes the outermost layer or position is highly desirable, whereas only the last function makes a bid for the inner berth. The arrangement chosen was the best compromise.

There are many disposition problems in the design of cars, for example, in components of the engine and in the engine itself, in the constant velocity joints and in the general layout. In the design of the suspension, it is not just the total volume occupied by the parts which matters, but their position relative to the wheel and what other uses compete for that particular space. The success of the Issigonis Mini design was due largely to its bold and effective solution of the disposition problem. Such underused space as there is in the traditional arrangement of a car, with the crankshaft lying fore-and-aft, is alongside the engine, because the engine is rather long and narrow. By arranging the engine with its crankshaft transverse, i.e., lying across the car, Issigonis was able to shorten the engine space and hence the whole car, producing a design with the maximum passenger space for its size.

The Mini also illustrates another aspect of design, in that it emerged at the ideal time (1963) to take advantage of an ingenious new form of suspension, the Hydrolastic system, which gave a much improved ride in a small car and also fitted well into the small space available, contributing further to the excellence of the solution of the disposition problem.

Whenever the abstract version of the design problem takes a standard form, like matching or disposition, then the experience of many past problems of the same kind can be drawn upon to help solve it.

10.9 Further examples and principles

10.9.1 An application of analogy

To see a simple use of a maieutic, suppose we wish to enable one cable to pass through another at a particular point, as in Fig. 10.11(a). The black square is to be a device permitting cable 2 to pass through cable 1, both carrying a load throughout the process.

An obvious (and very close) analogy occurs where a man drawing a boat along a quayside by means of a painter comes to an obstruction on the edge of the quay, a mast, say (Fig. 10.11(b)). Suppose the painter is in his right

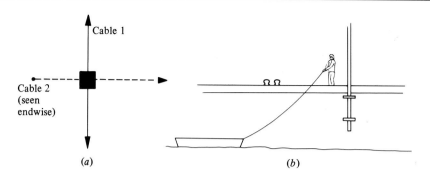

Fig. 10.11. Cable problem and an analogy.

hand: when he is opposite the mast he reaches round the far side of it with his left hand, passes the painter from one hand to the other, and proceeds. the mast is analogous to cable 2, the painter and the man's body to cable 1, and his arms and hands to the 'black box' of Fig. 10.11(a).

Now we have to embody this 'hand-changing' mechanism in some relatively simple hardware, and here the repertoire can help. We often have a choice between an alternating or a rotating system for producing a succession of events. For example, table-light switches of the press-button type can function on an alternating, see-saw type of mechanism or by the continuous rotation in one direction of a little shaft; lifts can either go up and down or always up one side and down the other, as in the paternoster kind. Even closer to our problem, consider double doors used in foyers to exclude draughts. You can either have two sets of doors, of which the first closes (ideally) before the second opens, or revolving doors where one leaf comes in behind you before the one in front opens on to the foyer or the street. In moving the boat past the obstruction, the man's arms work rather like double doors, but the revolving doors may be easier to embody in a mechanism (Fig. 10.12). Imagine that you, passing through such doors, are cable 2, the wall of the hotel is cable 1 and the black box is the revolving door: essentially, all we need add is some sort of sliding joint between the curved partitions AB and CD and the edges E of the door leaves, so that a tension

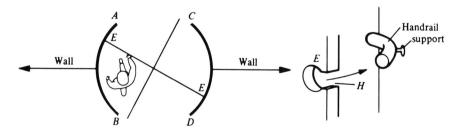

Fig. 10.12. Revolving doors.

in the 'hotel wall' can be carried right through. Such a sliding joint could be made by fitting a round handrail to the curved partitions, passing through holes H near the edges of the door leaves: a slot would have to be made from each hole through to the edge to allow the handrail supports to pass through.

The essentials of a scheme are now complete: we throw away inessentials, and are left with Fig. 10.13. One side of the revolving door can be replaced with a simple yoke, and guides (not shown) must be provided to ensure that cable 2 passes into the 'door' at the right place, so it does not jam against the yoke or in the corner K (incidentally, can you see why the number of spokes must be increased?).

In Gordon's terminology, this is an example of the use of Personal Analogy, leading to a Direct Analogy: Personal Analogy is a 'harder' maieutic because we can always try it but Direct Analogy depends on hitting on a suitable analogue. I did not hit on the swing-door analogy until I came to put this problem in this book, and the personal one should be enough for an experienced designer with a well-stocked repertoire.

10.9.2   The steam catapult

A very bold and elegant design is that of the steam catapult for launching aircraft from naval vessels, due to Mitchell.

A tank of very hot water under pressure acts as a spring, because if the small amount of steam over it escapes it is immediately replaced by spontaneous boiling. Such a tank can be drawn on for large volumes of steam with only a slow falling off of pressure, whereas a tank full of steam would soon be exhausted.

The steam from this tank or steam accumulator (which contains mostly water, not steam) is admitted to a very long cylinder in which is a piston carrying a hook engaged with the aircraft towing gear (Fig. 10.14(a)): the

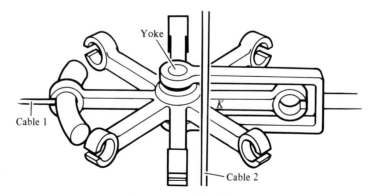

Fig. 10.13. Essential contents of black box in Fig. 10.10.

piston is driven down the cylinder, accelerating the aircraft rapidly to launching speed.

Now in Fig. 10.14(*a*) the connection to the aircraft is made by means of a very long piston rod, so that if L is the travel of the piston, the overall length of the device at the end of launching is rather more than twice L. A naive suggestion would be to make a slit in the cylinder wall and hook the piston directly to the aircraft, reducing the length overall by roughly L, which would be a very important advantage indeed.

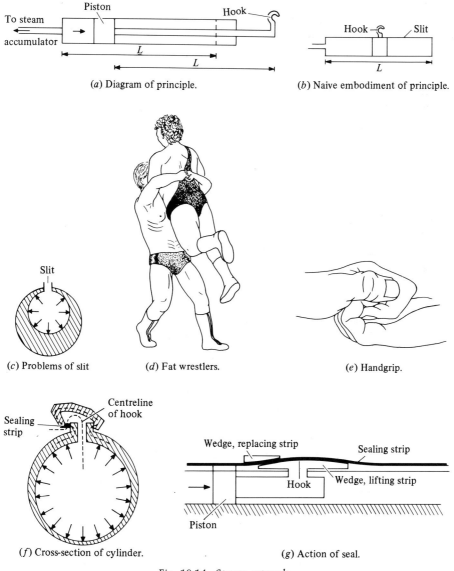

(*a*) Diagram of principle.   (*b*) Naive embodiment of principle.

(*c*) Problems of slit   (*d*) Fat wrestlers.   (*e*) Handgrip.

(*f*) Cross-section of cylinder.   (*g*) Action of seal.

Fig. 10.14. *Steam catapult.*

A practical man might pour scorn on this suggestion, for two reasons; what the designer should do is to note the problems but defer judgement, according to Osborn's precept. The two objections are, that steam would escape from the slit, and that the cylinder would burst by opening up of the slit. Can we overcome these objections by finding effective means of performing the functions of *sealing* and *retaining the pressure*, of preventing *leaking* and *bursting*? (Notice how like the problem of the gear-pump this abstract formulation is: perhaps in the last analysis there are only a few design problems, just as it used to be said that there were only seven jokes in the humorous magazine 'Punch'.)

We could solve the pressure problem by making the cylinder wall very thick, especially at the bottom where it is subject to the most severe bending moment. But the result would be bulky and objectionably heavy, and the slit would still open slightly, making the sealing problem more difficult.

Consider the fat wrestlers in Fig. 10.14(*d*). The arms of the man in the shorts will be much better able to prevent the other bursting out of his grip if he can lock his hands together as in (*e*): notice that his arms are then in tension, but his finger-tips are in compression and being forced together.

This principle is followed in the catapult, the cylinder of which has a cross-section of the form shown in Fig. 10.14(*f*). The slit is formed in a joggle in the cylinder wall and sealed by a steel strip. The action of the steam pressure throws the joggled part of the wall into compression like the wrestler's finger-tips, and clamps the strip firmly: the cylinder is now a much more efficient structure than the form in Fig. 10.14(*c*) (though still not nearly as good as a plain cylinder). The hook has to have an S-bend in it in order to pass through the slit, and there are also important details to observe in the longitudinal sense, which are shown very diagrammatically in Fig. 10.14(*g*). In front of the hook and travelling with it is a wedge which forces the sealing strip out of the slit, and behind the hook a further wedge replaces the strip. The piston must be even further behind, since behind the piston the sealing strip is firmly clamped by steam pressure and could not be moved by a wedge. Also, the slit must be properly sealed before the piston and the steam arrive.

10.9.3 Clothes-pegs: specification and classification

In this problem, the last, the specification of the device to be designed bulks large. I have left it till last, because it has defeated me, but, nevertheless, a good solution is probably possible and would be a benefit to many people. I mean to have another try when this book is finished.

There is another reason why it is not so illogical to deal with specification in the last problem. There is a general expectation that specifications should come first, but that is often the wrong way in design. A market study may prompt a company to decide that a certain type of car would be popular and a tentative specification may be drawn up. But as the design work proceeds, it may appear that some features would add excessively to the cost, while others, not included in the draft specification, can be added relatively cheaply, improving the value for money the company can offer the prospective purchaser.

It could be maintained that clothes-pegs are already satisfactory and that there is no scope for improvement. We can test this proposition by devising an ideal specification and seeing how the real peg compares.

Let us begin by drawing up an analysis of function for a clothes-peg, labelling the functions F1, F2, etc.

F1 secure peg to line
F2 disconnect from line when required

F3 secure clothes – light and heavy, thin and thick
F4 free clothes easily when required

F5 withstand shear
F6 move easily along line when required.

Most of these require little comment. It is convenient to have pegs which remain on the line when not in use, which is why F1 is needed, but they need to come off easily for storage, washing the line, etc. F5, resist shear, means that the peg, when retaining clothes, must resist forces tending to slide it along the line, otherwise heavy blankets may behave as in Fig. 10.15. However, when not holding clothes the pegs should readily slide along the line to where they are needed (F6).

As well as the analysis of function, in this case it is useful to consider some desirable features, or desiderata, (D), which will go towards specifying more fully the peg we seek to design.

D1 The pegs should not be 'handed'. By handed, the engineer means coming in left- and right-hand kinds – thus, ears and gloves are handed, but socks are not. Nor should they be handed in another, weaker sense; it is possible to design pegs so that although they are all the same, they are asymmetrical and must have a particular side towards the clothes they secure. The wheels

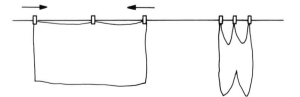

Fig. 10.15. Shear in clothes-pegs.

of a car are handed in this sense, in that they are all the same, but face in opposite directions in use, because they must have a particular side on the inside. With either kind of handedness, the pegs have to be in pairs of left and right, left and right, along the line (see Fig. 10.16). This in itself is a slight nuisance, but it is a worse one when articles fall out of step with pegs, e.g., when odd numbers of articles requiring only one peg are picked out of the laundry basket.

D2 Pegs should be readily operable with part of the hand, e.g., the thumb and forefinger of a hand holding a garment. The other hand is holding a basket, which leads to the important

D3 which could be expressed cryptically as F4 without F2. With present designs, you must pull the peg off or squeeze it to release the garment, remove the garment from the line and finally replace the peg on the line or put it in a container. If the peg were to remain on the line when the garment was released, the operation might be easier and quicker and better suited to the one pair of hands which are all we have.

D4 When the wind wraps the garments over the top of the pegs, tangling and catching should not occur.

The list of desiderata could be extended, with freedom from rust, corrosion, rot and staining, and so on, but these four are those which are likely to exercise an important influence on the design concept. D3 is not met by any present designs and those which meet D2 (with steel springs) do not meet D4 and often corrode and stain, so that there is room for improvement.

The desiderata form a sort of outline specification, lacking quantities such as maximum opening of jaws, gripping force exerted, and so on, which would be added in a proper study, which this is not.

Since I have not yet succeeded in designing a good clothes-peg, I can only indicate lines on which a solution might be found. The crucial function is F3, securing the clothes, which is most likely to be done by gripping them between jaws. These jaws must exert a consderable closing force, and they must do so over a wide range of gap between them. Let us suppose that the jaws open by a maximum of 10 mm, and that the gripping force F they exert is an average of 15 N (see Fig. 10.17(a)). If we pull them apart with our fingers, we have to exert a force averaging 15 N through a distance of 10 mm, which means doing work = force × distance = 15 N × 1/100 m = 0.15 J (see Chapter 2). The energy we expend in opening the jaws is transferred to the steel spring of the peg. The ability of the peg to function as

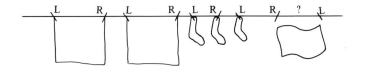

Fig. 10.16. Handedness in clothes-pegs.

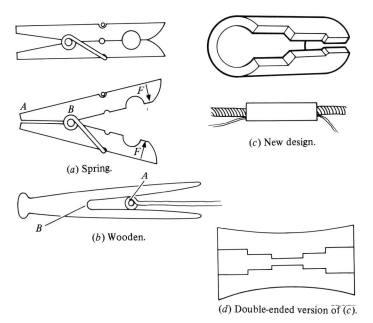

(a) Spring.

(b) Wooden.

(c) New design.

(d) Double-ended version of (c).

Fig. 10.17. Clothes-pegs.

a gripping device is entirely dependent on this stored energy. The same is true of bolts, which are also gripping devices, although the force is usually very much higher and the distance very small: in the very long bolts that run from top to bottom of large marine engines, however, the distance, the stretch of the bolt, may be larger than the opening of a clothes-peg.

Now when the clothes-peg is used, the user has to do some small amount of work, about 0.15 J in our example. But the force required is considerable – the arm AB is usually only about half as long as BC and so the *average* force the finger or thumb at A must exert is about 30 N, and the *maximum* force is about 40 N: moreover, an average force at the jaws of 15 N is rather low for a satisfactory grip – about 25 N would be better. The lever arms AB could be made longer, but would then tangle more with clothes blown over the top of them. Is there any way the energy storage requirement can be reduced?

There is, and it can be seen in the wooden clothes-peg of Fig. 10.17(b). In this case, the garment, wrapped round the line A, slips into the slot of the peg until it reaches a point where it fits. As it is forced beyond that point, it wedges the jaws apart a small amount, developing a large gripping force in doing so because the peg is very stiff (in the sense of Chapter 3). The work done is smaller because in the product force × distance the force may be rather larger than in the spring peg but the distance is much smaller, perhaps a millimetre (this is as well, because the energy storage capacity of wood is small when it is used inefficiently, as in this case where it easily splits from point B). Thus we have two kinds of clothes-peg, those like the wooden one which have a means of fitting the garment-plus-line and small energy storage, and those with large energy storage and no fitting.

The idea of matching is important in clothes-peg design; the spring peg of Fig. 10.17(a) is well-matched, because the wider the jaws are spread the tighter the grip becomes, and thicker garments or blankets need a greater force to hold them. The wooden peg of Fig. 10.17(b), on the other hand, is badly matched, because the thicker, heavier load will not go so far into the slot and so, being better placed to lever the jaws apart, will be less tightly gripped.

We have now assembled a few ideas which will help in approaching the design of a clothes-peg – a functional analysis, the outline of a specification in terms of desiderata, the beginnings of a classification based on function F3, securing the clothes – large energy storage, small energy storage plus 'fitting' and others which could be added. We have recognised that there is a matching aspect.

While none of the clothes-pegs I have designed meets the ideal specification in all points, the best may be a significant improvement on existing types. I will not disclose my most promising lines of attack, but Fig. 10.17(c) shows one which is interesting but which is unlikely to lead to a winning design. It consists of a short length of plastic tube tapered towards one end (the right-hand end in the figure) with a slot along one side: the slot tapers towards one end.

This design fails to meet D1 because it is handed in the weaker sense (i.e., all pegs are the same but they must be put on the line pointing alternately left and right). A double-ended version (Fig. 10.17(d)) meets this difficulty, but is clumsy. The classification is 'small energy storage plus fitting', and the matching is excellent, because the thin end of the tube is made smaller in wall-thickness. It is elegantly simple. D3 and D4 are met very well, but D2 is not – instead of the ideal 'half-hand' operation, it needs the help of a second hand to secure a garment: it is worse even than the wooden peg in this respect. The spring peg comes nearest to the half-hand ideal, and it does so by using one of the key features in the evolution of man, the opposable thumb. The peg is forced open, not by the co-operation of two hands, but by the finger and thumb of the same hand. Here is an important clue in the search for a better design: the collector has only one pair of hands, but two thumbs, two index fingers and six other fingers, all capable of some degree of independent action. In using the spring peg the ability of thumb and index finger to co-operate in opposition while moving as a whole in relation to the rest of the hand, and the ability of the three remaining fingers to grip the line or a garment in addition, all come into play: any successful design of clothes-peg is likely to use the versatility of the human hand more fully than the one in Fig. 10.17(c) does. Of course, the problem would be simpler if we had hands designed with two opposable thumbs.

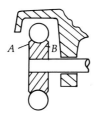

Fig. 10.18. Wheel of toy car.

It would be possible to write a book on this relatively simple design problem, introducing many aspects and many possibilities not touched on here – for example, are separate pegs necessary or desirable? could we have a combined line and pegs which secured the clothes to itself? what materials are suitable? Moreover, a relatively abstract and intellectual approach is possible and may be fruitful, using concepts like energy storage, matching, exploitation of the versatility of the hand, and so on. Design is a rich study.

10.9.4 Avoiding arbitrary decisions

Ideally, the designer should have a reason for every decision he makes, however trifling it may seem. For example, if two circular flanges are to be bolted together by six bolts, it might seem natural to space the bolts equally round a circle, at 12 o'clock, 2 o'clock, 4 o'clock, and so on. But by spacing the holes slightly unequally, it can be ensured that they can only be fitted together in one way, which may help to avoid mistakes in assembly.

Figure 10.18 shows a section through a wheel of a toy car: should the two flanges A and B which retain the tyre be made of the same diameter or not? If we think long enough, we see that by making the outer flange A of a larger diameter, we can make it almost certain that if the tyre comes off the rim, it will do so on the inner, or B side, and so be retained on the axle and not lost.

Figure 10.19(a) shows a plastic grommet for reinforcing holes in fabric, e.g., where a tent-pole fits: it consists of two rings C and D glued one either

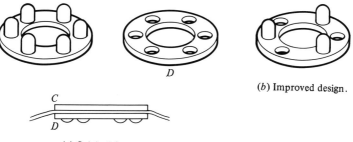

D

(b) Improved design.

C

D

(a) Original form.

Fig. 10.19. Reinforcing grommet.

side of the fabric round the hole. The protrusions on *C* are forced through the fabric and into the holes in *D*. Is any improvement possible in this very simple fitting?

Figure 10.19(*b*) shows a change which means that one kind of ring only need be made and supplied, saving on tooling, maintaining of stocks, and so on.

10.9.5   Clarity of function

There is one respect in which nature may not be a good example for the human designer. As a rule, and one which has few exceptions, good design by man is characterised by clarity of function: it is very clear how every essential function is performed, usually in a simple, direct fashion. Trousers are kept up by belt or braces, not by belt and braces, with a few tin-tacks to be on the safe side. Structures of any importance can generally be described in terms of one or two simple elements or sets of elements of the kind studied in Chapter 5. The I-beam in Fig. 5.20 has flanges to take bending and a web to take shear.

Function in nature is much more confused and complex, with a tendency for several components of different kinds to combine in performing one task. An example is the running of the cheetah, in which the whole body flexes and straightens to increase the distance covered at each stride so that every muscle of the body contributes the most it can to the speed of the animal. Sometimes this will give greater economy or higher performance, no doubt, but in engineering such complexities do not seem to pay.

10.10   Designing and inventing: summary

The last few sections should have given the reader some idea of the kind of thinking that goes into elegant functional design and the sheer intellectual delight it can afford. There is no single best method of tackling a design problem – each must be tackled on a catch-as-catch-can basis devised to suit both the cast of thought of the practitioner and the particular nature of the case. Remember always to watch for the emergence of a design philosophy, what Osborn has called the 'autonomy of the object'. Think as abstractly as seems fruitful in the beginning, use analogies from as wide a field as possible and look for indicators of a powerful sort, like the six times greater diameter of the water turbine as compared with an air turbine for a WEC. Use functional analysis and any useful elaboration of it, like classification, and draw lots of rough schemes. Seek the elegance that is usually found in good design even if it is restrained and appeals only to the educated eye. Be ruthless (and this is hard) in sacrificing the pretty but flawed idea for a sound but more prosaic one when you have tried, but failed, to eliminate the flaw. Above all, be greedy: let us not be satisfied with internal combustion engines which are quiet *or* economical *or* clean – let us seek to change the 'ors' to 'ands'. For a designer, it is legitimate to seek to have his cake and eat it too.

10.11   The engineering designer, society and the future

For a century and a half at least, there has been concern about the pressure on the earth's resources of the human species, with its rapidly-increasing numbers and its increasing material expectations from life, and recently such concern has become widespead. Crises of many kinds are foreseen, and two at least are real enough dangers. The first is the increasing population, which it requires only a responsible attitude on the part of ordinary people to halt. The second is the shortage of available energy which threatens to come in a couple of decades or so. It seems likely that one day we shall have virtually unlimited energy from fusion reactors, but it is not sure and in any case, the prospect is too remote for any complacency.

Those who blame our troubles on technology have much justice on their side. After all, the 'population explosion' is the result of reducing natural mortality rates. This is a proud achievement of medicine, agricultural technology, municipal engineering and other technologies, but it has brought its own troubles. Fortunately, the birth rate has dropped to near replenishment level in most of the developed world and is dropping in many other places. Nevertheless, it will be necessary for us to moderate the demands we make on our resources, and the designer has an important part to play in enabling us to do so with as little apparent loss as possible. For example, we must design means for large populations to retain freedom of individual movement and a measure of privacy when needed, while maintaining a pleasant and healthy environment. Above all, we must conserve energy (available energy, strictly, since energy conserves itself, as was seen in Chapter 2) and develop new sources of power to fill the gap before we have fusion power.

11

Some case studies

11.1 Introduction

It may be a good way to illustrate the absorbing intellectual problems presented by engineering design to recount some of the more interesting
examples that have come my way, and so I have added this new chapter to
the second edition. Several themes emerge, for instance, that there is nothing new under the sun and that most, if not all, invention reduces to the
adapation of old ideas to new circumstances or new functions. Another is
that contemplating a problem in a ruminative frame of mind may lead to a
feeling that there must be a better way than those so far thought of. Known
solutions sometimes have a hint of the digging of holes simply to fill them
up again, which may be a pointer to such a better way, if we can only spot
what it is.

The spotting of a case of digging a hole to fill it up is one example of the
important insights which mark the progress of much design. Often it is
only after a lot of apparently fruitless work that there suddenly dawns a
great light, and we know where we are going. There is then rapid progress
for a while, but alas, there are often further struggles before we achieve a
satisfactory conclusion, if indeed we ever do.

11.2 End-balancing of gas turbines

Figure 11.1 shows, very diagrammatically, the compressor and turbine of a
simple gas turbine. Air enters the compressor at the left and is raised to a
high pressure, passes through the combustion section (not shown) where it
is raised to a high temperature by the burning of fuel, and then enters the
turbine, still at high pressure. It expands through the turbine, doing work,
about half of which goes directly to drive the compressor via shafting which
is not shown. The other half constitutes the useful power output of the
whole engine. The air and products of combustion leave the turbine,
roughly at atmospheric pressure. The compressor and turbine rotors run in
ball and roller bearings, represented in the figure by small squares.

At the delivery end of the compressor and the inlet end of the turbine,
seals are needed to prevent leakage where the shafts penetrate the casings,

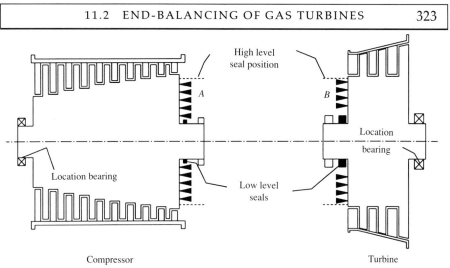

High level
seal position

A

B

Location
bearing

Location bearing

Low level
seals

Compressor   Turbine

Fig. 11.1. End loads on compressor and turbine of a gas tubrine.

and this example is concerned with those seals and the problems associated with axial pressure loadings on the compressor and turbine rotors. Because the engine rotates at a high speed, the seals cannot contact the shafts and so there must be a small gap, or clearance, and the seals must leak.

These leaks result in a serious loss of power, because the high-pressure air and gases which escape do not pass through the turbine and so do no work. If the losses total 2% of the intake air, then there will be a 2% drop in total turbine power. However, there will be no drop in the power demanded by the compressor, so that if this is 50% of the total turbine power, the useful power output will be reduced by $2 \times 2\%$, or 4%. It is highly desirable to keep down these seal leakages.

Now the leakage from such a seal will be much less if it is of small diameter. Not only will the circumferential length of the leak be reduced, but the clearance itself, the radial gap through which the leakage occurs, can also be reduced if the diameter is smaller, for reasons which will become apparent. It would therefore be desirable to make the seals at a small diameter (at a low level) as shown in the figure, were it not for the problem of axial pressure loads which would arise.

In a typical case, the bearings in which the compressor rotor runs might be one roller and one ball bearing. The ball bearing *locates* the rotor axially in the casing, while the roller bearing exerts no axial constraint, a classical arrangement in full accord with the Principle of Least Constraint (Section 7.4.2): if both ends ran in ball bearings and no special provision were made, then differential thermal expansion could lead to the shaft exerting large equal and opposite forces on the bearings and destroying them. In the classical arangement, the ball bearing is called the 'location bearing', and in the figure the location bearings are indicated by a diagonal cross.

11.2.1   End loads

With a small diameter or low-level seal at the delivery of the compressor, the pressure in the space A (see Fig. 11.1) exerts a large forward force on the compressor rotor, indicated in the figure by arrowheads. In the same way, with a low-level seal, the pressure in the space B at the front of the turbine gives rise to a large rearward force on the turbine location bearing. Because these forces were unacceptably large, it was usual to provide large-diameter seals and put up with the power losses. The challenge to the designer, however, is to have the cake and eat it – to reduce the seal leakage without exerting large loads on the bearings. The art of avoiding large axial pressure loads on rotors such as those of compressors and turbines is called 'end-balancing', so that in effect, what we would like is good end-balancing without high-level running seals.

11.3   A first solution to the end-balancing problem

This problem involves the 'digging-a-hole-just-to-fill-it-up' effect. The compressor rotor is subject to a large forward force due to unbalanced pressure loads, and the turbine rotor is subject to a similar, roughly equal, rearward force. With low-level seals these forces are balanced out, one against the other, via the location bearings and the external engine casings, involving the transfer of large loads from the rotating to the stationary parts and back again, an expensive thing to do. Why not transfer the load directly through the rotating parts? This could be done simply enough with a long whippy tiebolt running down the centre, and the replacement of one of the locating bearings with a non-locating bearing. The tiebolt would have to bend easily to cope with inaccuracies in alignment of turbine to compressor, in accordance with design principles (Chapter 7).

   This idea was rejected because it was felt that it was important to retain a separate location bearing for each rotor. I then proposed that we should insert a pneumatic cylinder in the tiebolt to fix the load in it while allowing it to lengthen and shorten freely under that load to cope with relative movements axially between the two location bearings.

   This version is illustrated in Fig. 11.2. The cylinder is rigidly attached to the turbine rotor behind the rear bearing and the piston which slides in it is connected by a long flexible tiebolt to the front end of the compressor rotor. Air at compressor delivery pressure is admitted to the cylinder and exerts a large forward force on the cylinder, and hence on the turbine; the piston is subjected to an equal and opposite rearward force which is transmitted to the compressor. As a result, the resultant axial pressure loads on the location bearings are both made acceptably small. Because the piston does not rotate relative to the cylinder the seal between them can be made

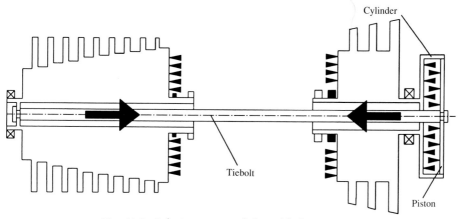

Fig. 11.2. Solution using tiebolt and balance piston.

near perfect, so leakage is restricted to that of the two low-level seals, which is quite small compared with that of high-level seals.

This account has been somewhat simplified, particularly as regards the turbine, but it retains the essential feature, namely, the transmission of the balancing force between the two rotors without going through the stationary components. Notice how obvious this approach seems looking at Fig. 11.1, with the complementary unbalance forces, forward on the compressor and rearward on the turbine, with the loop needing to be closed as directly as possible, that is, via the rotating parts.

This invention was greeted with enthusiasm in the company, which was Napiers, and patented. Figure 11.3 shows a refinement which was introduced

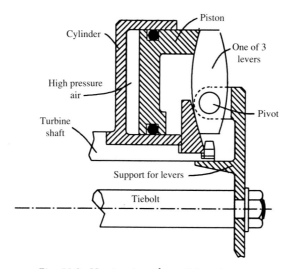

Fig. 11.3. Version to reduce piston size.

in order to improve the overall arrangement. The diameter of the cylinder in Fig. 11.2 was inconveniently large, so I designed a lever mechanism to enable a smaller cylinder to do the job. The smaller load from the smaller cylinder was applied to the outer end of a number of radial levers which multiplied the force by about 2.5 times before applying it to the tiebolt, so that a cylinder of only about 60% of the diameter sufficed. These levers and their bearings posed another interesting design problem because of the large centrifugal field in which they operated, but that is another story.

11.4   Further ideas

I then had a much better idea, but unfortunately, the first had so taken hold that I was never able to get a proper hearing for the second. In effect, it was the simple tiebolt scheme with only one location bearing, as described at the beginning of the last section, but with one crucial refinement, as shown in Fig. 11.4. To understand this it is necessary to know something more about the problems of gas turbines.

The blades of compressors and turbines work by virtue of a pressure difference between the two sides which ideally is maintained right along their lengths. There is therefore a tendency at the tips of the blades for a flow to take place over the end, from the high- to the low-pressure sides, and if this flow is large, it will involve large losses (Fig. 11.5(a)). The loss can be reduced by fitting shrouds on the tips, as in Fig. 11.5(b), thus shutting off the leakage route, or, alternatively, by making the clearance between the tips of the blades and the wall of the casing enclosing them very small, so that only a small leakage takes place through it. The shroud solution works well in preventing loss, but is expensive and heavy: keeping the clear-

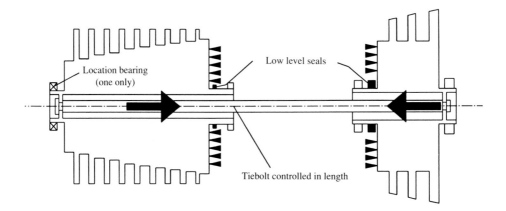

Fig. 11.4. Simple solution with tiebolt of controlled length.

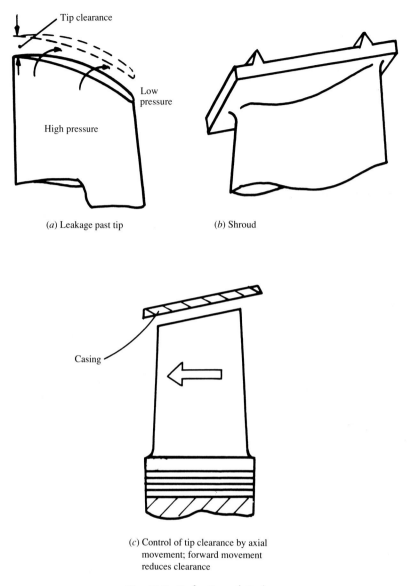

(a) Leakage past tip   (b) Shroud

(c) Control of tip clearance by axial
    movement; forward movement
    reduces clearance

Fig. 11.5. Reduction of tip losses.

ance small works well but is made difficult in the turbine by thermal expansions in the rotors and casings.

11.4.1   Clearances during transients

When a turbine starts up from cold, the casings warm up and expand much faster than the turbine rotor. The tip clearance over the turbine blades will thus increase to begin with, and then decrease as the rotor slowly follows the casing in expanding. If the turbine is now shut down, the casings will contract faster than the rotor, and may actually tighten over the blades: the engine cannot be restarted in this condition. It is therefore necessary to make the clearance big enough in the first place not to disappear under these circumstances, which means large tip losses. The condition which produces the minimum tip clearance will be called the *critical transient*, because normally it will be a transient state. Because of this difficulty of the critical transient, it is usually impossible to avoid large tip clearances in the steady running state, and so it is usual to fit shrouds, but the large centrifugal load they impose on the blades means the blades must be thicker and heavier, and so on through the discs and the shaft, increasing the overall weight of the engine substantially, not to mention the cost.

Now because of the conical form of the turbine casing, the design of Fig. 11.4 enables the tip clearance in the turbine to be reduced or increased by moving the turbine rotor axially forward or backward respectively, so maintaining it at a small value at all times. The means of effecting this is by controlling the length of the tiebolt, which can be done mechanically with a small actuator or by heating and cooling it, say, by making it hollow and flowing hot or cool air through it as required. Although this is a slow means of control, it can still be made faster than the changes for which it compensates, and so quite fast enough.

This design offers low seal loss, low blade tip losses, low cost and low weight. It is also mechanically simpler than other schemes, but subtler, which is often the way with progress in engineering. It all hinges on the observation that the two large unbalanced forces on the rotating parts ought to be balanced out via the rotating parts, not via a transfer from rotating to stationary and back again, a digging and refilling holes operation.

Modern aircraft gas turbines have two or three fans or compressors and as many turbines to drive them, interconnected by two or three shafts running one within another, and it would be very difficult, at the least, to apply these ideas to them.

11.5 Desalination of seawater using ocean temperature differences

This example also shows the avoidance of unnecessary transformations, the digging of holes simply to refill them.

In deep seas in the tropics, the surface water may be at about 24 °C, while below 3000 m in depth the temperature is always close to that at which water has its maximum density, which is about 4 °C. For almost a century proposals have been made to use this temperature difference to generate power, a source of energy now usually known by the acronym OTEC

(Ocean Thermal Energy Conversion). Most modern proposals are based on using the warm water to boil ammonia, expanding it through a turbine and condensing it with the cold water: this is just like the ordinary steam power cycle except that ammonia is used because of its lower boiling-point. The weakness of this proposal lies in the low overall temperature drop available, which means, for instance, that the ammonia boiler will have to have a very large surface area to avoid using up too much of the precious temperature drop to transfer the heat across it, from the warm seawater to the liquid ammonia. Similarly, the condenser must be very large, and both condenser and boiler must be able to withstand salt water and be kept free of fouling: these considerations tend to make the cost of using this 'free' energy very high.

It occurred to me that some way not involving heat exchangers (boilers, condensers and the like) should be found. Also the fact that the source energy was low-grade thermal (Section 2.5) suggested that some useful task which could be done using such energy directly, without conversion into mecahnical work or electrical energy, would be more likely to result in a viable device, thus avoiding the digging of holes simply to fill them up again. The task must also be a large one, and the only one that came to mind was the desalination of seawater. One way of removing the salt is to freeze some of the seawater, giving crystals of ice that are ideally free of salt, which becomes concentrated in the remaining liquid. To do this requires low-grade thermal energy in the form of refrigeration, which can be derived from a small temperature difference by an absorption cycle, in a way which will now be explained.

In my invention, liquid propane is injected into some cold seawater. The drop in pressure as it is injected causes the propane to boil, extracting heat from the seawater and producing a cloud of ice crystals, which are subsequently washed free of salty water and melted to give relatively salt-free water. Figure 11.6 illustrates the plant, and this first step of propane injection takes place in the reactor, in the top right corner. However, we now require to restore the vaporised propane to the higher pressure and liquid state so it can be used again; the first step is to absorb it in liquid butane, which is less volatile than propane. However, the absorption liberates a great deal of heat, so liquid butane and cold water are sprayed into the propane vapour, which is absorbed in the butane to give a liquid mixture of propane and butane, the heat of absorption being given up to the cold seawater; neither liquid propane nor butane, nor a mixture of them, mixes with water. This process takes place in the absorber, in the top left of the figure. The mixture is then pumped by the pump at the bottom of the figure to a higher pressure, at which propane is liquid at the temperature of the cold water. Because of the small volume of the liquid, the pumping work is not a serious

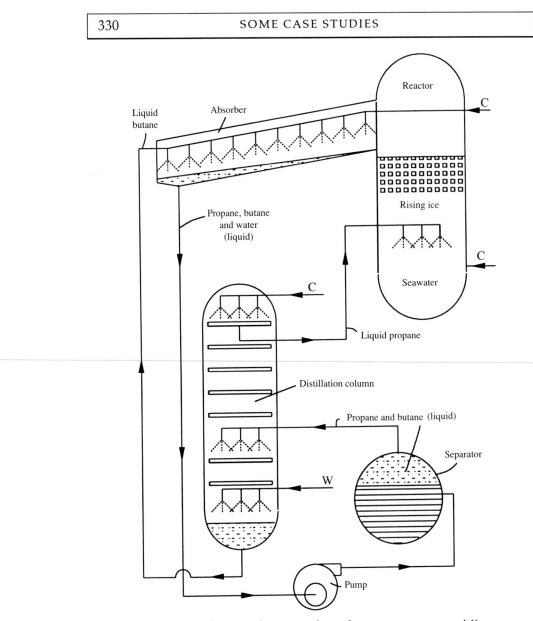

Fig. 11.6. *Desalination of seawater driven by ocean temperature differ-ences: C, cold water from 3000 m down; W, warm water from near surface.*

problem; it is a small term in the energy balance. The water is separated out in a separator, leaving the liquid solution of propane in butane which goes on to the last step.

11.5.1 Generating the liquid propane

The reader may have noticed that no use has been made so far of the warm surface water, and the available energy must be associated with the extraction of heat from the warm water and the rejection of some of that heat to the cold water; have faith, the former is now going to happen (the latter took place in the absorber).

The last step in the process is to separate the propane from the butane, and this is done by distillation, just as alcohol is separated from water in the making of whisky. The liquid mixture of propane and butane is pumped into a vertical *distillation column* (in the centre of the figure) which is heated at the bottom by warm surface water injected directly into it, and cooled at the top by sprays inside the column of cold water from the depths. All this is illustrated diagrammatically in Fig. 11.6. The interior of the column is full of vapour, and contains also a system of plates with liquid dripping down through them, which improves the efficiency of distillation, in a way very akin to the regenerative systems looked at in Chapter 7.

The higher boiling-point butane separates at the bottom and the lower boiling-point propane at the top, both as liquids. The liquid propane is injected into cold water in the reactor, and so the process goes on. The ice crystals are allowed to float upwards in water in another column (not shown) and melted, some of the fresh water being allowed to run back through the column to wash down the salty water which would otherwise contaminate it.

11.6 Construction of the desalination plant

All this is difficult to understand, even for someone familiar with thermo-dynamics. The important things to note are that the whole process is just an adaptation of a known invention, the absorption refrigerator, and once started, an engineer competent in the field could easily complete the whole scheme; there are no boilers or condensers, just large vessels containing sprays, like ordinary showers; the process is driven almost entirely by the low-grade thermal energy available in the difference in temperature between the two levels in the sea, with only a little mechanical power required for the pump and a few other minor purposes. If such a plant were built on land, then the chief cost would lie in the large pressure vessels, able to withstand a pressure of several atmospheres, such as the reactor, the distillation column, and so on.

However, this plant would be in the sea, so that by arranging the various parts at the right depths, the pressure of the sea outside could be made equal to that inside, and the vessels could be made of heavy proofed fabrics of the kinds used for inflatables. The reactor vessel would be nearer the surface than the distillation column, because it is at a lower pressure. The pump is pumping the mixture of butane, propane and seawater to a higher pressure but also to a lower depth, so it only has any work to do because the density of the mixture is rather less than that of the surrounding sea-

water, and has to be pushed down against its own slight buoyancy. By
taking advantage of the sea, which provides an environment of different
pressures at different depths, the structural demands (Section 5.2) become
small and cheap to provide, and even the pumping requirement is greatly
reduced.

It must be confessed that there are serious practical problems. The butane
and propane are slightly soluble in water, and would tend to be lost with
the seawater leaving the plant; the depths at which the plant would need to
be placed are not sufficient entirely to protect it from storms, and making
it strong enough to resist them would add greatly to the otherwise low cost;
above all, it is difficult to find anywhere where there are the right conditions, of deep water close inshore, a sufficient demand for fresh water and
warm surface water. It is a pretty concept, but not, I fear, a really practical
one. However, it presents another interesting aspect which is worth looking at.

11.7   Raising the cold water

A plant of this kind or any OTEC installation needs a vast flow of cold
water from depths of around 3000 m. The power required for this purpose
depends on the size of the pipe through which it is raised, and in the design
I looked at this was 25 m for a pumping power of 1500 kW. Now to make
such a huge pipe, 3000 m long and 25 m in diameter, and to install it is
clearly very expensive, so that it was natural to think of making it also of
fabric. However, if suction were applied to the top, such a tube would collapse under the external pressure, so the pump must be at the bottom, so
as to pressurise it (Fig. 11.7). This presents some difficulties, but the fabric
tube with the pump at the bottom seems to me likely to present the best
solution. There is a problem in making the pump itself cheaply, because it
needs a huge, propeller-like rotor running very slowly. It could be made of
reinforced concrete, however, and driven by a massive belt, and running a
suitable electric motor at that depth should not be difficult.

11.8   Ocean transport of liquid natural gas

The late 1950s saw the beginning of preparations for the transport by sea of
liquid natural gas (LNG), which consists mainly of methane. Unlike liquid
petroleum gases (LPG), methane, which is much more abundant and one of
our most important fuels today, is too volatile to be carried as a liquid under
pressure at ambient temperature. At atmospheric pressure it boils at
−160 °C; the idea was to carry it as a cold liquid at that temperature and at
very slightly above atmospheric pressure, so that gas would tend to leak out
of the tanks rather than air leaking in, which would be much more danger-

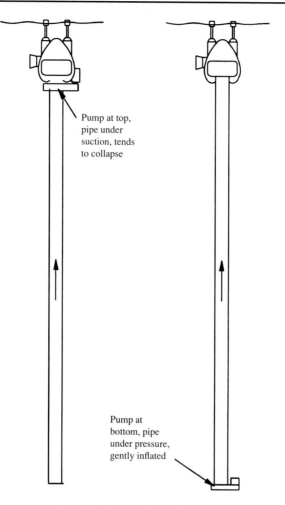

Pump at top,
pipe under
suction, tends
to collapse

Pump at
bottom, pipe
under pressure,
gently inflated

Fig. 11.7. Raising the cold water.

ous. If enough air leaked into a tank to form an explosive mixture with the
methane vapour above the liquid, it would form a potential bomb.

The tanks must be insulated, to reduce the heat leaking in from the surroundings: nevertheless, some heat will always leak in, so that the cargo is
always simmering, as it were, and evolving a flow of gaseous methane which
is used as part of the ship's fuel, or if there is too much for any reason,
burn safely at a stack.

I was employed to look at this subject in 1959 and 1960. I made a systematic study of the possibilities, which simply endorsed the lines already being
followed but showed up nothing very new. The general idea was to have
large holds, with an insulating lining, probably of balsa, faced with a secondary barrier to contain any leakage from the tank proper, which in turn fitted

inside the lining (Fig. 11.8(*a*)). The problem lay in the word 'fitted', because if the tank fitted when it was warm, it would no longer fit when it was made cold and shrank accordingly. If an aluminium alloy tank 40 m long is cooled from 20 °C to −160 °C, it shrinks in length by about 140 mm, which is a lot of slack. If such a tank were placed in a hold which it fitted well when warm, and then cooled by adding liquid natural gas, the weight of the liquid would burst it, unless it were made very strong. Now at the temperature of liquid natural gas, ordinary steel is very brittle, so that some more expensive material must be used. To make a *free-standing* tank in such a material which is strong enough to withstand the internal pressure due to the weight of the LNG is a costly option.

The free-standing tank option is another design which smacks of digging holes to fill them, as a consideration of Fig. 11.8(*b*) will reveal. The LNG presses outwards on the walls of the tank: a fraction of a metre outside the tank is the ship's hull, with the sea pressing inwards upon it, with nearly the same total force, though that force is differently distributed, as the arrows in the figure show. If the tank could just lean on the lining of the

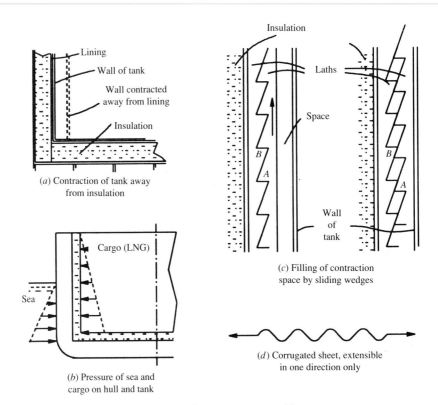

(*a*) Contraction of tank away from insulation

(*b*) Pressure of sea and cargo on hull and tank

(*c*) Filling of contraction space by sliding wedges

(*d*) Corrugated sheet, extensible in one direction only

Fig. 11.8. The contraction problem.

insulation, then it could be relatively lightly constructed and very much cheaper and the hull would be relieved of some load also.

11.8.1 A naive idea

I had a naive proposal to overcome this problem, which did not find favour, however. This was simply to let the tank contract away from the lining when it was cooled prior to adding the cargo, and then slip something into the gap to fill up the space. One element of the something I proposed is shown in Fig. 11.8(c): it is just a pair of wooden laths with interlocking sawtooth profiles. A pair of laths A and B are placed as shown, with their teeth fitting into each other. Sliding lath A upwards causes its teeth to ride up on those of B, taking up the gap.

The objection to this simple scheme was that when the tanks had been emptied of cargo and started to warm up, if one lath A had inadvertently been left in the 'up' or 'thick' position, then the freedom of that tank to expand would be impaired and it would be seriously damaged. I thought this was hardly a good enough reason to abandon a cheap and effective solution, and that good safeguards could be provided to prevent such an accident, with horns blaring and lights flashing all over the place and spare lath pairs ready to slide down beside any that failed to work, but they would not have it.

11.9 Membrane tanks

Another way was to make a stretchy tank which would stretch under the pressure of LNG to fit the hold tightly and lean on the walls. If we had been able to use leather, which was the material used in some early trials of London buses running on LNG, then a leather tank would easily have stretched to fit. But a huge tank holding 20 000 tonnes is a different matter from a bus fuel tank holding a fraction of a tonne, and the registration authorities demanded a metal tank, quite rightly. So we set out to devise a 'membrane', as we called it, a stretchy metal wall made of thin sheet corrugated in some cunning way so that it could stretch uniformly in its plane by the required amount without being too highly stressed. If the sheet were flat and it were stretched the full amount of the contraction, so it remained the same size, then call the stress which would be produced in it f. Then we needed to make a corrugated sheet such that if it were stretched the same amount, the maximum stress in the metal would be only about one-fifth as much, or $0.2f$.

Consider a case where the sheet was only required to stretch in one direction. Here the answer is a simple corrugation, like corrugated iron. Such a sheet is quite stretchy at right-angles to the corrugations (see Fig. 11.8(d)), but not at all stretchy along them. What we needed was a sheet that would stretch uniformly in all directions. I do not think I have ever tackled a problem where it was so difficult to see how to start, but in the end I found a solution.

I deduced certain things about a solution, as follows.

(a) It would contain no long unbroken straight lines (very easy, that one).
(b) It would only stretch in all directions at once, i.e., if you stretched it in one direction it would automatically extend at right-angles by the same amount.
(c) It would have a repeat pattern, like wallpaper, which would be like a chess-board and not, say, like a honeycomb.
(d) In the chessboard pattern, there would be elements that rotated in the general plane of the wall, one way in the white squares and the other way in the black squares, as in Fig. 11.9(a).

(For engineers, (b) is equivalent to an apparent Poisson ratio of −1, but when the sheet is bent out-of-plane it has an equal anticlastic curvature, as if it had a Poisson ratio of 1).

11.9.1 A solution

Using these and other deductions about properties that any solution must, or would be likely, to have, I eventually arrived at the solution shown in

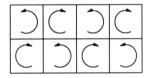

(a) Chessboard pattern of rotations

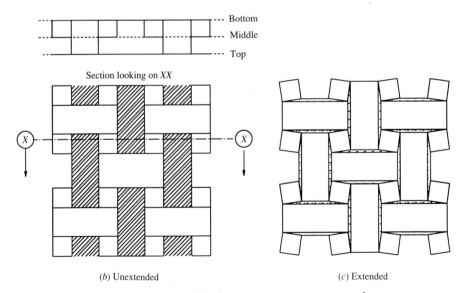

(b) Unextended   (c) Extended

Fig. 11.9. Extensible sheet (movements exaggerated).

Fig. 11.9(*b*). Imagine three planes, all parallel to the general plane of the sheet, a top plane, a middle plane and a bottom plane, as shown edgewise at the top of Fig. 11.9(*b*). The sheet is built up from rectangles lying in these three planes and walls connecting the rectangles to form an unbroken surface, all the walls being perpendicular to the planes. View (*b*), looking directly on the planes, shows the rectangles, consisting of the little squares, all of which lie in the middle plane, and two sets of rectangles, the shaded ones, lying in the bottom plane and the others in the top plane. When the sheet is stretched, the squares all rotate, alternately clockwise and anti-clockwise in a chessboard pattern, and the walls distort, as in Fig. 11.9(*c*), in which the stippled areas are the walls. The whole sheet constitutes a mechanism, with all the curious kinematic properties which I had predicted. This approach of studying what properties a solution would have is sometimes known as Pappus's principle, and it can be very useful to engineers as well as to mathematicians like Pappus.

11.9.2 Practical forms

The shape shown in Fig. 11.9 is not easy to make and we worked on it to produce more practical forms. My colleague John Petty observed that the rectangles can be done away with if two sizes of square are used, one turning clockwise and one anti-clockwise, as in Fig. 11.10, a very ingenious idea. Such a tank was built and tested, and then a small ship, the M.V. *Findon*,

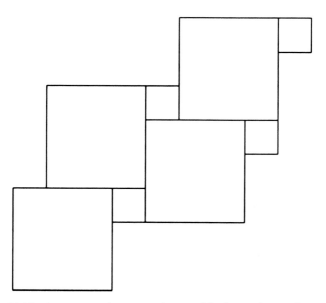

Fig. 11.10. Asymmetrical version of extensible sheet. The small squares are all on one level, and turn one way. The large squares are all on another level, and turn less and the other way.

was fitted with one and tested, but the idea was not proceeded with further. I think now we went astray with the asymmetrical design, and we should have produced a practical version in quite a different way. In the meantime, I had a more promising idea, the hybrid tank.

11.10   Hybrid tanks

The concept of the hybrid tank stemmed from a consideration of how a weak unstretchy tank would fail. The sides would bulge out until they rested against the insulation and be happy, as it were; it is the corners which would suffer as the pressure of the LNG tried to iron them into the corners of the hold, from which they would now be separated by a gap which might be almost 100 mm. Figure 11.11 shows a horizontal section through such a weak unstretchy tank, with the contraction greatly exaggerated: the broken line shows the tank cooled and hence contracted, before the cargo is pumped in, and the bulged shape caused by adding the LNG is shown in solid. Everywhere except in the corners the tank is well-supported, so that if some arrangement could be made to support this narrow band round each wall, A in the figure, the tank would be entirely satisfactory. I produced a number of designs of this type, based on supporting the walls at A by short beams, like floor joists structurally, but of metal, some of which were patented.

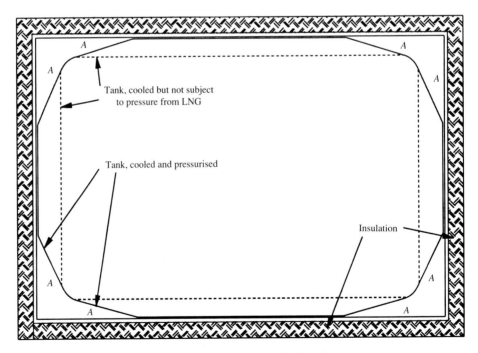

Fig. 11.11. Principle of hybrid tank.

Some designs had no moving parts, and supported the 'joists' by flexing elements. I am sure these solutions would work very well, but they would be rather expensive, and possibly more expensive than some of the alternatives. Other ideas were simpler, but had undesirable features like pivots instead of flexures; it is not a good idea to put pivots where they cannot easily be got at. I would like now, 30 years or so later, to try to design something working on the same principle, but rather less elegant and a great deal cheaper: I am sure it should be possible.

11.11   Nitrogen backhauling

It requires large expensive plants using considerable power to liquefy the natural gas at the port of supply, so that the 'cold' in the LNG represents a lot of money: 'what use could be made of it?', I was asked.

Any liquefied gas is a potential source of energy – indeed, at Lancaster University we built a go-cart that ran on liquid nitrogen, at a time when the National Coal Board were considering the use of this source of energy in the mines, for safety reasons. But the amount of energy from the surroundings made available by the cold sink provided by LNG (see Section 2.5) was not enough to bother with: it was only as 'cold' that the 'cold' was valuable, and only one industry used such large quantities of such deep cold. It was, of course, the LNG industry, so the logical thing to do was to return the cold to the supply port, where it could be used to liquefy more LNG, and the tankers would have a return cargo. The cold had to be in some cold substance which must be either a genuine cargo in itself, and there was no such thing, or else in something which was virtually free and no trouble to dispose of. The only answer was to ship the cold back in the form of liquid nitrogen obtained from the air. I was very pleased to hit on this answer myself, although I was not the first to do so as it appeared later.

Two new kinds of plant were needed: at the gas field end, a plant which liquefied natural gas, using liquid nitrogen in the process and returning warm gaseous nitrogen to the atmosphere, and at the other end, a plant which vaporised and warmed up the natural gas while separating nitrogen from the air and liquefying it ready for transport back to liquefy more natural gas (Fig. 11.12). The total energy requirements could thus be reduced by about half. There were other interesting possibilities, such as the use of the nitrogen at the supply end, together with more gas, to make artificial fertiliser. Nitrogen had to be used rather than liquid air because the oxygen in liquid air makes it very dangerous near methane.

Had the boiling-point of liquid nitrogen (about 74 K) been nearer that of methane (113 K), the scheme would have been much more attractive, but the large gap of nearly 40 K meant that quite elaborate plant was needed,

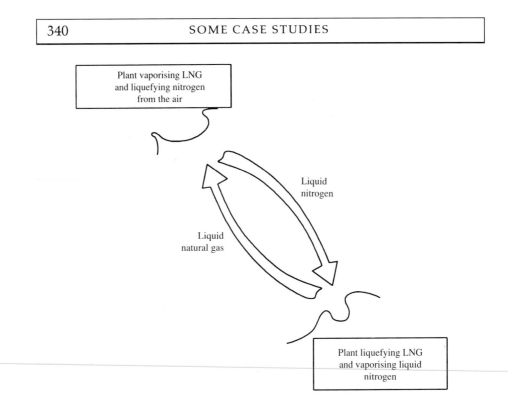

Fig. 11.12. Nitrogen backhauling.

of high capital cost. It was very interesting devising these plants, some of which were patented, but their complexity took a lot of the gilt off the gingerbread of this nitrogen backhauling concept. The chief principle I used was that of matching, for matching the nitrogen and the methane was the essence of the matter: bridging the 40 K gap was really a matter of heat-pumping

We did not use nitrogen backhauling, for reasons which have to do with matching, strangely enough, the matching of the supply of gas to the demand, which is very uneven. A plant which vaporises liquid gas while liquefying nitrogen needs to work at a steady rate. It is easy, however, to produce large flows of gas from the liquid merely by heating it, in a relatively cheap plant, and as only one-tenth of the supply in Britain was to be from liquid imports it made sense to keep it for topping-up purposes at times of high demand. In a country like Japan, which imports a great deal of LNG, it might be that nitrogen backhauling would pay.

11.12   Air-cooled turbine blades

The power delivered by the turbine of a gas turbine engine is roughly proportional to the absolute temperature of the gases entering it, so that raising

that temperature (inlet temperature) was an important engineering objective in the early 1950s. At Napiers we were the first to make a significant
step forward, with an increase on an experimental engine equivalent to an
increase of turbine inlet temperature of 150 K. If we had been able to do
this without introducing other losses, this rise of about 12% would have
increased the turbine output by 12%, but since about half the original
power went to drive the turbine, the engine output would have gone up by
about 24%.

The difficulty in raising the temperature was the limited heat resistance
of the materials, most importantly, that of the turbine blades. Brilliant work
by the metallurgists on the Nimonic series of alloys had pushed the art
almost to its limit, and new materials were then a long way off. Any big
improvement had to come by cooling the turbine blades. The most favoured
plan was to make the blades hollow and pass relatively cold air direct from
the compressor through them, via passages designed to ensure the material
was kept everywhere sufficiently cool to survive the stresses imposed upon
it. The principal difficulty lay in designing such a blade so as to be manufac-
turable, and the two philosophies available were, broadly speaking, to make
passages in a solid blade or to fabricate the blade from sheet. For a company
with small resources and if early success was required, the fabrication route,
though seen as beset with awful difficultes, seemed the best hope. In the
outcome, this hope proved justified, though it did Napiers no good in the
end.

One of the difficulties, which turned out to be trivial, was that of centrifugal loads. To reduce centrifugal loads in solid blades, they are tapered
strongly, so the root cross-sectional area which takes the load is much bigger
than the average cross-sectional area, to which the load is roughly proportional. With a sheet metal blade, the only way to achieve low centrifugal
stresses is to taper the thickness of the sheet, from thick at the bottom to
about 60% less at the tip (see Fig. 11.13(a)) and scepticism had been widely
expressed about the practicability of making such sheet. We rolled pieces of
sheet round mandrels and turned them down in a lathe, not just to a uniform taper but with a change of taper part-way up, which gave a very good
approximation to the ideal (Fig. 11.13(b)). I believed this could be done, but
I went to the people in the shops and asked them very diffidently if they
thought they could manage it: they were sure they could and they proved
it, and that was one difficulty out of the way.

Unfortunately, we were unable to obtain sheet in the most heat-resisting
alloy, Nimonic 90, and the best we could get was Nimonic 75, nominally
only suitable for temperatures of about 150 K less.

The cooling air had to flow through the blade in a way which ensured
every part was adequately cooled. We devised a simple way which might
work, with an internal divider which led the air up the most severely-heated

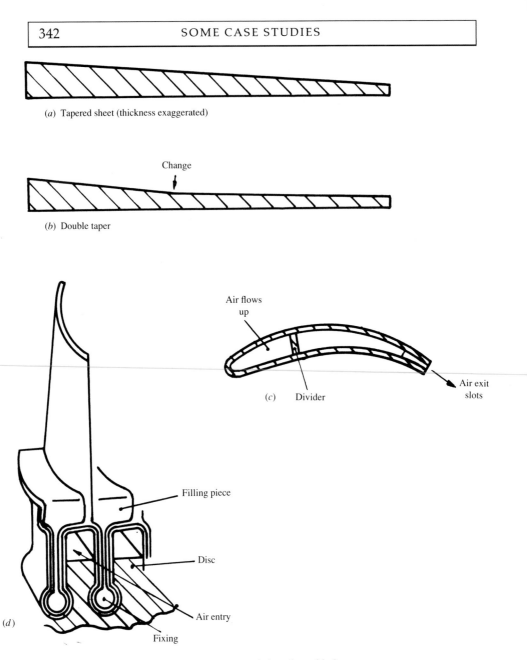

(a) Tapered sheet (thickness exaggerated)

Change

(b) Double taper

Air flows
up

(c)   Divider

Air exit
slots

Filling piece

Disc

Air entry

(d)

Fixing

Fig. 11.13. Air-cooled turbine blade.

part, the leading edge, and back down the centre and trailing edge (Fig.
11.13(c)). On the way back, it was exhausted through a series of small slots
in the trailing edge. We had devised more refined designs as well, but we
tested this simple idea on static rigs and after a little adjustment it worked
well. Then in 1951 we made a set of blades and ran them in an engine with
an inlet temperature of 1300 K, which would have been equivalent to 1450 K

with Nimonic 90 blades, and they worked, surviving a severe test without damage. The final step would have been to replace the Nimonic 75 with Nimonic 90, but that should have presented no major difficulties. It was a triumph technically, but it was not proceeded with.

11.12.1 Performance

To keep the blades cool enough required so much air tapped off from the compressor delivery that the efficiency was not significantly increased: the extra power obtained was almost matched by the extra fuel burnt. However, with development it would probably have been possible to reduce the amount of cooling air, and in any case, the power/weight ratio would have much been much improved. The hollow cooled blades were much lighter than solid ones, but that meant the discs in which they were mounted could be lighter in roughly the same proportion; there would have been weight savings on the shaft as well. So the engine with air-cooled blades would not only have been more powerful, it would also have been lighter. A final bonus was that the new blades, in spite of their complexity, were cheaper to make than solid ones. The version shown in Fig. 11.13(*d*) is not the one tested, which would not have been so cheap, but is based on a more advanced design by my colleague, Len Dennis.

The basic idea of a cooled blade made in sheet metal was not new, and during the war the Germans had made hollow cooled blades. Much development of the manufacturing processes was needed, particularly to ensure that the metal emerged from all the operations it went through without loss of properties. A great deal of credit should go my colleague Brian Butler and his team, who did the testing quickly and thoroughly with very limited resources. Unfortunately, the Ministry of Supply lost interest, and air-cooled turbine blades had to wait for heavier and much more expensive kinds which were also less extravagant in cooling air.

11.13 Continuously-variable transmissions (CVTs)

An important matching problem, as was noted in Chapter 6, is that between a vehicle and its engine, which requires a change in gear ratio according to the speed of the vehicle and the gradient. In a manual gear-box there are a number of fixed gear ratios, first, second, top, etc. The imperfection of matching associated with these steps wastes energy, and ideally the conventional gear-box should be replaced by a stepless or *continuously-variable* one. Much ingenuity has been expended on devising such continuously-variable transmissions, or CVTs (Section 2.9.6). Savings of between 10 and 20% of fuel consumption are to be expected from CVTs as compared with manual boxes, and considerably more compared with most automatics, which have fewer steps and extra losses.

A few small cars are obtainable with CVTs of the belt drive type, the essence of which is shown in Fig. 11.14. Two shafts A and B each carry a pulley, and the two pulleys are connected by a belt. If the A pulley is bigger than the B pulley, then the A shaft will drive the B shaft faster than itself, and vice versa. The pulleys are made of variable diameter so that A can be made smaller and B larger in such a way that the belt always remains tight. As this change is made, the gear becomes lower and lower. The variable diameter is achieved by constructing each pulley of two inward-facing cones C, one of which can slide on the shaft, and giving the belt a trapezoidal cross-section, as shown in Fig. 11.14. Moving the cones together forces the belt out to a larger radius, so increasing the effective diameter of the pulley, and vice versa. Because belts of fabric and rubber are of limited durability, great ingenuity has been expended in making steel belts, consisting of blocks threaded on multiple steel ribbons.

Other kinds of CVT have been designed which do not use belts, but while many of these are successful for other uses, none is currently used for cars. Most of them use a direct friction drive between rotating bodies pressed hard together. Figure 11.15(a) shows the working principle of one such 'contact' CVT, the Beier gear. Two shafts A and B carry stacks of discs D

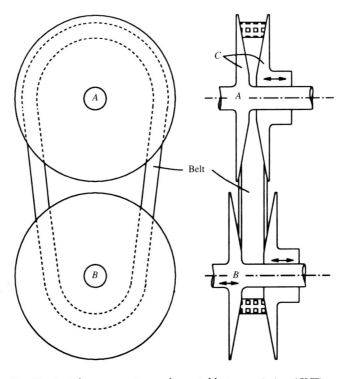

Fig. 11.14. Belt-type continuously-variable transmission (CVT).

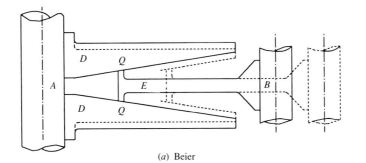

(*a*) Beier

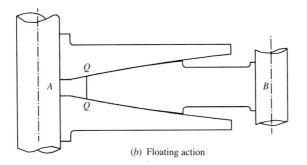

(*b*) Floating action

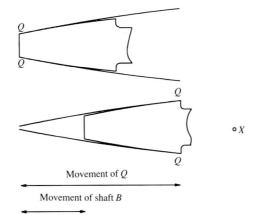

Movement of *Q*

Movement of shaft *B*

(*c*) Floating movement

Fig. 11.15. Disc-type CVT elements.

and E respectively, and the two stacks are interleaved. In the figure there are just two D discs sandwiching one E disc, the simplest arrangement possible, but there might be up to about 10 of each. The discs D are tapered or conical in form, while the E discs have T-section rims. Rotating shaft A causes discs D to drive discs E (in this case, the disc E), so driving shaft B in the opposite direction to shaft A. The speed ratio depends upon the radii of the two kinds of disc at the points of contact Q between them. Moving the shaft B to the right in the figure increases the radius to Q on the D discs while that on the E discs remains unaltered, so increasing the speed of B for a given speed of A.

In the actual gear-box, there were several shafts B and many discs in each stack. The shafts B moved on swinging links and drove a common output shaft via fixed-ratio gearing. Perhaps the largest CVT ever built was used in a remarkable diesel engine/gas turbine compound, the Napier Nomad, which was possibly the most efficient heat engine ever in its time. In effect, it was a turboprop engine in which the combustion chambers were replaced by a diesel engine so that power was produced by both the engine and the gas turbine. To combine the two output powers and yet enable their relative speed to be changed, which is necessary for the best matching, a CVT was used. This Beier-type gear had three swinging B shafts each carrying 10 discs, so that the drive was transmitted through 60 contact points.

It seemed to me that the Beier form might be improved as follows: if the contact points Q moved inward on the E discs as well as outward on the D discs when shaft B moved, a greater range of ratio could be had, or, alternatively, the same range of ratio with a larger minimum radius on either set of discs, or any combination of these advantages. The way of achieving this *floating contact point*, as I called it, is shown in Fig. 11.15(b). The discs D and E have bowl-like forms, with radial sections which approximate to circular acts, with the radius on the E form only about 60% as large as that on the D form. When the discs are closest together, contact occurs at the inner radius of the D form and the outer radius of the E form (see Fig. 11.15(c)), while when the discs are farthest apart, contact is between the outer radius of the D form and the inner radius of the E form. As the discs move apart, the points of contact Q move away from the A shaft two and a half times as fast, so that the motion required from shaft B is only 40% of the radial extent of form D.

Figure 11.16 shows the overall form of the CVT I designed; it can be arrived at by *bending* the arrangement of Fig. 11.15(b) through 90°. Leave shaft B where it is and in your imagination move shaft A through 90° anti-clockwise, pivoting it about the point X in the figure below, and bending the working part of the discs to suit: you will arrive at something like the left-hand side of Fig. 11.16. Mirror the shaft A and the discs D in

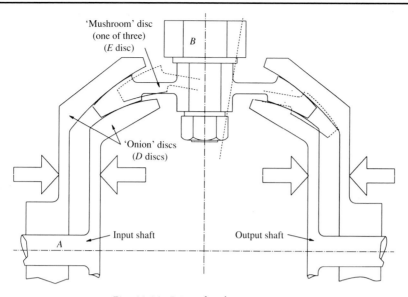

Fig. 11.16. Principle of 'onion' gear.

the axis of shaft *B*, add two more *E* discs to make three equally spaced round the circumference of the *D* discs and you have the whole thing.

In Fig. 11.16 the *D* discs have been called 'onion' discs because they resemble the fleshy layers of an onion and the *E* discs 'mushroom' discs because this is what they look like. When we were working on it, we called the whole thing the 'onion' gear, because of a strong resemblance in the early versions, which had more layers.

The onion gear is thus like a two-stage Beier gear, with the power flowing through two sets of contacts in series. The gear is varied by altering the angle of the *B* shafts to the main axis, and because of the two stages and the floating contact points a large range of ratio is obtained with only a small change of angle (in the example shown, only ±9.5° for 1 to 7).

11.13.1 Critique of the 'onion' gear

Although it was patented by a major British company, I think they were right not to proceed with it, for reasons which became apparent as a result of research which was in progress at the time. One advantage of my design seemed to be that the contacts were very favourable in that they fitted together rather closely, and that would mean low contact stresses and a robust CVT capable of transmitting a large power.

Now in such drives there is an oil film between the parts, though a very thin one, and the torque has to be transmitted through that oil film, the very device we saw used in bearings (Chapter 7) because it is virtually frictionless. It might appear that the whole notion of contact CVTs is doomed to failure, but at very high pressures, oil behaves less like a fluid and more like a plastic solid, like warm toffee, say, and will transmit a significant force parallel to the surfaces it is squashed between. Special oils have been formulated which do this more readily than ordinary ones, but it still requires very high contact stresses between the discs.

Another problem lies in the sharing of load and the matching of the ratio between the three mushroom discs, but that can probably be solved well enough by the application of the sort of principles discussed in Chapter 7. For instance, if one of each pair of onion discs is able to tilt slightly relative to the shaft, this will share the loads equally between the three pairs of contacts, just as a three-legged stool centrally loaded will carry an equal load on each leg.

In its favour, the onion gear is simpler and more compact than other designs, and it is capable of the high range of gear ratios which is necessary if the matching is to be really good and the fuel economy high. The high loads required are obtained by squeezing the onion discs together, which is done by hydraulic cylinders not shown in Fig. 11.16, and there is no need to transfer any load between rotating and stationary parts. Perhaps with the increasing interest in CVTs someone will take it up.

11.14 Conclusions

None of these ideas has been commercially successful. The end-balancing arrangements for turbines, the plant for desalination by OTEC, the hybrid LNG tank, the backhauling of liquid nitrogen and the 'onion' CVT have never been put into practice, though any of the last three might still be some day. The air-cooled gas turbine and the membrane tank for LNG were built on an experimental basis but were not adopted. Nowadays the first has been superseded by better designs and the second went astray when the asymmetrical version was adopted. Of my wave-power inventions, the one with air bags I now think was not a good idea, although it has been taken up by others who are still working on a variant. The latest, however, is really promising and I nurse high hopes for it. Nevertheless, it will not proceed unless there is a will to do something about wave power, and currently there is virtually no support for renewable energy in the UK.

You might think it must be very disappointing and a sign of incompetence to have so many failures. The work has been mostly a joy, however (and how many people are as lucky in their jobs as that?), and I have had successful inventions, none of which I have mentioned. Moreover, it is the fate of inventors to have few of their brainchildren survive in the marketplace, and that is how it must be. It is not enough for an idea to be good and sound, it must also be timely and find the right channels to exploitation.

There is nothing new under the sun, and most of my ideas have been adaptations of old ideas from other fields or in new forms. Several have depended on spotting the element of digging holes just to fill them up. Matching, in one form or another, has been a common element, and other design principles have been important.

There are several other inventions I mean to have a try at, like a good clothes-peg, an improvement to coolers for instruments in satellites and a wave energy converter even better than P. S. Frog. One of these, which I shall look at as soon as this manuscript is away, is a little toy bear, in a tutu perhaps, Tumbelova of the Bearshoi ballet. At rest she will stand up straight on her hind legs, but given a tap of just the right force between her shoulderblades, she will execute a neat somersault and regain her original pose. It is quite a difficult task, much more so than was the cylinder I still sometimes show in lectures. About 100 mm in diameter and 250 mm long, it lies on its side on the table and then suddenly jumps up on end, leaping slightly into the air as it does so: it is just a manifestation of Newton's Laws of Motion, having in it a flywheel which is rapidly spun up to speed by a spring and then stopped even more quickly. I would also like to understand in detail the remarkable arrangement of the muscles of fish, not just the broad explanation given in Section 6.6.2 but the geometry of the flakes in canned tuna, and many other aspects of natural design which we only half-appreciate.

The design of functioning things is a fascinating subject, whether studied in nature or practised in our clumsy human way. Have a go!

Questions

Here is a collection of short questions arranged by the chapters to which they refer. Many are not explictly numerical, but it is good practice in design to introduce numerical values and calculations whenever possible, even if these are very rough. Many are such that tutors can readily change the subjects to suit their own course or to introduce particular areas of engineering which the students may currently be studying. Many of them are suitable for group solution or discussion in class.

There follow some notes on the lines which might be taken in answers, and numerical answers where appropriate.

Q1.1 Write short notes on the proposition 'engineering innovation is slowing down again, because we have done all the easy things'. Suggested illustrations are the steam engine, railways, automobiles, heaver-than-air flight, television, the gas turbine, computers, genetic engineering and fusion power.

Q1.2 Vertebrates use blood to carry oxygen to the tissues that need it, while insects transport it directly as air. It is suggested this may be because transport in blood requires less pumping power. Assuming the density and viscosity of blood to be similar to those of water, estimate the minimum fraction of oxygen by mass in arterial blood if this is to be true.

Q2.1 The height a certain species of flea can jump is 200 mm. Using the data given in Table 2.1, calculate the minimum fraction of its body weight which must be resilin.

Q2.2 A battery electric car of mass 1000 kg is to accelerate from 0 to 30 m/s in 6 seconds. Using the data in Tables 2.1 and 2.2, and ignoring losses, find the minimum mass of the lead–acid accumulator required (a) on the basis of power, and (b) on the basis of energy stored. In (a), assume constant power, i.e., a perfect CVT.

Q2.3 In the study of economical speed of ships in Section 2.7, no account was taken of turn-round time in port. Decide, by reasoning or by mathematics, whether the effect of including turn-round time is to increase or decrease economical speed, or perhaps leave it unchanged.

Q3.1 List the advantages and disadvantages of the following materials for use in car-bodies: mild steel sheet, aluminium alloy sheet, glass-fibre reinforced polymer, carbon-fibre reinforced polymer, plywood. Note any advantages there might be in combining any pair of these.

Q3.2 Discuss the properties desirable in materials for the following purposes:

(a) an energy-storing flywheel (as in Section 2.9.6);
(b) an energy-absorbing fender for quays, to take the slow but heavy impact of a ship coming alongside;
(c) the two parts (ball, to replace the end of the thigh-bone, and cup, to fix in the pelvis) of a human hip joint replacement;
(d) a liquid for use in vehicle engine cooling systems.

Where you can, suggest actual materials.

Q4.1 Discuss the degrees of freedom necessary in an automatic bowling or pitching machine for training in cricket or baseball. Outline a design, noting the relevance of nesting order and parts of Section 4.7.1.

Q4.2 List the degrees of freedom in commercial robot arms and machine-tools you have seen, such as lathes and milling machines, and their nesting orders. Try to think of some other combinations you consider might be useful for particular purposes.

Q4.3 Several devices which extend the reach of the human arm – apple pickers, long-reach pruners, 'long arms' for retrieving light objects from high shelves – require a jaw-like action at the far end from the hand. Discuss different mechanisms which can be used for the purpose, and their suitability in different circumstances: are they required to hold something, once engaged, when the hand is relaxed? how strong a grip or cutting force is needed? is catching on branches a problem, favouring smooth outlines?, and so on. Include some rough calculations if you can.

Q4.4 In a patented arrangement of transmission for a four-wheel drive vehicle, the engine drives directly into one shaft of a differential. The other two shafts connect, one each, to the two axles, which have their own differentials as is usual. The connection to the front axle is direct, but that to the rear axle goes via a clutch and gear-box. Will it work? and if it will, how will its behaviour differ from that of the conventional arrangement with the clutch and gear-box between the engine and the first (or middle) differential?

Q5.1 Bridges may be roughly classified into beam, arch, suspension, cantilever (like the Forth railway bridge) and stay-supported types (like the one in Fig. 5.1).

Try to classify the examples you can think of, and to list the types in order of spans for which they are suitable.

Q5.2 Consider the problem of prestressed concrete beams such as that shown in Fig. 5.36. If the load is removed, the tension in the cables at the centre, which produces a hogging moment in the beam there, will produce tension in the top of the section where there is no steel to take it. What consequences does this have for the sequence of operations to be followed?

Q5.3 Carry out the crude experiment on twisting a long thin square-section box which is described in Section 5.4.3. Instead of a box, a square tube formed by folding up a rectangle of card, say, 12 cm by 40 cm, will do very well. You will find that it is not necessary to tape the whole length of the open edge to develop most of the potential stiffness. Find experimentally the spacing between short lengths of tape to which you can go before the stiffness begins to fall off rapidly, and express the result in a simple form useful to designers.

Q6.1 Buildings are often decorated with hanging wire baskets filled with flowering plants which require watering every few days. Outline a number of possible systems for doing the watering economically, ranging from hand methods to fully automatic ones, and estimate costs as best you can.

Q6.2 Discuss matching in the following cases:

   (a) force required at the foot to air-pressure in the cylinder in a foot pump for inflating tyres;
   (b) petrol (gasoline) flow to air flow in carburettors and in fuel injection systems;
   (c) oxygen supply to the muscles in a human at rest and at various levels of physical activity.

Q6.3 Some work has been done on automatic braking systems for automobiles cruising on major roads, to prevent them running into the vehicle in front. Make preparatory notes for a specification for such a system. In particular, address this question: in Fig. 6.18, if Y is the pressure applied to the brake cylinders, what should be the input X, that is, what kind of function of the data provided by the sensors?

Q7.1 A centrifugal pump is required to supply 0.1 m^3/s of water at a head of 1 m. Using Fig. 7.16, estimate very roughly the possible range of overall diameter of such a pump.

Q7.2 Consider the stiffness of a hydrostatic journal bearing such as that shown in Fig. Q7.2, with oil at high pressure admitted to four pockets spaced equally round the central shaft. If the shaft moves to the left, the forces on it to the right from the left-hand pocket, and to the left, from the right-hand pocket, remain nearly unaltered and balanced, and do not tend to return the shaft to the centre. To confer stiffness, it is necessary for the pressure in the left-hand pocket to rise

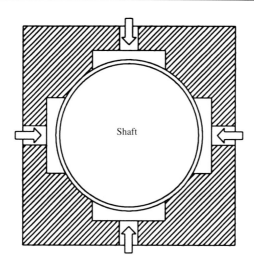

Fig. Q7.2. Hydrostatic journal bearing with four pockets supplied with high-pressure oil (clearance exaggerated).

and that in the right-hand pocket to fall. We need to arrange that the pressure in any pocket rises as the shaft moves closer to it, and vice versa. Can you see a simple way of achieving this?

Q7.3 The heart of a giraffe has to be able to pump blood to its head, 3 m higher, and its veins will collapse if the pressure in them falls below that in the surrounding tissues. Consequently, the lower, or diastolic, blood pressure of the giraffe has to be at least 3 m of blood, or 220 mm of mercury, about three times that of a human. The head of a brachiosaurus, a dinosaur with a very long neck, may sometimes have been as much as 8 m above its heart, raising the diastolic pressure to at least 590 mm of mercury, or about seven times that of a human being, imposing a terrible load on that muscular pump. Suggest a means whereby nature might have relieved that load, other than making veins that would not collapse if the pressure in them was less than that in the surrounding tissues. Hints: the load on the heart muscle is determined by the difference between the pressures inside and outside; note Fig. 7.7(a).

Q9.1 Examine some common manufactured objects and try to decide what material each part is made from and how it was produced. Suitable subjects are a stapler, items of cutlery, a garden spade, a bicycle brake, a light fitting and the bulb, shoes of various kinds, kitchen pans and other equipment, a door lock. Note the large number of sheet metal pressings and polymer mouldings, and the various surface finishes.

Q9.2 Do a library search for information on these relatively modern processes: ultrasonic welding, laser welding, rapid prototyping techniques, such as stereography, injection mouldings with foam fillings, 3D woven reinforcements.

Q9.3 Write notes on the evolution of the design of some product in recent years. Choose one with which you are familiar or about which you can readily obtain information. Note which changes had long been considered but were not adopted earlier, which depended on new ideas and which stemmed from progress in other fields – thus in car engines, multiple valves fall in the first category, toothed belts in the second and fuel injection in the third.

Q10.1 List the functions involved in a can opener and the means which might be used to perform them, and construct a table of options like Table 10.1. Note which means are used in examples known to you, and note advantages and disadvantages of each means.

Q10.2 Repeat the previous example for some other simple and familiar device.

Q10.3 Can you recognise this simple and well-known object from its functions? These are, store fuel as a solid, melt fuel progressively, retain liquefied fuel in a small reservoir, transfer fuel at the required rate from the reservoir to the combustion zone? This is a case of complicated functions embodied in a very simple form.

Q11.1 A satellite contains a spherical tank filled with liquid oxygen, which presents the same problem as the much larger LNG tank, in that it shrinks when filled, although there is no problem with strength. It is supported and located by a number of thin metal straps fixed to its skin and to its warm surroundings. The straps are flexible in bending, but stiff lengthwise. How should they be placed to enable the tank to contract without stretching them?

Answers

A1.1 The case can be argued, particularly on the length of time between first demonstration and widespread appearance of major innovations. Development often takes much longer than it might have done in the nineteenth century. It could be held on the other side that the important ideas come closer together.

A1.2 This argument can be very complicated. The simplest approach is to imagine two animals with the same length, area and number of passages, which can be justified on the grounds that most of the drag will occur in the smallest ones, and their dimensions are dictated by the area necessary to diffuse the oxygen into the tissues. Such passages are so small that the flow will be laminar, and determined entirely by the viscosity, to which it will be inversely proportional. If we increase the viscosity m times and the velocity n times, the pressure drop will increase $m \times n$ times and the power will increase as the pressure times the volume flow, that is, by $m \times n^2$ times. For the same power, $m \times n^2$ must be equal to 1.0. That is, n must fall inversely as the square root of m. The viscosity of water is about 36 times that of air, so that its speed will be only one-sixth, but as its density is about 900 times as great, its mass flow will be 150 times as high. It follows that as oxygen is about 21% by mass, to need the same pumping power the oxygen content of the blood would have to be 21%/150 or 0.14%. But this is a higher concentration than in pure oxygenated haemoglobin, so the original argument falls.

A2.1 For simplicity, assume the mass of the flea is 1 kg – it will make no difference and save dealing in powers of 10. It will weigh 10 N and it will take two joules to raise it 0.2 m. From Table 2.1 it will need a minimum of (2/1500) kg of resilin to store 2 J, so the answer is 1/750.

A2.2 (a) At 30 m/s energy is 0.45 MJ, so power required is 75 kW, needing 417 kg of accumulator. (b) However, to store 0.45 MJ only needs 5.6 kg.

A2.3 First, divide the B-costs into B_1-costs, those of the engine, and B_2-costs, those of fuel. By mathematics, proceed as in Section 2.7, but this time journey distance, D, must be taken into account. If, as before, x is the speed, the overall journey time will be $t + D/x$, where t is the turn-round time. Now differentiate the cost of a journey with respect to x. It will be seen that if the old optimum value of x, that

355

made the differential zero, is inserted, there remains a positive term, one in B_1, the others cancelling out. This means that at the old optimum x, dC/dx is now positive, so that the new optimum x must be lower.

To see this by reasoning, note that the A-cost for the turn-round is a fixed sum, but the B_1-cost for the turn-round time rises with v^3, so here is a variable B-cost unmatched by any corresponding variable A-cost, which must reduce the economic speed (since without variable A-costs, the economic speed would be zero!).

A3.1 Important properties are manufacturability into the required forms, freedom from corrosion, rot, insect attack or deterioration due to ultra-violet radiation, energy-absorbing capacity, ease of repair and for most markets, the ability to take a hard polished finish. Some deficiencies, like corroding, can be corrected by surface or other treatments.

A3.2 For the flywheel, strangely enough, density is a disadvantage, because the energy which can be stored per unit volume is proportional to the allowable stress. For the energy-absorbing fender, do not overlook the possibility of using a fluid material. For the coolant, there is a long list of requirements, which require a mixture usually, and then not all of them are fully met, like non-toxicity, not supporting combustion, not damaging various materials with which they may come in contact, either inside or outside the engine, and so on.

A4.3 Note that using rotary motion to transmit the effort will generally be weak and 'spongy', for fundamental reasons. For many applications a string and a return spring is the best approach, with the string running inside a hollow handle if a clean exterior form is important.

A4.4 Yes, it will work: when the clutch is disengaged, there is no drive to either axle. As the gears are changed upward, the proportion of torque going to the front axle increases. This brilliant idea (Jaguar) has advantages, but also disadvantages.

A5.3 The result is most simply presented as the ratio of the distance between tapings (the pitch) to the side of the square at which stiffness begins to fall off rapidly.

A6.1 The solutions can range from piped systems, with sensing of basket moistness, to a man with a watering can and a step-ladder. Do not overlook solutions in which the basket is lowered, but these need careful design and special equipment.

A6.3 The input X should be some function of the distance between the vehicles and the relative speed, i.e., the speed of closing. It should increase with both, but be zero when the distance is large. Various mathematical expressions combining the two can be proposed and considered.

A7.1 Calculate c (4.4 m/s) and note that in Fig. 7.16 the values of p/c for centrifugal pumps range from about 0.006 to a little above 0.06. Calculate the corresponding values of p, and hence, from the required flow, the extreme diameters (about 0.7 m to 2.2 m).

A7.2 If the shaft moves to the left, the flow from the right-hand pocket will increase, since there is a freer leakage path, and that from the left-hand one will decrease, because the leakage path will be narrower. If a restriction is placed in the inlet to each pocket, then increased flow will drop the pressure, and this effect will provide stiffness. To obtain substantial stiffness, the restriction must be severe enough for a large part of the total pressure drop from supply to exit to take place across it, leading to a disposition problem (Section 10.8.2) in which the commodity to be disposed of is pressure drop.

A7.3 One solution might be a strong bag round the heart, maintained at about the diastolic pressure. A duct full of lymph, say, leading from the space inside the bag and around the heart to a tank in the head, would serve to maintain the right ambient pressure, with a small amount of work involved in shunting lymph up and down the duct at every heartbeat. It is not suggested that large long-necked dinosaurs were so provided, indeed, it is difficult to see from what structures such a system could have evolved.

A10.1 A suitable list of functions is, provide power (2), cut tin (3), guide cutter (3), retain lid (2); the numbers in parentheses are the number of likely means you might suggest.

A11.1 When the tank contracts, the attachment of a strap moves radially inwards. If the strap remained unchanged in length, and it were tangential to the tank, it could follow this motion without stretching. However, it will itself contract thermally and shorten. If the strap is of the same material as the tank, and assuming its mean temperature drop averaged along its length is half that of the tank, a little geometry shows that the perpendicular bisector of the length of the strap should pass through the centre of the tank. Structurally, the strap should be tangential (Section 7.4.9), but then it is impossible for the condition for no change of length to be met, so a compromise is necessary.

Suggestions for further reading

Engineering Design

Svenson, N. L. (1981). *Introduction to Engineering Design.* New South Wales University Press.

Pahl, G. and Beitz, W. (1984). *Engineering Design.* London: Design Council.

French, M. J. (1971; 2nd ed. 1985). *Conceptual Design for Engineers.* London: Design Council.

French M. J. (1992). *Form, Structure and Mechanism.* London: Macmillan.

Slocum, A. H. (1992). *Precision Machine Design.* New Jersey: Prentice-Hall.

Most simple books on engineering design are less concerned with design itself than with the context in which it takes place: Svenson's short book is of this kind. *Conceptual Design* is strictly about design rather than its context, and how to use ideas like matching, nesting and disposition to develop good schemes. It is strong on principles. Pahl and Beitz scarcely touch such matters, but are long and deep on formal methodology generally, including their version of morphological charts or tables of options, and deal thoroughly with some contextual aspects. *Form, Structure and Mechanism* is an easy book to read, dealing mostly with mechanical engineering design and nearer the detail end than *Conceptual Design*. Slocum's book is a great expensive volume and limited in its scope, but it is full of lovely stuff, including beautiful examples of disposition, of error in this case, developed mathematically.

Design in Nature

D'Arcy Thompson, W. (1917). *Of Growth and Form.* Abridged version edited by J. T. Bonner. (1961), Cambridge University Press.

Paturi, F. R. (1976). *Nature, Mother of Invention.* London: Penguin.

Hertel, H. (1966). *Structure, Form, Movement.* New York: Reinhold.

Alexander, R. M. (1968). *Animal Mechanics.* London: Sidgwick & Jackson.

Duddington, C. L. (1969). *Evolution in Plant Design.* London: Faber & Faber.

Wainwright, S. A. *et al.* (1976). *Mechanical Design in Organisms. Studies in Biology* (series). London: Arnold.

Gray, J. (1968). *Animal Locomotion.* London: Weidenfeld & Nicolson.

D'Arcy Thompson's justly famous book should be read and reread by anyone interested in the field. The abridged version contains most of the best stuff, although enthusiasts may wish to dig in the original, especially for the discussion of phyllotaxy. The book is delightfully written, in an ornate but lucid and easy-paced style, and has a rare distinction in such works that it is 'hard to put down'. He ignores genetics, and puts altogether too much importance on physical forces such as surface tension as determinants of natural form, but the genius of the book transcends such faults.

Alexander's book lacks the poetry of Thompson's, but is another joy. Brisker in pace, it is still easy to read and is fascinating almost from cover to cover and full of interesting examples. On the design of plants, Duddington has provided the nearest parallel I know to Alexander's book. Wainwright *et al.* are chiefly concerned with material properties: Gray's book on animal locomotion is excellent and comprehensive. Paturi and Hertel are rather similar in approach, both strong on the superiority of living organisms over engineering products, to the extent of exaggeration in the case of Paturi, but both are instructive and interesting. Paturi deals only with plants, and Hertel chiefly with animals.

Energy, Materials

Gordon, J. E. (1968). *The New Science of Strong Materials.* London: Penguin.

I know of no work on energy I can recommend in this context. Books on the subject abound, but they are either too difficult or not related to design, or both.

On materials, Gordon is most readable. He seizes the essence of his examples and expounds it simply and with humour.

Mechanism

Chironis, N. L. (1965). *Mechanisms, Linkages and Mechanical Controls.* New York: McGraw Hill.
Molian, S. (1982). *Mechanism Design.* Cambridge University Press.

Chironis has edited a collection of articles on mechanisms of different kinds, grouped together in sections each having a common theme, most of which originally appeared in the journal *Product Engineering*. It gives some idea of the great variety of mechanisms which have been designed and some indications of how to find one to suit a given purpose. In his early chapters Molian demonstrates simple but powerful means of synthesising mechanisms to perform specific tasks.

Structures

Gordon, J. E. (1978). *Structures.* London: Penguin Books.
Cowan, H. J. (1971). *Architectural Structures.* New York: Elsevier.
Morgan, W. (1964). *Elements of Structure.* London: Routledge.

Torroja, E. (1962). *Philosophy of Structures* (translated from the Spanish, *Ser y Razón de los Tipos Estructurales*). Berkeley, University of California Press.
Hilson, B. (1972). *Basic Structural Behaviour via Models*. London: Crosby Lockwood.

Here are riches. Gordon is stimulating and interesting. Cowan and Morgan are excellent and systematic within their fields, but do not treat structures outside them, naturally enough. Torroja is perhaps rather for designers than beginners, but is deep, fascinating and concerned with aesthetics. Hilson shows how to gain insight into structural behaviour by handling simple models in household materials, a royal route to understanding.

Form

Schaeffer, H. (1970). *The Roots of Modern Design*. London: Studio Vista.
Whyte, L. L., ed. (1968). *Aspects of Form*. London: Lund Humphries.

The collection of articles edited by Whyte is full of interesting observations about biological forms. Schaeffer deals with the relation of the aesthetic and the functional, with many convincing illustrations from the last century.

Designing and Inventing

Osborn, A. F. (1953). *Applied Imagination*. New York: Scribners.
Gordon, J. J. (1968). *Synectics*. New York: Collier.
Von Fange, E. K. (1959). *Professional Creativity*. New Jersey: Prentice-Hall.

The first two books are briefly discussed in Chapter 10. There seems to be little since which adds to them on the 'soft' side, but Von Fange is interesting because he introduces the 'hard' side as well.

Aesthetics and Invention

Wechsler, J., ed. (1978). *On Aesthetics in Science*. Massachusetts: M.I.T. Press.

This excellent collection of essays uses the word 'aesthetic' in Poincaré's sense, and is rich in good things but is concerned with science, not design.

Education of Inventors

De Simone, D. V., ed. (1968). *Education for Innovation*. Oxford: Pergamon.

These essays are well worth reading, for good examples of invention, for accounts of educational experiments and for interesting points of view. Particularly interesting (and irritating) is one by J. Rabinow.

Psychology and Creativity

Barron, F. (1969). *The Creative Person and the Creative Process*. New York: Holt.
Hudson, L. (1967). *Contrary Imaginations*. London: Penguin Books.

Lytton, H. (1971). *Creativity and Education*. London: Routledge.

These books give data on creative people from a variety of sources and refer to various experiments. Hudson originated the 'converger–diverger' classification.

Index

abstraction, 291, 307, 309
acceleration, 24
aesthetics, 15
  based on function, 235
  of bow, 250
  of mass production, 249–50
air, 76
air-cooled blades, 340–3, 348
aircraft
  evolution of propulsion,
    280–1
  style, 249
  undercarriage, 146
  variable geometry, 230–1
allometry, 275
analogy, 296–7, 310–12
analysis, 286
  of function, 300–3, 315
  morphological, 300
animal, 4
anisotropy, 60
appearance and rightness,
  240–1
apple, 24
aqueduct, 6
aramid, 74
arch, 115–18
  masonry, 118
Archimedes, 287
architect, 2, 295
architecture, 5
arm
  degrees of freedom, 97
  design of, 322–3
attribute listing, 298
avoiding arbitrary decisions,
  319–20

bacterium, 106–7
badger, 80

balance, 236
ball joint, 99
ball, bouncing, 29
balloon, 145
band-width, 171
beam, 126–9
  concrete, 142
  economical design, 131–2
  I-, 126–7, 320
  prestressed, 142–4, 352
  twisting of, 132
bearing, 79–80, 182–7, 348,
  352
  car wheel, 183
  flexural, 88
  instrument, 183
  location, 323
  machine tool, 183
  porous, 187
  range of, 186–7
  rolling element, 183, 185,
    186
  tilting pad, 187
beauty, 222
  derived from function,
    235–43
  of components, 250
  of mechanism, 230
'belt and braces', 200–1
bending, 126, 146
  in connecting rod, 17
  moment, 130
Bernal J. D., 286
bicycle
  characteristic, 151–2
  manufacture, 277
  wheel, 139–40
biologist, 2
bird
  brain cooling, 202–3

difficulty of design, 230
  flight, 33–7, 39–40, 41–2
blade, fluid machine, 196
  dovetail fixing, 206–8
blood, 76, 149
  in tall animals, 353
boat, asymmetrical, 236–7
Bondi, H., 291
bone, 71
bottle, 253
bow, 4, 47–9
  aesthetic aspect of, 250
  late invention, 178
bowling machine, 351
brain, human, 169, 173–4, 175
brainstorming, 297
braking, regenerative, 44–6
bridge, 351
  arch, 115–18, 246
  bridle-chord, 141–2
  Salginatobel, 116–18, 200,
    245
  Saltash, 119, 131, 245
  soldiers crossing, 66
  suspension, 67, 113–15
  Tacoma, 67
  Waterloo, 245–6
brittleness, 58–60
brontosaurus, 119, 131
Brunel, I. K., 119, 246
Brunel, M. I, 277
buckling, 64–5, 145
Budal, K. and Falnes, J., 213
bursting and leaking, 3, 305
Butler, F. B., 343
buttercup, 1, 15
butterfly valve, 80–5

cable, 114–15, 310–12
cake, cutting of the design, 177

camel, 239–40
can opener, 354
cantilever, 127–9
car, 68, 310
  braking, 352
  electric, 350
  in collision, 58
  roll of, 87
  style, 247
carbon-fibre reinforced plastic
  (CFRP), 68
carnivore, 97
case studies, 322–49
casting and moulding, 255–6
  ceramics, 255
  die, 255
  glass, 253, 255
  sand, 255
cathedral, 6, 108, 119
cats-eye, 355
cell, 20
  division, 265
cellulose, 70, 74–5
chain molecule, 69–70
chair, 55
characteristics, 151–8
  bicycle, 151–2
  cyclist, 151–2
  pump and fountain, 153–4
chassis, 133
chatter, 65
chimera, 237–40
chlorophyll, 20
chromosome, 265
classification, 303–4
closure, 81–5
  force, 96
clothes-peg, 314–19
Cockerell, C., 302
combinations, 302
communications, 170–3
complexity
  cost of, 19
components of systems, 149
composite, 70–5
composition, 235–6
compression,
  strength in, 60–1, 64–5
computer, 13, 170, 174, 178, 254
  learning by, 178
concrete, 71, 192
  prestressed, 142–4
  reinforced, 142
conformity, 96–7
connecting-rod, 17
  big end, 209–11
continuously-variable
  transmission (CVT), 45–6,
  343–8
  Beier type, 344–6
  floating contact in, 346

control systems, 167–70
  augmenting stiffness, 176
  positional, 166–7
  thermostat, 166, 178
  tie-bolt length, 328
  wing load, 177
cow, 97
crack stopping, 69
craftsmanship, 4
creative people, 284
creativity, 283, 286
  measures of, 284–5
cross-linking, 70
curves, regular and organic,
  232–5
cylinder
  autofretted, 190
  filament-wound, 191–3
  lazy layers, 188
  slit, 313–14
  that stands up, 349
  wire wound, 190
cylinder, high pressure, 188–92

dam, 216–18
  arch, 216–17
  buttress, 217–18
  forms, 233–4
Darwin, C., 18
data-processing, 173–5
deformation, plastic, 57–9
degree of freedom, 86, 95–104,
  351
  balancing, 101–4
  in arm, 97–100
  in characteristics, 154–5
  nesting order, 98–100, 281
Dennis, L., 343
desalination, 328–32
  construction of plant, 331–2
  raising cold water, 332
design
  aesthetic basis, 251–2
  aids to, 295–312
  and drawing, 11
  elegance in, 15–18
  evolutionary, 4, 8, 276–81
  fail-safe, 169
  functional, 3, 5
  in genetic engineering,
  281–2
  in nature, 19
  in the schools, 289, 292
  intellectual nature of, 3
  nichewise evolution in,
  276–7, 354
  non-evolutionary, 10, 282
  of assembly, 13
  relation to science, 15
design philosophies, 199–211
  dams, 216–18

eagle, 220–1
  gas-cooled reactor, 218–20
  grass, 221
  suspended roofs, 216
  wave energy, 211–16
design principles, 199–211
  avoid arbitrary decisions,
  319–20
  biasing, 199
  cascade, 206–9
  clarity of function, 320
  combination of functions,
  305
  direct closure of loops, 325
  kinematic design, 200–1
  least constraint, 199–201
  matching, 209–11
  regenerative, 201–5
  separation of functions,
  305–7
  tangential support, 211
  thermodynamic, 205–6
design-build-test, 293
designer, 283, 287–9
  and energy conservation, 321
  dispersion of, 288–9
  higher education of, 295
  society and the future, 321
designing, 283
  summary, 320–1
desoxyribonucleic acid (DNA),
  2, 264–8
differential, 102–4, 115, 351
digging holes to fill them, 322,
  324, 328, 329, 334
dinosaur, 20
disposition, 309
distillation, 331
divergers and convergers, 285,
  287
division of labour, 3
Djinn's dilemma, 38, 274–5
dome, 138
dovetail fixing, 206–8
drag, 33–8
drawing, 10, 12
  for thinking, 292, 293–4
ductility, 57–60

ecology, 2
economy
  of material, 15
  of production, 254, 264
  of ship speed, 37–9, 350–1
  of scale, 254, 264
Edison, T., 13
efficiency, 36–7
  of joint, 122, 207
Einstein, A., 290–1
elbow, 164
electric motor, 157

elegance, 15, 85, 320
energy, 23–7
    available, 32
    conservation, 42–6, 321
    conservation of (law), 27–8,
        36
    convertibility of, 27–8
    forms of, 26–7
    from ocean temperatures,
        328–32
    kinetic, 26
    of deformation, 58–9
    specific, 52
    storage, 45–52
engine
    mechanism, 88–9
    multicylinder, 182
    rotating piston, 278
    steam, 3, 276–7
    Trojan, 15–17, 19
    Wankel, 278–9
engineer, 1, 295
enzyme, 104, 106–7
epiphyte, 226, 278
equilibrium, 24–5, 112
evolution, 4, 18, 20, 268–73
    extinction of, 281
    hill-climbing analogy, 272–3
    limitations of, 22, 272, 275
    nichewise, 270–3
    of man, 271–2
exoskeleton, 21, 90
extensible sheet, 335–8
extrusion, 261–3
eye, 173
    educated, 242

familiar, making strange, 296,
    297
family tree, 291
feather, 134–6
feed-back, 147, 148, 166
femur, 55
film,
    hydrodynamic, 184–5
    hydrostatic, 185
    lubricating, 183–5, 210
    solid, 185
    synovial, 183–4
fish, 21
    form, 230
    muscle, 163, 349
    speed of, 37–9
flange, 126, 134–6
flea, 47, 350
Fleming, A., 106
flexibility, 68–9, 115
flight, 33–7, 177
    insect, 89–90
floor, 134
flowers, 22

fluid
    machines, 194–8
    materials, 75–7
    structural, 144–6
flywheel, 45–7
force, 23–5
    components of, 50, 112
forging, 256
form
    and aesthetics, 235
    of animals, 229–32
    of trees, 225–6
    related to function, 15, 20
Fox, Uffa, 13
friction, 183
function
    analysis of, 211–12, 300–3,
        307, 354
    conflicts between, 108
    critical, 303
    dictates form, 234
    separation of, 3
furniture, 246–7
fusion power, 26, 193–4

gas holder, 242
gas turbine, 150, 280–1
    air-cooled blades, 340–3, 348
    end balancing, 322–8, 348
    tip leakage, 326–7
gas-cooled reactor, 218–20
gear-pump, 305–7
gearing
    use in matching, 155–7, 158,
        160
genetic engineering, 281–2
genetics, 18, 265, 266–8
giant elk, 275
glass-fibre reinforced plastic
    (GRP), 60, 73, 74
glycogen, 40
Gordon, W. J. J., 296, 312
grommet, 319–20
group, advantages of, 296
growth, 19, 265
    gnomonic, 273–4
gun barrel, 162
    evolution of, 190–1

Hadamard, J., 286
Hardy, G. H., 252
heat, 23
    convertibility of, 29–31
Heisenberg, W., 286
helicopter drive-shaft, 71–2,
    310
hinge, 79–80, 117
    flexural, 101
    in bridge, 144, 201
    in reactor duct, 218–19
hoe, 101–2

homo faber and jabber, 289
horse, 47, 101–2
    motion, 231
Hugo, V., 11
hydrofoil, 231

ideas, stimulating, 295–312
image intensifier, 263
Industrial Revolution, 20
information flow, 170–1, 290
insect, 21
    flight mechanism, 41–3,
        89–90
insight, 85, 244–6, 293–4, 297,
    303
instantaneous centre, 87, 91,
    93–4
intuition, 244–6
inventing, 283
    stages of, 286, 296
    summary, 320–1
invention, 20
    adaptation of old ideas, 322
    aids to, 295–312
    of nature, 20
    systematic, 51, 300
inventor, 283, 287–8
invertebrates, 20
Issigonis, A., 310

jaw, 80, 96–7
joint
    efficiency, 122
    synovial, 183–4
    three kinds, 300
joule, 25–6

karabiner, 294
kelvin, 30
keratin, 136
kinematic design, 200–1
    in reactor duct, 218–19
kinematics, 84–98
kinematics-statics relation, 103,
    294
kinetic fluid machines, 194–5
    parametric plot, 196–8
knee, human, 90–4
knee, in ships, 9
known means, 181, 187, 194

lamp, 'Anglepoise', 160
lathe, 65, 166–7
leaf, 15
lift, 33–6
ligament, cruciate, 90–4
lines, ship's 10
liquid natural gas tankers,
    332–9
    membrane tanks, 334

naive design, 335
 shrinkage problem 334
lock, 79
 ancient, 104–6
 canal, 203–5
 Yale, 106
locomotive, 1, 15
lubrication, 76–7

maieutics, 295–312
 hard, 298–307
 soft, 296–8
Maillart, R., 116, 118
masonry
 arches, 118
 tension in, 119
mass
 apparent, 43
 effects on form, 230
matching, 148, 340
 driver to driven, 151
 electrical, 157
 in big end, 209–11
 in bows, 164–5
 in car jack, 161
 in fish muscles, 163
 in flapping-wing flight,
  158–60
 in gas turbine, 150
 in powder grains, 161–2
 in reading lamp, 160
 in ships, 156–7
 in windlass, 149–50
 muscle-wing, 158–60
 of components, 149
 pump-fountain etc., 153–5
 rider–bicycle, 151–2
 use of gearing, 155–7
 in human muscle, 164
materials
 appreciation, 54
 choice of, 54–67
 fluid, 75–7
 for telescope, 176
 properties, 55–64
mathematicians, 286
 algebraists and geometers,
  291
Maudslay, H., 288
Maxwell, C., 124
mechanism, 79–107
 clock, 291
membrane tank, 335–9, 348
 corrugated forms, 335–8
 hybrid forms, 338–9, 348
Mendel, 18
mine, 60
Mini, 310
missile, throwing, 49–51
Mitchell, C., 312
molecular structure, 69–70

molluscs, 20
 gnomonic growth, 273–4
moment, bending, 130
 hogging and sagging, 130
morphological chart, 303
morphology, 300
muscle
 in fish, 163, 349
 in worm, 144
 insect, 90
 red and white, 39–40
mutation, 267
myoglobin, 40

Napiers, 325, 341
natural selection, 4, 18, 22,
 268–73
 limitations of, 122, 272, 275
neural networks, 175–6
Newcomen, 3
newton, 23
Nimonic, 341–3
nitrogen backhauling, 339–40,
 348
nothing new under the sun,
 349
nutcracker, 294

ocean temperature differences,
 328
 energy from (OTEC), 328–
  32
one-over-N rule, 205, 207, 209
orchid, 270
orthographic projection, 12
Osborn, A. F., 296, 300, 320

pastry, 261
patents, 180–1
pattern, 240
pelvis, 146
pendulum, 28–9
 apparent mass, 43
Pepys, S., 9
philosophy, design, 211–21,
 303
physicists, 286
pick-up 55,
pigeon-hole, 304
pitching machine, 351
pivot, 79–80, 182–7
 flexural, 101, 115
plant, 20
 paper-making, 148
 process, 147
plant forms, 223–5
 in forest, 225–6
pliers, 241
ploughing, 102, 103

Poincaré, H., 287
 mathematical invention,
  251–2
 sieve, 295–6
pollination, 22, 268–70
 avoidance of self-, 269–70
polyhedra, 298–300
polymers, 69
Pont du Gard, 6–7
power, 23, 32
 peak, 39, 41–3. 45–6
 specific, 52–3
practice, 180–199
pressing, 57–8
prestressing, 139–44
princess and the pea, 210
problems
 disposition, 309–10
 standard, 309–10
process, 147
production, 19, 55, 57–8, 253–
 64, 353
 casting and moulding, 255–6
 effect on style, 246–7
 etching, 257, 258
 explosive forming, 263–4
 extrusion, 261
 knife-and-fork, 253–4
 of VLSI chip, 258–60
 photograhic, 258
 planar-replicative, 257–60
 pressing, 256–7
 printing, 257
 replicative, 253–60
propellant, 161–2
psychologists, 283, 285
pulley block, making, 277
pump, 194–8
 displacement, 196–7
 gear-, 305–7
 kinetic, 194–5, 352
 other, 195, 198

quickness, 47–9

radio, 147–8
radio-telescope, 176
railway, 44–5
 system, 147–8
reaction, 24
redundancy, 169
refrigeration
 absorption, 329–30
 cascade, 208
 mixed refrigerant, 209
regenerative
 feed-heating, 203, 206
 tax evasion, 202
regenerative principle, 201–3,
 209
 gas liquefaction, 208

regenerative principle—*cont.*
  side ponds, 203–6
  steel-making, 203, 206
reliability of systems, 168–9
repertoire of known means,
  198–9
replicative production, 253–60
reproduction, 19, 107, 264–70
resolution of forces, 50, 112
revolving door, 311–12
rhythm, 240
robot, 98, 178, 223
rock, seaside, 262–3
rocket, 41, 146
  cascade principle, 209
roll centre, 87
Roman aqueducts, 6
ruby, 185
rudder, 9–11

Salter, S. H., 301
Samson, 109–10
satellite, 354
scale effects, 35, 37
scientists, 2
scythe, 4–5, 241
servo, 10
  in telescope, 176
  in tools, 176–7
  positional, 167–7
shape memory, 77
shark, 20
shear, 60–2, 125, 129
  in twisting, 132
  strength, 60–1
  stress, 61–2
shell, 80
Sherlock Holmes, 214–15
ship, 136, 280
  economical speed, 37–9, 350–1
  Egyptian, 130–1
  form, 234
  lines, 10
  sailing, 8–11
  standardisation, 181–2
  steering, 9–10
signal-to-noise ratio, 170
silk, 70
sketch, 12
sledge, 25–6
slider-crank chain, 88–90
sling, 50–1
sonar, in whales, 178
spatial relationships, 5
spear-thrower, 50
specific speed, 197
specification, 315
specificity, 104
spider, 55–6
  leg, 145–6
  silk, 70

squid, 20
stability, in tension, 116, 138–9
standardisation, 181–2
steam catapult, 312–14
steam engine, 3, 276–7, 279–80
  triple-expansion, 280
  uniflow, 279–80
steam turbine, 189, 277
steel, 70, 72
  mild, 57
steering, 9–10, 86
stem, 15
stereospecificity, 79, 104–7
stiffness, 64
  replacement by control,
    176–7
Stradivarius, 8
strength, 56
  compressive, 60
  of composites, 74
  shear, 61–2
  specific, 62–3
  tensile, 56–7
stress
  hoop, 137–8
  in tree trunk, 228–9
  longitudinal, 137–8
  shear, 61
  tensile, 56
structural element, 108, 146
  arch, 115–18
  beam, 126–9
  cantilever, 127–9, 134
  dome, 138
  economy in, 122–5, 146,
    334–5
  spoke, 139–40
structure, 108–46
  allied with control, 177
  minimum material, 124
  strut and tie, 111–13, 120–5
  suspended roof, 2, 216
  tiered, 134
strut and tie, 109–12, 120–5
  kinematic test, 111–12
  stability of, 109–11, 138–9
style, influence of production,
  246–9
submarine, 139
superconducting wire, 263
suspended roofs, 2, 216
suspension
  vehicle, 68, 85–8, 96, 294
  Hydrolastic, 310
  nesting order, 100
  strut, 85–7
  wishbone, 88
suspension bridge, 67, 113–15,
  116, 201
swan, 156
  feet, 201–2

Swift, J., 4
swingletree, 101–2, 103, 115
sword blade, 70, 261
symmetry, 1, 236–7
synectics, 296
synthesis, 286
system, 147–9
  aircraft control, 148
  analogue and digital, 169
  branching, 223–5
  hydraulic, 149
  in human body, 149
  in nature, 177–9
  quality of, 176
  service, 147, 148
  watering, 352
system for assembly design
  (SIMAD), 13–14
systematic design
  of sling, 51
  of spear-thrower, 51
  of valve, 82–4

tab-washer, 249–50
table of options, 302–3
technology in schools, 289
television, 170
  signal rate, 171
  band-width, 171–3
temperature
  control, 21–2
  thermodynamic, 29–30
tensile strength, 56–7
texture, 240
thermodynamics
  first law, 27
  second law, 31–2
thought, visual, 289–92, 293
  and verbal, 289–92
  capacity of, 290
  proprioceptive link, 290
time, direction of, 29
tomography, 174–5
Torroja, E., 220
transfer function, 168
transformer, 157, 160
transient, 327–8
tree, 132
  as engineering design, 227–9
  forms, 225–6
  stress in trunk, 228–9
  'structural', 229–30
truss bridge, 242–3
tuning fork, 43
twisting, 132–3, 352
typewriter
  evolution, 277
  manufacture, 281
tyre, 141, 146

vertebrates, 20, 21, 350